THE GREAT
SCHOOL WARS

Diane Ravitch

THE GREAT SCHOOL WARS

NEW YORK CITY, 1805-1973

A History of the Public Schools
as Battlefield of Social Change

BASIC BOOKS, INC., Publishers · New York

for Stevie

MAY 11, 1964–DECEMBER 14, 1966

Contents

BETWEEN THE WARS
1896–1913

THIRD SCHOOL WAR
The Crusade for Efficiency

BETWEEN THE WARS
1920–1954

FOURTH SCHOOL WAR
Racism and Reaction

Preface

A NEW WAVE of educational historians has in recent years accused the public schools of implicitly accepting inequality in institutional arrangements while mouthing empty democratic rhetoric. The schools, they charge, have pursued a common culture based on a white Anglo-Saxon-Protestant model, ignoring the richness and potential of other cultures; they have perpetuated self-congratulatory legends about American society instead of exposing the persistent malfunctions of the American social and governmental system; they have exalted conformity in the name of harmony; they have prepared lower-class children to take their place in the lower class; in sum, they have served the needs of an unjust class structure instead of promoting equality.

This angry reinterpretation of American educational history is to a large extent a reaction to the one-sided, over-idealized histories which viewed the development of American education as an unfolding series of triumphs, symbolizing the victory of democracy and modernity over aristocracy and error. The revisionist perspective reflects a particular historical moment in which awareness of past injustice to American minorities is painfully acute. But some revisionists tend to be as guilty of excess in their interpretations as were the eulogistic historians of the past, especially when they confuse motivations with consequences. Too often, the revisionists presume that proving a desire for social control is the same as proving the existence of social control. History is rarely so simple. For example, the genteel aristocrats who established New York's earliest free schools may have intended to mold their lower-class pupils into compliant citizens and thereby to secure the status quo, but it didn't work. The status quo did not hold: the patricians of the early nineteenth century were superseded in the city's political and industrial leadership by a new class, some of whom had learned to figure and spell in the repressive atmosphere of the public schools. What seems to bind together the extreme revisionists and eulogists is a common belief that the public schools, apart from any other factors and forces, have the power to transform society.

But despite its flaws, the new criticism of the schools has stimulated a much needed reconsideration of scholarly and popular attitudes towards education. For the fact is that the lofty, rhetorical version of American educational history bears slight relation to the reality of American public schooling. Nowhere is this more true than in New York City.

New York's public school system has a present reputation as a system that was once great, but which has fallen onto bad times. This notion that public schools are not as good as they used to be is fairly widespread throughout the country. If there is in fact a decline in the quality of public schooling, then it is important to know why; if it is not so, then it is equally important to understand why there is a misconception. Have the schools changed for the worse? Or have standards of judgment and levels of expectation risen?

It first occurred to me to read about the history of the New York City public schools during the acrimonious, discordant months of the 1968 teachers' strikes. The strikes left many people, myself included, perplexed about the direction of public education. It seemed to be common knowledge that the public schools had worked wonders for poor white immigrant children in the early decades of the twentieth century. Yet it was clear that black and Puerto Rican children were not succeeding in the present-day system. Why had the schools "succeeded" in one period and "failed" in another? Was the system hopelessly racist? What had caused the centralization of a massive system with more than one million pupils? Did the schools have better teachers fifty years ago? Had community control of public schools ever been tried? Would community control solve the academic problems of the minority pupils?

In my search for answers, I discovered that the only histories of the school system in this century had been written under the auspices of the Board of Education. One was written in 1905; the other spanned the period from 1898 to 1948. Neither answered my questions. The first was inadequate because it surveyed the history of the school system, including the initiation of centralization, without explaining the political and social context in which the changes were made; the second began after centralization was a fact and ended before the problems of integration and community control arose.

After additional research, I learned the reason why there has never been a comprehensive history of public education in New York City: the secondary material about the schools is thin, while the primary materials are voluminous. How should one organize the history of an institution fraught with controversy into a meaningful form? The two earlier accounts conveyed the history on a decade-by-decade basis. This, however, seemed arbitrary, since not all decades were of equal importance and interest. Some recent books about the public schools have analyzed a single issue, but that approach leaves aside the evolution of the entity of public education.

At the end of several months' research, I began to see a pattern, to realize that each major reorganization of the school system was the result of intense political struggle, and that each of these battles coincided with a huge wave of new immigration. Though the issues in each instance were different, the public school was the battleground where the aspirations of the newcomers and the fears of the native population met and clashed. I realized then that a history of the New York City public schools had to be a political history, for no other perspective was broad enough to capture the colorful controversies that have shaped the schools through the years.

The struggle in 1968 between the teachers' union and militant blacks was but one more of these head-to-head, highly emotional contests for control of the school system. Following the phraseology of Nicholas Murray Butler, a prominent school reformer of the 1890s, I have called them "great school wars." In each case, the role and purpose of the public school were bitterly disputed by intense and hostile factions. The political system of the city and state, which usually aims to compromise differences and pacify discontent, on each occasion engineered a political solution which satisfied both wants and fears, terminating the contest.

The issues at stake in New York's school wars were not purely local questions. Just as the Scopes trial represented far more than an interesting episode in Tennessee school history, so New York's major school controversies dramatized significant national issues. New York's crises anticipated the emergence of similar trends in other parts of the country. The problems of New York City—integration, decentralization, community control, the church–state issue, equality of opportunity—are the problems of other American cities on a magnified scale; because the stage is larger and the confrontations are more vivid, the issues appear in sharper focus than elsewhere. The participants are New Yorkers, but the issues are American.

Not only is each issue of national interest, but each is a recurrent problem in American education. Though specific political contests were resolved by political settlement, the underlying philosophical disputes remain and recur at different times and in different parts of the country. The problem of whether to grant public funds to religious schools, for example, is as controversial today as it was in the 1840s and as relevant to the rest of the nation as it is to New York City. Each school war is characterized by the combat of principle against principle, of one set of rights against another, of the strongly held interests of one group against those of another. For these reasons, none of these issues is ever permanently settled.

The first great school war grew out of the Catholic clergy's desire to get public funding for their schools. In the city of New York in the 1840s, the great question of the proper relationship between church and state was hotly disputed. Abstractions gave way to personalities as the tough and aggressive bishop of New York allied forces with the idealistic governor of New York to destroy the upper-class corporation which managed the public schools. When the struggle was concluded, the city had a public school system, publicly funded and publicly controlled by elected local school boards, an outcome which none of the participants had sought. The Catholics won their battle to disestablish the private society which monopolized public education funds, but lost their war for a share of the money and determined to build a separate system of their own.

Half a century later, the public schools were once again plunged into controversy as school reformers demanded modernization, centralization, and professionalization of the system. The reformers crusaded to eliminate local lay boards and to replace them with a strong centralized bureaucracy. Centralization was initially construed as a means of reforming the schools by getting the politicians out of education and putting educational experts in charge. But, as usual, more was involved than educational reform. The reformers were white Protestants reasserting their dominance over the city's educational system, seeking to shape the rude immigrant masses in their own image. In the pursuit of assimilation and Americanization, alien languages and cultures were disregarded, even disparaged.

Another aspect of the centralization movement was its emphasis on businesslike efficiency. Growing out of admiration for the accomplishments of American business, the "cult of efficiency" developed as a national movement on its own and returned full-force to New York City in the second decade of the twentieth century. The cause of the efficiency expert was taken up by upper-class reformers, who lobbied energetically for the adoption of the "Gary plan" in the public schools, a space-saving, money-saving plan, which had been praised by John Dewey and other progressives. The Gary plan, also called the "platoon system," appeared to be destined to transform the American public school until it came to a disastrous end in New York City. Suspicious immigrant parents, unimpressed by its glowing prospectus, were repelled by the reformers' portrayal of the school as an efficient factory; they helped to oust the reform mayor who had imported the Gary plan.

By the 1940s, the city's schools were populated by second- and third-generation children of European ancestry; the schools had not conquered the problems of first-generation immigrant pupils but rather had survived them. Not realizing this, educators imagined that the schools were at the top of an evolutionary spiral of social progress, since the crises associated with immigrant absorption were in the past. It came as a rude surprise, therefore, when a new immigration crowded the schools in the decades following World War II.

The first response to black immigration was the school integration movement, which was first stymied by white resistance to forced transfers, then stopped dead by the fact that black and Puerto Rican children had become a majority of the school register. The next effort to make a place for the new immigration in the system was the militant blacks' community-control movement. Once again the "outs," now blacks, made war on the "ins," who were school professionals. The blacks' demand for community control echoed the separatism of the Catholics in the 1840s. The "great school war" of the 1960s came to an end in 1969 with the passage of a law decentralizing the public school system.

While New York City learns to live with its decentralized school system, the rest of the country continues to debate integration, decentralization, community control, and the education of poor children, issues which were fully aired in New York's last great school war. These questions were not conclusively answered, any more than earlier school wars settled the religious issue or the tension between centralization and decentralization. The issues live on, because the tensions that produced them are woven into the fabric of American political, social, and cultural life.

Acknowledgments

ONE of the lessons I learned from writing this book is the importance of inadvertence and unintended consequences. A series of accidents and chance encounters carried me from my original intention of writing an article about the genesis of centralization in the 1890s to the eventual five-year labor that resulted in the present work. I am indebted to many who assisted in that process: to Peter Shrag, who as an editor at *Change* magazine encouraged my desire to convert what began as an article into a larger study and introduced me to Harris Dienstfry, who as an editor gave me moral support during the first years of my research; and next to Margaret Mahoney and Barbara Finberg, both then at the Carnegie Corporation, for directing me to Lawrence Cremin at Teachers College. Dr. Cremin, I soon discovered, was not only a great and humane scholar of American educational history but a man who confounded me by his unstinting generosity, interest, and commitment.

During the writing of this book, there were many people who kept me going: Rita Kramer, the first person to read the manuscript, many drafts ago; Dorothy Dubin and Anita Shefrin, who minded the store; Grace Hechinger, whose sense of balance often propped me up; Catherine Lochart, Marian Hudgins, Agatha Joseph, Barbara Stonecipher, Maria Fuentes, and Judy Skaggs, who at various times kept my home and family together while I was in the library.

I owe special thanks to those who opened their files and correspondence to me and to the many people who permitted me to interview them, as well as to those city and state officials who turned over to me confidential documents—documents which should never have been confidential in the first place, since the usual reason for the secrecy was that the papers contained unpleasant truths or unpopular opinons.

I am grateful to the librarians of the New York Historical Society, which was my haven for many months, the New York Public Library, the New York Society Library, the Columbia University Library, and the Teachers College Library. I owe thanks to Judith Klugman, for her sensitive and intelligent photo research; to Dr. Edward Solomon of Morningside Heights, Inc., for providing me access to the Columbia li-

xvii

braries; and to my friends and associates at the Institute of Philosophy and Politics of Education, Teachers College, whose perceptions helped to broaden my understanding of American education.

For their humor, patience, and sharp criticism, I especially thank Paul Neuthaler, my editor, and Erwin Glikes, my publisher.

I am most deeply indebted to my husband, Richard, who has been an invaluable source of encouragement, wise counsel, and sharp insight; I have been strongly influenced by his profound knowledge of the workings of the political process. And I thank my sons, Joey and Mike, who shared their mother's time and attention with more understanding than I ever expected. For their love and their willingness to bear with me during these years, I am grateful.

DIANE RAVITCH

FIRST
SCHOOL
WAR

Public Schools or
Catholic Schools?

CHAPTER 1

Early New York:
Social Conditions
and Schools

In the beginning of the nineteenth century, New York was a bustling, thriving seaport of 60,000 people. Almost all its inhabitants lived south of Grand Street, near the docks and wharves which were the source of the city's vigor. North of Grand Street were farms, scattered houses, rocky ledges, rolling hills, swamps, fields, and forests. Almost every year the town's northern boundary was pushed outward by its expanding population, which by 1805 had grown to 75,000. As the town became a city, the natural formations which characterized it were one after another eliminated by "the leveling hand of improvement." A contemporary guide book observed that "hills have been dug down, and swamps have been filled up. Knolls have been pared away, and gullies brought to a level." Within a few years it would be forgotten that the irregular terrain of southern Manhattan had once included a large fresh water lake and numerous hills.[1]

Because of its superb harbor, New York was the leading commercial center of the new American nation. The city's economic vitality permeated its social life. The aristocracy, such as it was in a republican city, included the wealthy descendants of old Dutch and English families, as well as merchants who had recently made their fortunes in shipping, importing, and exporting. Many of the town's prominent citizens were not natives at all, but had migrated from New England. Though travelers and seamen from all over the world gave the town a cosmopolitan air, New York had not yet experienced any massive foreign immigration. The large-scale influx of poor Irish did not begin until 1818.

Until then, New York's population was comparatively homogeneous, being mostly native-born white Protestants of Dutch or English descent. Of some 4,000 blacks, almost half were slaves in 1800 (slavery was gradually abolished between 1799 and 1827 in New York State).

Even after New York had developed into a sizable city, there was little notion of community responsibility for vital public services; many critical public needs were either ignored or left haphazardly to private enterprise. Except for the main thoroughfares, the streets were cleaned only once a month. Because of the infrequency of garbage collection, many city dwellers dumped their garbage into the roads. Most foreign visitors to the city wrote disapprovingly of its filth and of the pigs which roamed about at will.

Inadequate sanitation was not only an esthetic problem but a serious health hazard, which contributed to the frequent epidemics of yellow fever and cholera. The spread of these dread diseases was also caused by the city's abysmal water supply. The city's water was drawn from public pumps in crowded neighborhoods and, according to contemporary reports, tasted terrible. Most of these wells were probably polluted. A private corporation received a charter in 1799 from the legislature to supply the city with pure water, but supplied only a few hundred houses with what was still, even then, considered poor water. Not until the Croton Reservoir was opened in 1845 did the city have good and sufficient water.

The unreliable supply of water impaired the city's ability to fight fires. Fires and epidemics were the greatest threats to the life of the city and its people. Almost every night there was a fire, and every few years a disastrous fire consumed major business and residential sections of the city. The city's defense against fire was weakened by its reliance on volunteer fire-fighters. Men joined volunteer fire companies for social and political reasons. The companies competed fiercely to arrive first at a fire and to gain possession of the hydrant; very often the firemen battled each other while a building burned down. Not until 1865 was a professional fire department created, despite the community's desperate fear of fire.

Although the port city had many rowdies and toughs, many of them organized into vicious gangs, police protection was still primarily voluntary. Each of the city's wards elected two constables annually, who served during the day; at night, a citizens' watch was relied on to maintain order. When brawling gangs took over the streets, the mayor had to call on the military to restore peace. A professional police department was at last established in 1844.

The city was governed in the early years of the century by an appointed mayor and an elected legislature, the Common Council. A Council of Appointment, consisting of the governor and four state senators, selected the mayor and other leading town officials until 1834, when the mayoralty became an elective office. The Common Council was composed of aldermen and assistant aldermen; it passed laws, set taxes, laid out roads, granted franchises, and approved contracts for city services. All free male inhabitants over twenty-one who either owned property or paid taxes were permitted to vote in elections for aldermen, tax assessors, constables, and tax collectors; the property restrictions on voting were removed in 1821.

For the wealthy, early nineteenth-century New York was a beautiful and elegant city. Broadway and the Bowery, the two finest avenues, were lined with fine houses and shops. The handsome red brick houses of Broadway were owned by the gentry, who emulated their European counterparts in manner and dress. British visitors wrote that the shops which lined Broadway displayed as much variety in their wares as could be found in London. In times of epidemic, the well-to-do moved out of the city until the threat had passed.

For the poor, living in New York was an altogether different experience. The city had a conscience, but it was a private conscience, not a public one. Public services for the poor were minimal. True, the city provided an almshouse for the poor; but the commissioners of the almshouse were empowered to compel its several hundred inmates to work as apprentices or servants. A debtor, no matter how small his debt, might be sent to jail to languish at the whim of his creditor. The jail provided neither food nor heating fuel to its prisoners; that charitable duty was left to such organizations as the Humane Society. Poor widows with children could apply to the Ladies' Society for the Relief of Poor Widows with Small Children; these benevolent ladies would investigate the applicant's character and morality and grant her assistance only if she agreed to put her small children in a charity school and send her older children to work.

Slum areas began to develop as early as the 1790s. Though the town of New York had not yet experienced any large-scale immigration, native New Yorkers already associated immigration with crime and poverty. There was a reality to their attitude, since newcomers were more likely to be poor than the native-born, and most criminals were reported to be foreigners. There is no more constant theme in the unfolding of New York City's history than the reciprocal relationship between the native and the immigrant: the immigrant arrives poor, lives in crowded slums with others like himself, suffers discrimination and terrible living conditions, and (as a group) produces a disproportionate number of criminals and paupers; the native blames the immigrant for bringing crime, poverty, and slums to the city, discriminates against him, and wonders whether *this* particular group can ever be assimilated into American society. With each major wave of immigration—Irish, Italians, Jews, blacks, Puerto Ricans—the scenario has been replayed. And in each instance, the cultural clash of the old and the new has occurred in and around the school.

SCHOOLING in turn-of-the-century New York City was altogether in keeping with the private philanthropic tradition of minimal public services. Since 1664, when the English gained control of the town, there had been no provision for free public education. Several churches maintained charity schools for the children of their congregations. Families with means hired tutors or sent their children to private pay schools. Private-venture schools were patronized by the middle class as well as the rich. Some pay schools were within the means of workingmen, but schooling arrangements were uncertain and haphazard. Those children whose parents lacked either means or church membership received no schooling. The transiency of formal schooling was not a critical problem, because most education was then informal: many children learned to read at home by the fireside and learned a craft as apprentices.

Two private efforts to provide schooling for some of the children who were left out of existing arrangements were launched during this period. The city's first nonreligious free school was opened in 1787 by the Manumission Society on behalf of black children. The Manumission Society was formed by such prominent men as John Jay and Alexander Hamilton, for the purposes of "mitigating the evils of slavery, to defend the rights of the blacks, and especially to give them the elements of education." The school began with twelve pupils and grew steadily each year; in 1794 it was incorporated as the African Free School. Then, in 1801, a second free school was opened, a school for poor white girls established by the Female Association, a group of Quaker women. Both these schools were private attempts to fill what was not yet recognized as a public need.[2]

That New York City did not have free public education at this time was in large measure due to the religious heterogeneity of the populace, who belonged to many differing, mostly Protestant, sects (in 1800, New York was more than 95 percent Protestant). Education had been traditionally associated with religion; where there was great religious diversity, there was no impulse towards establishing communal schools for all. By contrast, the New England states, which were settled primarily by one sect, the Calvinists, made an early start on communally supported schooling. The Calvinists, or Puritans, were zealously devoted to general education as an article of their faith; in Puritan Massachusetts, laws were passed in 1642 and 1647 which required that all children learn to read and that all towns of 100 or more householders establish elementary and secondary schools. Thus, when the New England states passed laws requiring the maintenance of tax-supported district schools soon after the American Revolution, they were confirming established practice. New York State had no such tradition to build on.

The task before the state of New York at the opening of the nineteenth century was the creation of a mechanism to encourage the development of public education. The state began to lay the groundwork for its common school system in 1795, when the legislature appropriated $50,000 annually for five years as matching funds for towns which set up their own schools. Within a few years, more than 1,000 schools with almost 60,000 pupils had opened across the state. In 1805, the state earmarked a permanent school fund to support common schools; the first apportionment from this fund was made in 1815 when sufficient monies had accumulated. With a relatively small investment at the outset, the state had nudged localities into setting up free public schools.

In New York City, however, the state money from the Act of 1795 was not used to create free public schools, but was released instead to the support of the ten existing church schools and the African Free School. The state and city evidently felt that it would be too expensive to create a whole new network of schools for its largest city. Thus, while other parts of the state were making a start on their public schools, New York City was not. The size and heterogeneity of the city, as well as its constantly shifting population, diluted any sense of community responsibility for education. Parents determined how much schooling their children received. Carl Kaestle has estimated that 52 percent of the city's children attended "some kind of school for some time" in 1795–96.[3]

Because of the limited availability of free education, many poor children were not receiving any schooling and, after the distribution of the state funds among the existing schools, it was clear that no improvement in the situation was in prospect. Illiteracy among the poor was nothing new. Like other cities, New York always had a large number of illiterates in its population. But in the early nineteenth century, a new urgency was attached to the spread of education. Since the Revolution, leading statesmen had expressed their belief that illiteracy was a threat to the high ideals of the new nation, that a democracy with broad suffrage must have an enlightened electorate. George Washington, in his farewell address as president, had urged the people to "promote . . . as an object of primary importance, institutions for the general diffusion of knowledge." Thomas Jefferson advocated the universal extension of education at state expense, in the hope that an aristocracy of intellect would replace an aristocracy of wealth. John Jay, James Madison, John Adams, and others wrote forcefully of the importance of popular education in a republic. Others sought to end illiteracy as a means of stamping out vice and poverty and generally improving social conditions. Men with strong religious convictions advocated the spread of education in order to enable all men to read the Bible and attain salvation.[4]

A combination of these attitudes set the stage for the first broad-scale effort at popular education in New York City. In 1805, John Murray, Jr., and Thomas Eddy, both wealthy Quakers, agreed that something had to be done about the large number of children in the city who were not receiving any schooling. Inspired by the success of the free school for girls established by the Female Association, Murray and Eddy wanted to create a similar school for poor boys. Both were men with established fortunes who devoted their leisure time to humanitarian and civic causes. Murray's family owned Kips Bay farm on Murray Hill; his father and uncle were the chief ship-owners in the colonies. John Murray, Jr., was active in the Manumission Society and the Humane Society, served on New York Hospital's Board of Governors, and participated in the relief activities of St. Patrick's Society, an organization that aided poor Irish immigrants. His wife too was an active Quaker, and it was in their home at 335 Pearl Street that the Female Association had been organized. Like Murray, Thomas Eddy was a conscientious citizen, interested in reform of prison conditions and the penal code, an officer of the Board of Governors of New York Hospital and a commissioner of the city almshouse.

On February 19, 1805, Murray and Eddy invited ten other men to a meeting at Murray's home on Pearl Street to discuss the need for a free school for poor boys. Among them were Matthew Clarkson, William Edgar, and Thomas Pearsall, three of New York's wealthiest men. Clarkson, a neighbor on Pearl Street, was president of the Bank of New York, a regent of the State University, president of the Board of Governors of New York Hospital, a trustee of the New York Society Library, a director of the City Dispensary, and a layman of Trinity Church. Eight of the twelve men at the meeting were part of the city's business elite;

four of them were Quakers. (Though there were only 160 Quaker families in New York in 1805, their influence in the business sector and humanitarian activities far exceeded their numbers.)

The dozen men assembled at Murray's home agreed to support this new philanthropy; the group applied to the state legislature for an act of incorporation, which was promptly granted, recognizing them officially as "The Society for establishing a Free School in the City of New York, for the education of such poor children as do not belong to, or are not provided for, by any religious society." The new organization was soon renamed the Free School Society. Religious domination of schooling was so openly conceded that the Society asked permission to educate *only* those who were not in a religious school. The Free School Society's first president was DeWitt Clinton, then mayor of New York, a gesture which signaled the Society's semipublic nature. As in the Manumission Society and the Female Association, the dominant influence in the Society was Quaker, which explained both its social concern and its emphatic nonsectarianism.

The Society's nonsectarianism proved to be its greatest strength and its fatal flaw. Because of its nonsectarianism, the Society was able to make a stronger case for public funds than the church schools and eventually to establish its schools as *the* public schools. Yet its nonsectarianism, which was in reality nondenominational Protestantism, repelled conscientious Catholics, especially the clergy. At that point in American history, to offend a man's religion was to offend his entire culture and identity; to Catholics, nonsectarianism was *by definition* Protestantism and necessarily offensive. Since the great immigration of the 1820s and 1830s consisted mostly of poor Irish Catholics, the Society had the least success in drawing the poorest class into its schools. Thus, the irony: The institution created to eliminate illiteracy, criminality, and poverty could not bring under its influence that group which most needed its benefactions. As the Society became more assured and the Irish grew in number, it became inevitable that there would someday be a confrontation between the insiders and the outsiders, the natives and the immigrants.

The Society was a private organization composed of upper-class philanthropists, men who shunned partisan politics but helped to found New York's great libraries, hospitals, and museums. Because of the trustees' social and political connections, they were able to convince the city and state governments to assume financial responsibility for the education of all children; because the trustees were themselves prominent citizens, they could persuade the business community to accept taxation for

public education. But because of the trustees' social prominence, the Society never developed strong popular support. Historians Kaestle and
Michael Katz have described it as a waning elite using its schools to control the lower class.

In its early years, the trustees saw schooling as a bounty which they
were graciously making available to unfortunate members of the community, just as other boards had undertaken to provide charity hospitals. Their overt condescension and their patronizing concentration on
the morals and habits of their pupils gave way over the years to a
different conception of schooling. In time, the trustees came to assert
that lack of education kept some men ignorant and poor and that
schooling was therefore a right of free men. This transformation in their
rhetoric caused the trustees eventually to espouse common schooling
for all children, rich and poor, and to demand full public support for
their schools. The evolution of the Society's rationale for schooling
reciprocally affected public attitudes; the idea, nourished by the Society,
that education was a democratic right and a *public* responsibility eventually contributed to the Society's downfall.

THE FREE SCHOOL SOCIETY's first appeal to the public for funds stressed
the duty of society to remove children from "the contagion of bad example," so that instead of becoming "the burden and pests of society," they
would grow to be useful members of the community. The Society believed that it was better to fight crime and irreligion "by the salutary
discipline of early education than to make laws punishing crime. Their
proposed Free School would permit the "opulent and charitable" to employ their "influence and aid . . . to diminish the pernicious effects resulting from the neglected education of the children of the poor," which
was attributed to the "extreme indigence" or intemperance of parents.[5]

The first public school in New York City, then, was conceived of as
an antipoverty program, an institution that would directly counter the
bad influences of the poor home. The Free School Society felt that the
community had to educate poor children in order to safeguard its own
welfare. Its trustees did not want New York to develop a large pauper
class, as was common in big European cities. They hoped to abolish
poverty and crime by making education universal. Their aims, on the
one hand, were conservative, because of their clear desire to preserve
and stabilize society; yet at the same time the benevolent founders of
the Society sought to better the lot of the poor by eliminating the cause
of their poverty. The trustees did not view poverty as endemic or the re-

sult of innate inferiority, as did many patricians of their generation, but viewed it rather as the consequence of ignorance, which was remediable. Given the customary public indifference to the squalor and illiteracy of the poor, the Society's concern for them, whatever the motive, represented an ameliorative impulse.

Implicit in the proposal of the Society was the European two-track system of education, one for the rich, another for the poor. Still, a two-track system was better than a system which made no provision for the poor at all. As a private philanthropy acting on behalf of the public weal, the Society had no difficulty getting financial aid from city and state governments, as well as private contributions.

The thirteen trustees of the Free School Society had no ambitions beyond a single school, but their small school for poor boys was destined to be the forerunner of the New York public schools. The first school opened May 19, 1806, in a small apartment on Bancker (now Madison) Street. Attendance in the first year was seventy. The Society adopted the Lancasterian, or monitorial, system, which one of its trustees had observed in operation in England. This system had achieved fame in England for its economy and efficiency, as well as for its nonsectarian approach to religious instruction. The system enabled one teacher to conduct a school of 1,000 students by using students as monitors.

In little time, the Free School Society succeeded in committing both the city and the state to the support of its program and in expanding the number of pupils eligible for schooling by the Society. In 1807, the trustees received funds from the state legislature, which was doubtless delighted to see a private group providing schooling for the poor in the city. The trustees agreed to educate fifty children from the city almshouse; in exchange, the city gave the society the funds to convert a building next to the almshouse into a school. In 1808, the legislature enlarged the Society's scope to include all poor boys, not just those who did not belong to a religious society.

The first new building of the Society was erected on land donated by the city, at a total cost of $13,000. It was a two-story red-brick building, capable of accommodating 500 children in one classroom on the second floor. On the first floor were an apartment for the family of the teacher, a meeting room for the trustees, and a classroom for another 150 pupils, which was used by the Female Association's girls' school.*

---

* The schools of the Society were for boys only until 1845, when the Society absorbed the schools of the Female Association; for many years afterwards, most classes were segregated by sex because of parents' objections to coeducation.

The school's public dedication, on December 11, 1809, was a festive event. The main speaker was the distinguished president of the Society, DeWitt Clinton. Clinton lavishly praised the Society and the system it introduced to the United States:

> When I perceive that many boys in our school have been taught to read and write in two months, who did not before know the alphabet, and that even one has accomplished it in three weeks—when I view all the bearings and tendencies of this system—when I contemplate the habits of order which it forms, the spirit of emulation which it excites, the rapid improvement which it produces, the purity of morals which it inculcates—when I behold the extraordinary union of celerity in instruction and economy of expense—and when I perceive one great assembly of a thousand children, under the eye of a single teacher, marching, with unexampled rapidity and with perfect discipline, to the goal of knowledge, I confess that I recognize in Lancaster the benefactor of the human race. I consider his system as creating a new era in education, as a blessing sent down from heaven to redeem the poor and distressed of this world from the power and dominion of ignorance.[6]

A few years later, DeWitt Clinton, then governor of the state, compared the Lancasterian system to labor-saving machinery, both in its efficiency and in its revolutionary impact. In a time when universal education seemed still to be a radical, visionary scheme, the system made it appear attainable. Private-venture schools continued to enroll a far larger number of children than did the schools of the Society, but it was the Society that was making education universal. As the Society increased the number of its schools, thousands of poor children received an education under its auspices. In 1817, annual attendance exceeded 1,000; in 1834, it reached 12,000; by 1839, more than 20,000 children were enrolled in the Society's schools. It was the Lancasterian system that made this "remarkable progress" possible, not only in New York City, but in England, Scotland, Ireland, and other European countries; furthermore, groups in other American cities followed the lead of the Free School Society and opened Lancasterian schools.

What was so amazing about this system which caused statesmen to consider it a heaven-sent blessing? Two things: first, it was a fully packaged system that was known to work; and second, it was cheap. Its efficiency made the system attractive; its economy made it irresistible. The system left nothing to chance in the operation of the school. It was not only analogous to an industrial machine, it *was* a machine, the school machine. Each part of the machine had a role to perform, which was spelled out in minute detail in manuals. Joseph Lancaster, the English Quaker who developed the system, was responsible for the mottoes "A place for everything and everything in its place" and "Let every child at every moment have something to do and a motive for doing it." ° The boys moved about the school in a prescribed, measured pace. To speed the morning roll call, the students were each assigned a number which was posted on the wall; every morning, each boy stood under his number, enabling the attendance monitor to ascertain with a glance who was absent.

The teacher of the school was almost a silent bystander. There was little for him to do "except to organize, to reward, to punish and to inspire." The running of the school was parceled out to various monitors, who were the oldest students. Monitors assigned new students to their class; taught reading and arithmetic lessons; took attendance; gave examinations; promoted deserving students; ruled the writing paper; took care of pens, paper, slates, and books. And there was a monitor who looked after all the other monitors.[8]

Though the overall system was highly mechanistic, the plan of grading and promotion was flexible. Pupils were classified separately for reading and arithmetic and promoted separately in each subject.† There were eight reading classes, representing specific levels of ability. Each class had a clearly defined block of material to master. The first, or youngest, class learned the alphabet; the second class learned words and syllables of two letters; the third class learned words and syllables of three letters; the fourth class learned words and syllables of four letters; the fifth class studied reading lessons of one-syllable words; the sixth class studied reading lessons of two-syllable words; the seventh class studied the Testaments; the eighth class studied the Bible. The most advanced class was also permitted to read books from a list approved by the Society's Board of Trustees.

Arithmetic was similarly divided into ten classes, each learning a single rule of arithmetic. The first class learned to make and combine fig-

° An English biographer of Lancaster wrote that he "was a man who, if he had only found out for himself the advantage of tying a knot on the end of a thread, would have proclaimed aloud that he had made an original discovery destined to regenerate society, and would have elaborated a complete scheme of knots for different threads or different kinds of sewing." [7]

† Years later, when the Lancasterian system was abandoned, this practice was replaced by the graded system, in which an entire class was taught all subjects on the same level; in the mid-twentieth century, educational reformers urged the adoption of ungraded classes, in part a reversion to that aspect of the Lancasterian plan which advanced pupils according to proficiency in separate subjects, rather than age.

ures; the second learned simple addition; the third learned simple subtraction; the fourth learned simple multiplication; the fifth learned simple division; the sixth learned compound addition; the seventh learned compound subtraction; the eighth learned compound multiplication; the ninth learned compound division; and the tenth learned the higher rules of arithmetic.

Since virtually all the teaching was in the hands of the monitors, a manual described each detail of the teaching process precisely, leaving nothing to the imagination of the monitors or pupils. Imagination, in fact, was actively discouraged, since it might cause a pupil to offer answers that were not prescribed. Submission to authority was a necessary part of the machine, which worked perfectly only if all the cogs played their preordained parts. Learning was synonymous with memorizing, and teaching was little more than mastery of the material to be memorized.

Three teaching techniques, all based on memorization, were used to teach reading and spelling. The first of these was dictation. The pupils sat in long rows called forms, which corresponded with the eight reading classes; each form had its own monitor. The monitor in charge of dictation, appropriately called the "dictator," would move from form to form, giving each class a syllable or word on its level of ability. The monitor of each form would repeat the word for his pupils, who then wrote it on their slates. After the dictator gave each form six words or syllables, he would then order all eight classes to prepare for inspection; at that signal, each boy would put his hands behind his back, and the class monitors would inspect their pupils' work. The dictator would then give an order to clean slates, the monitors would check again to see that the slates were indeed clean, and the process of dictation would begin again.

After dictation, teaching would proceed by a method called "reading from boards." Groups of eight or nine pupils would stand in semicircles around lesson charts suspended from the wall. These groups were called "draughts," because children were "drawn out" from their classes for this activity. The monitor would direct his pointer to the printed lesson, and the pupils would read in turn until a mistake was made. The next pupil would correct the mistake and change places with the previous reciter. The second boy would read until corrected by the third, and so on. When a student mastered the lessons of his reading class, his monitor would recommend him for promotion to the next form.

The third method of teaching described by the manual was the spelling test, conducted by the draught monitor with the lesson board out of view. As in reading from the boards, the students competed with each other and changed their relative positions in the group according to the correctness of their answers.

The youngest children were taught the alphabet by drawing the letters in sand. A monitor would stand before them with an alphabet stand, or wheel, on which appeared all the letters. He would say to the pupils, "Prepare," to get their attention, then point to the letter A and say, "Print A." He would then say, "Hands behind," and examine their work, correcting mistakes and smoothing the sand before them for the next letter. Beginners were seated next to students who had already made some progress.

Arithmetic too was taught by dictation and draughts. The monitor would dictate several numbers, then tell the boys how to add them. He might say, "eight and four are twelve, and six are eighteen and five are twenty-three," and so on. At the draughts, the boys dictated to themselves, writing down examples, which were then examined by the monitors ("If a boy commit an error, or deviate from the prescribed words in the smallest degree, his error is corrected, and he forfeits his place"). The boys would write and repeat equations over and over, until they could recite them by reflex. The monitors would quiz the boys in draughts by asking questions to elicit precise answers: "One and one are?" "One and one are two." "Two and one are?" "Two and one are three." At the end of each day's draught in arithmetic, the monitor general presented each first boy with a reward of one ticket.

The system of rewards in the Society's schools was based on a currency of pasteboard tickets with a value of one-eighth of a penny. They could be redeemed for toys. Tickets were distributed daily to all of the monitors for performing their duties; the highest-ranking monitor, called the monitor general, received eight each day, while other monitors received fewer. Each day, the head boy of each draught was rewarded with a ticket. On promotion to a superior class, a scholar received twelve tickets. Each boy who attained the eighth class won fifty tickets. Each teacher had discretionary power over 1,000 tickets per month as special rewards for the deserving.

To discourage corporal punishment, which the Society strongly disapproved, students were fined tickets for various forms of misbehavior. The list of punishable offenses, like everything else, was explicit: [9]

OFFENSE	FINE IN TICKETS
Talking, playing, inattention, out of seats, etc.	4
Being disobedient or saucy to a monitor	4
Disobedience of inferior to superior monitor	8
Snatching books, slates, etc., from each other	4

OFFENSE	FINE IN TICKETS
Monitors reporting scholars without cause	6
Moving after the bell rings for silence	2
Stopping to play, or making a noise in the street on going home from school	4
Staring at persons who may come into the room	4
Blotting or soiling books	4
Monitors neglecting their duty	8
Having dirty face or hands, to be washed and fined	4
Throwing stones	20
Calling ill names	20
Coming to school late, for every quarter of an hour	8
Playing truant, first time	20
Ditto, second time	40
Fighting	50
Making a noise before school hours	4
Scratching or cutting the desks	20

The system of rewarding and punishing by tickets was abandoned in the 1830s. The trustees came to feel that they were more often rewarding the cunning than the meritorious; some boys even resorted to stealing tickets. The trustees decided that the ticket system "fostered a mercenary spirit" and "engendered strifes and jealousies." In its place they issued a handsome certificate, testifying to the trustees' commendation of the pupil. These certificates, which were helpful in getting employment, were distributed semiannually "with appropriate formalities." Though the trustees frequently voiced their disapproval of corporal punishment, it was a traditional form of discipline which they were powerless to abolish by fiat. In the mid-1840s, the Society awarded testimonial certificates to principals who conducted their schools successfully for one year or more without resorting to corporal punishment. Only a handful of principals qualified.[10]

The pride of the trustees was the economy of their system. The Society was able to maintain the average annual expenditure per pupil at little more than four dollars over their forty-seven-year span of existence. When it was necessary to retrench and cut back on expenditures, the trustees cut teachers' salaries, restricted the use of the library to only the best scholars, eliminated assistant teachers, or withdrew the monitors' grant of board and clothing. To save paper, the youngest classes did their lessons in sand, and the intermediate classes used slates; only the most advanced students used paper. It was to save the cost of books that reading lessons were taught from a chart affixed to the wall. The trustees expected their low-paid teachers to be paragons of virtue and promised them that while they might or might not receive their "merited reward" from men, they would certainly receive "what is infinitely more valuable—the approbation of Heaven." [11]

Each year, enrollment in the Society's schools increased. After 1815, the Society began to receive its share of the state common school fund, which was proportionate to the number of its pupils. A large excess accumulated after the teachers' salaries were paid, since the system required so few teachers, and the surplus funds were spent to build new schools. The Society continued to receive special grants from the legislature and to get a warm hearing, because of its reputation for integrity and thrift. The trustees attended to their management of the schools with zeal. For many of them, the unsalaried position was a nearly full-time job. Each year the fifty to one hundred trustees reported the precise number of personal inspections of the schools in their care, and the aggregate visits usually numbered more than 10,000.

The Society's charter designated the mayor, the city recorder, and all members of the city legislature as permanent ex officio members of the corporation. Though the Board of Trustees was elected by the membership, and membership was open to anyone who could pay the membership fee, it was nonetheless a self-perpetuating board. From its inception in 1805 until its dissolution in 1853, there were thirty trustees who served an average of twenty-five years on the board, and another twenty-five who served an average of fifteen years. While the Society considered this constancy an evidence of selfless devotion, the same facts later lent proof to critics of the Society's undemocratic and monopolistic character.

In the early years of the Society, the trustees' public statements contained an overt tone of condescension towards the pupils and their parents. An 1819 address to the public was typical:

> It may not be improper to state to you, that the establishment of the New York Free School has been attended with much labor and personal exertions on the part of its friends and patrons . . . and as all this is done in order to promote the good of your children, and to improve their condition, you cannot but feel a weight of obligation to the friends and patrons of so valuable an institution. . . . There are diverse other things which we could enumerate as connected with the subject of this address; but it cannot be expected, in a communication of this nature, we should embrace every duty or point out minutely every thing which might have a bearing on your religious and moral character. [12]

The notice detailed what was expected of students in the free schools. They were required to appear promptly each day; the only excuse for

absence was sickness; six unexcused absences in a month would constitute grounds for expulsion; children who were not neat and clean would
be punished; children under six would not be admitted; children whose
parents could afford to pay for schooling would not be admitted; parents were expected to see that their offspring regularly attended some
place of worship.

This tone of reproval softened over the years. As the Jacksonian
rhetoric of democracy became popular, the notion of schooling as a
bounty of the rich became unfashionable. A change in tone was also hastened by the fact that the schools of the Society were operated with
public funds although they were controlled by a private board of trustees. Even so, the trustees never failed to blame ignorant, indifferent, and
neglectful parents for the truancy of their children. It never occurred to
them that the cold, military atmosphere of their schools might be in part
responsible for the refusal of some children to attend, or that some parents might refuse to be objects of charity and of priggish moral uplift.

The trustees believed that their nonsectarianism was synonymous
with democratic tolerance and was superior to the divisive sectarianism
of the church schools. They did not care what religion their charges adhered to, so long as they attended to their religious duties regularly.
Tuesday afternoons at school were set aside for instruction in the cate-
chisms of the various faiths, supervised by an association of female volunteers. On Sundays, the children were required to assemble at school
and were led by their monitors to the church where they belonged. The
trustees incessantly reminded the students that the good life was attainable only through proper moral and religious training.

Each day began with a Bible reading; after 1821, a volume of
"Scripture Lessons" and a "nonsectarian catechism" were included
among the approved schoolbooks. The daily Bible reading was followed
by a religious exercise, a responsive recitation between teacher and students. The recommended exercises were printed in the Society's official
manual. A typical portion was:

TEACHER: Children, who is good?
ANSWER: The Lord is good.
T. To whom should we be thankful?
A. Be thankful unto Him.
T. Whose name should we bless?
A. Bless his name.

A portion of the exercise recommended for younger children was
simpler but equally homiletic:

TEACHER: Therefore, my children, you must obey your parents.
SCHOLAR: I must obey my parents.
T. You must obey your teachers.
S. I must obey my teachers.
T. You must never tell a lie.
S. I must never tell a lie.
T. You must never steal the smallest thing.
S. I must never steal the smallest thing.
T. You must never swear.
S. I must never swear.
T. God will not hold him guiltless that taketh His name in vain.
S. God will not hold him guiltless that taketh His name in vain.
T. God always sees you. (*Slowly, and in a soft tone.*)
S. God always sees me.
T. God hears all you say.
S. God hears all I say.[13]

The trustees believed that scrupulous nonsectarianism, coupled with strenuous inculcation of what they took to be commonly accepted moral and ethical values, would preserve their schools' ability to teach children of all religious groups. They understood that the Free School Society, though founded with limited aims, had become far more than a useful and charitable association. Its schools were free of charge and open to all. Though controlled by a private board which was self-selected and accountable to no one, the Free School Society had come to realize that its schools served a public purpose. They were not the same as the publicly supported, publicly controlled common schools that were developing in other parts of the state, but they were the closest thing to "public schools" that New York City had.

It must be remembered, moreover, that during most of the 1820s, for every child in the Society's schools there were three in private schools and another three in no school at all. For all the trustees' rhetoric, their schools bore the stigma of their pauper origins. It was far easier to espouse democratic ideals, which the trustees learned to do, than to introduce democratic practices, which proved to be impossible.

CHAPTER 2

The Society Expands

IN THE FIRST HALF of the nineteenth century, religion was a vitally important force in American life, and, as such, was the source of bitter conflict among contesting sects. In New York City, where Protestantism was represented by numerous groups, and Catholicism was viewed with hostility, the spirit of sectarianism threatened the harmony of any institution which sought to mix religious factions. Nowhere was this more true than in the schools, since education and the inculcation of religious values were traditionally bound together. The trustees of the Free School Society hoped to stay out of contentious situations by respecting the general values of all sects.

But despite their desire to remain above religious disputes, the trustees were drawn into a series of rancorous controversies with religious organizations. In 1822 and 1832 the Society's monopoly of nondenominational free education was challenged by church schools. Then in 1840, the Roman Catholic Church directly contested the Society's control of public education funds, and began one of the bitterest debates in the history of American education. Though control of public education was the surface issue, the conflict was charged with undertones of religious, ethnic, and class animosity. A pattern emerged which characterized subsequent battles for control of the schools in New York City: since schools by their very nature deal in values, they inevitably became the focal point for cultural conflict, usually at a level of acceptable tension, but occasionally, as in 1840, reaching a point of irreconcilability and forcing profound changes on the school system.

THE FIRST OF THE RELIGIOUS CONTROVERSIES that challenged the Free School Society's monopoly of public education funds began when the Bethel Baptist Church opened a school for poor children of all faiths in

20

1820 and obtained a portion of the common school fund the following year. In 1822 the church trustees obtained the legislature's permission to apply any surplus funds to the erection of new buildings. This was the same privilege which had enabled the Free School Society to grow to an enrollment of almost 3,500 pupils by 1822. The Bethel Baptist Church, by underpaying teachers and overstating expenses, soon had enough public money to operate three free schools. Following their example, other churches opened free, nondenominational schools in competition with the Free School Society's schools.

The Bethel Baptist Church's expansion plans alarmed the Free School Society, which had become the major agent of popular education in the city. The Society tried to have the Baptists' special legislation repealed and urged that church schools be restricted to receive only the children of their own congregations. As the dispute evolved, the Society adopted the position that no public funds should go to *any* sectarian school. The trustees feared that schools built with public funds might later be diverted to sectarian use, unlike the schools of the Free School Society, which were forever dedicated to nonsectarian education and open to all. The Society contended that the common school fund was strictly for civil purposes and ought not to be applied to any religious endeavor. The issue was drawn. The legislature would have to decide whether the state would continue to grant public funds to charity schools operated by churches.

The legislature directed the city's Common Council to take charge of the division of the common school fund within the city and to appoint a commissioner of school money for each of the city's ten wards. Since the Common Council sided with the Free School Society in its bout with the Bethel Baptist Church, this decision was a victory for the Society. Until this time, the church charity schools had received a proportion of the state common school funds. But, in settlement of this dispute, the Common Council passed an ordinance in 1825 denying common school funds to any religious society.

Cutting the church schools out of the common school fund was a major change in the city's school policy. The Free School Society's principle of nonsectarian education received official sanction, and the Society itself gained official recognition as the city's foremost agency for free education. With this step, the concept of a common school, as it was developing in New York City, was clearly transformed: the common school was to be not only free and open to all, but was to be devoid of religious sectarianism.

Having achieved a new stature in the community, the Free School

Society began to think of ways to enlarge its scope and to fulfill better its role as the city's chief purveyor of popular schooling. A committee of the trustees surveyed the city's educational facilities and reported that there were some four hundred private pay schools, many of them kept in small rooms, without regard to proper lighting, adequate ventilation, or cleanliness. These schools were found to have teachers of questionable qualifications, dissimilar textbooks, and varying plans of study, inconsistencies which placed a burden on the family that moved to a different part of the city. The committee noted that those families who sent their children to pay schools were also supporting the common school fund with their taxes and yet derived no benefit from the schools of the tax-supported Free School Society. Believing that many citizens were too proud to send their children to free schools, the committee recommended that the Society increase the number of its schools and admit scholars on a paying basis.

In 1826 the legislature granted the organization a new charter, changing its name to the Public School Society and permitting it to accept all children, regardless of religion or economic level. The Society was allowed to charge a nominal tuition, based on ability to pay, but none were to be refused admission for inability to pay. The mayor and the city recorder were again made ex officio members of the Board of Trustees. The new charter directed that all the Society's buildings and real estate be conveyed to the city, which would then grant the Society a perpetual lease so long as the facilities were used for education. This reorganization transformed the Society into a quasi-public agency, with public funds, public facilities, and public duties, under the immediate supervision of a private, unpaid board.

As the Society's responsibilities increased, so did its ambitions. Counting on the success of the new tuition plan to attract children of the middle class, the trustees discussed among themselves the need for a central high school for the instruction of monitors and advanced pupils. For children over the age of three they set up infant schools which were not conducted along Lancasterian lines.

But their plans for the future were halted by discovery that the pay system was a failure; it had actually led to a decline in attendance. In search of an alternate method of support, the trustees urged the adoption of a real estate tax solely for the benefit of the schools. A careful study of the city's school-age population showed that there were 463 schools (eleven of them belonging to the Society) with about 25,000 pupils. The Society schools had 6,000 students, the charity schools had about 4,000, and another 15,000 were in private pay schools. In addition, there were about 12,000 school "dropouts" under the age of fifteen, and another 12,000 who were without any education whatever. The trustees calculated that the proposed real estate tax would permit them to open three new schools each year, to educate all classes of society without charge, and to open a high school and a teacher-training academy.

At about the same time, the relationship between public education and social equality was a subject of agitation by the radical Workingmen's Party, which was active in the late 1820s in New York City. Led by Robert Dale Owen, Frances Wright, and George Evans, the Workingmen's Party advocated universal education and other reforms to protect workers against the stratifying effects of industrialization. Party leaders feared that industrialization would harden class lines unless workers had equal opportunity to advance through education. The party reached a zenith in 1829 when it elected a carpenter to the state assembly; it lost support because of its leaders' outspoken agnosticism and radicalism, and because of the gradual adoption of many of its demands by the Equal Rights faction of the Democratic party. Robert Dale Owen wrote a widely read series of essays on the significance of universal education in promoting social equality and class mobility. He advocated that children of all classes be fed, clothed, and educated by the state in order to obliterate artificial class distinctions. Workingmen were not prepared to accept such radical egalitarianism, but his assumptions about the relationship between education and class structure found a receptive audience.[1]

Whereas universal education had previously been discussed as a good though visionary ideal, after the 1820s its attainment became inevitable: first, because it was politically popular with the newly enfranchised voters; and second, because it was a widely accepted idea that a society which gives all men the vote must educate them to vote wisely. The election of Andrew Jackson to the presidency in 1828 started the slow decline of the dominant aristocracy of wealth and social position. Political equality forced the emergence of new political patterns, and the schools, like other social institutions, began to adjust to the demands of the rising middle class.

In New York City, where the Workingmen's party had its greatest following, its intellectual leadership, and its newspaper, its impact on the Public School Society was almost immediately apparent. In the trustees' Annual Report for 1830, they responded directly to "the recent public remarks and attempts to create an excitement on the interesting subject of common school education." They defended their schools, saying that they were open to all the children of the poor and laboring classes; they continued to assert that the persistent problem of vagrant children was due to no fault of the Society, but rather to parental neglect.

By the late 1820s, the trustees' attitude toward schooling had changed. No longer did they describe education as a valuable gift which good-hearted patrons bestowed on the deserving poor. When they turned to the community for a real estate tax to subsidize better education, they pointed out that in the entire state, New York City was alone in making no legal provision for maintaining schools other than the state common school fund and the matching city funds. In 1828, the trustees issued an appeal which differed radically from the statement with which DeWitt Clinton launched the Society twenty years earlier:

> It is time for us to pause, and inquire whether this subject has yet received the consideration to which it is entitled, and whether our public schools occupy their merited station among our political institutions. . . . The want of knowledge is the most important of all wants, for it brings all others in its train. If education be regarded as a charity, it is the only one whose blessings are without alloy. . . . It may not be without just cause that, in some other countries, it is considered a dangerous thing to enlighten the people. But with us, the question of their political power is settled—and, if they are true to themselves, it is settled forever. We wish to keep that power in their hands, and to enable them to exercise it with wisdom.[2]

They urged that their schools have the financing necessary to transform them into truly *common* schools, schools which would enroll children of all backgrounds. "We desire to see our public schools so endowed and provided, that they shall be equally desirable for all classes of society." They recognized that to attract children who were not poor, their schools would have to offer the very highest quality education; to achieve this end, they urged an increase in teachers' salaries, to raise the dignity of the position. The enlarged and improved schools which they wanted "should be supported from the public revenue, should be public property, and should be open to all, not as a charity, but as a matter of common right." The ultimate object of the schools, they averred, was to "open to universal emulation the path to all public distinctions." As the decade of the 1820s closed, the trustees had arrived at the view that the public schools were the chief instrument for achieving equal opportunity in society. The fact that this principle was enunciated by a board of sedate aristocrats who were neither elected themselves nor appointed by an elected official is evidence that the idea of a distinctly American public school which served the needs of a democratic society had taken hold as an aspiration, though still far removed from reality.

THE SOCIETY'S REQUEST for a real estate tax was forwarded to the legislature in 1829 with the signatures of 4,000 "of the most respectable citizens" in the city. Since the trustees were themselves leading members of the propertied class, they were well situated to win the support of their fellow property-owners, who would be most affected by a new real estate tax. The legislature only partially filled the Society's request, giving them one-fourth the tax they had asked. Two years later the legislature raised the tax to the original request, but by 1831 the increase was no longer enough to permit the Society to embark on both a building program and educational reforms. The trustees were not able to open a high school or a teacher-training academy, but they did eliminate the pay system altogether. The new tax funds permitted the Society to open primary schools for the youngest children and to improve the curriculum of advanced students, introducing for the first time such subjects as algebra, trigonometry, geometry, astronomy, and bookkeeping.

In the year 1832, the Public School Society had every reason to anticipate that its future service to the community was assured. The trustees felt confident that their schools were held in high esteem by their fellow citizens. They believed deeply in the cause of popular education and were convinced that their schools were breaking down the caste spirit, that "deadly enemy of practical democracy." They had cooperated successfully with the city government and the state legislature. They had expanded even further by negotiating a merger with the schools of the Manumission Society, which had 1,400 Negro pupils enrolled in its six schools.°

° The Manumission Society schools remained separate as "colored schools." Segregated public schools were phased out in the city beginning in 1873, when a state law was passed stating that all public schools were open to all "without distinction of color." In 1883 the Board of Education decided to disestablish the three remaining separate colored schools. One of these schools was closed, but "colored citizens" successfully petitioned the legislature to prohibit the abolition of the other two; those two, which were designated Grammar School 80 and 81, were nominally open to all, but in fact continued to be colored schools. In its report for 1884, the Board of Education stated: "The causes which led to the establishment of colored schools having ceased to exist, except as a matter of history, all legislation with reference to the establishment and maintenance of such schools has thus at last been repealed, and the color line has finally and happily disappeared from our schools." The board noted that only "a small minority" of the system's colored children attended GS 80 and 81 and expressed the hope that the two schools would be assimilated "in practice as well as theory . . . at the earliest practicable date." The city of Brooklyn abandoned its policy of segregated schools in 1890.

After the metropolitan city was consolidated in 1898, the borough of Queens had the city's only colored schools, one in Flushing and another in Jamaica. Several Negro families, led by a Mr. and Mrs. Cisco, refused to send

Consolidation with the Manumission Society further strengthened the position of the Public School Society as the city's quasi-official agency of free education. Therefore, when a new religious controversy arose in the same year as the merger, the Society felt concerned, but not threatened. First the Roman Catholic Orphan Asylum and then the Methodist Episcopal Church school applied for a share of the common school fund. The Society argued against admitting either institution as a recipient in the fund. The trustees particularly objected to the sectarian nature of both applicants and felt that the entire question had been decided by the Common Council in 1825, when it ruled that public funds could not go to any sectarian school.

The Catholic Orphan Asylum maintained that the New York Orphan Asylum, which shared in the fund, was commonly called the Protestant Orphan Asylum and that the Catholics were entitled to receive an equivalent portion. It further contended that an orphan asylum was both home and school and that to deny it a share in the fund would be to withhold the benefits of education altogether. Despite the Public School Society's objections, the Common Council approved the Orphan Asylum's petition and admitted it to participation in the common school fund. At the same time, it denied the Methodist Episcopal request, which had come from a conventional church school. While the Public School Society was not pleased by the granting of public funds to a sectarian institution, it was not deeply disturbed: the money was going specifically to an orphan asylum, and there were no others to take advantage of the precedent. What the trustees did not know and had no way of foreseeing, was that the 1832 controversy with the Roman Catholics was but a shadow of the trouble to come. For, as many thousands of Irish immigrants entered the port of New York during the 1820s, a new and angry population was building which shunned their schools, despised their patronizing manner, rejected their Protestantism, and viewed them as an extension of the hated English masters from whom they had fled.

---

their children to the colored school in Jamaica and sought a writ of mandamus compelling the school authorities to admit them to white schools. All the courts, including the Court of Appeals, sustained the school officials who assigned colored children to separate schools. As a result, Governor Theodore Roosevelt had the legislature pass a law abolishing colored schools in the state. The schools in Flushing and Jamaica were probably the last colored schools in the state, with the exception of one in Hempstead, Long Island.[3]

CHAPTER 3

The Irish Arrive

RECURRENT FAMINE and political oppression made life unbearable in nineteenth-century Ireland for thousands of small farmers. What had been a trickle of immigration became a flood in the 1820s and 1830s, as economic disaster in Ireland coincided with a demand in the United States for raw muscle power to build canals and railroads.

The Irish became the single largest national group to settle in New York in the mid-nineteenth century. Those with funds and contacts tended to pass through New York and continue on to the interior, where they found jobs or bought land. Many who stayed in New York had exhausted their meager resources on the long sea voyage. Those who managed to arrive with some money were often robbed and fleeced by unprincipled compatriots who met them at the docks. Others were paupers who were exported from almshouses by the English authorities. Nativists complained that Europe was dumping its human refuse on America's shores.

While most immigrants moved through New York to other parts of the country, enough remained behind to alter radically the character of the city's population and institutions. In a relatively short period the quiet seaport town became a city with vast social problems. The population grew rapidly, from 120,000 in 1820 to 200,000 in 1830, to almost 270,000 in 1835, to more than 300,000 in 1840. (Immigration slowed down temporarily in the late 1830s because of a severe financial crisis in 1837.) The proportion of aliens in the population increased sharply from 11 percent in 1825 to 35 percent in 1845. By 1855, more than half the city's residents were foreign-born, and more than half the foreign-born were Irish.

New York was not prepared to absorb the sudden influx of poor Irish. Public services, especially for the poor, were minimal. Thousands of immigrants crowded into old housing in lower Manhattan. Well-to-do

27

families moved north, and once comfortable homes were divided into cramped quarters for numerous families. Once fashionable neighborhoods became Irish slums. Dilapidated buildings and flimsy shanties were converted into dwellings, side-by-side with factories. The first building designed as a tenement was built in 1833. It had two apartments on each floor; only the front and rear rooms had light and air— the rooms in the middle were dark and unventilated. Between and behind the tenements were irregular frame structures, jerrybuilt to house the overflow of people who could not find or afford tenement space. The streets and alleys which connected these dwelling-places were notoriously filthy, dark, and foul-smelling. Many of the old, twisting streets were without sewers. Garbage collection was irregular. Street cleaning was performed at will by private contractors who bought their contracts from corrupt ward politicians. Contemporary writers described ankle-deep garbage and open sewage in the streets of the worst slums. Most tenement houses had backyard, wooden privies which frequently overflowed and were a source of disease and foul odors. There were laws regulating privies and sewers, but they were rarely enforced; ill-paid health officers were usually incompetent, and respect for the rights of private property transcended concern for public health.

Life in the slums was a constant struggle against death and disease. There was no inspection of housing or sanitary conditions; from time to time, houses or factories collapsed or caught fire. Thousands of the poor lived in damp cellars. Tuberculosis, pneumonia, bronchitis, and scrofula were common, and medical care was scant. When deadly epidemics broke out, hundreds of people died in the slums, since they were too poor to leave town until the danger was past.

The heart of the Irish slums was an infamous section called "the Five Points" in lower Manhattan where the Criminal Court now stands. The Five Points grew up over the site of the Collect, originally a lovely pond, then a dreary swamp which was finally drained and filled in 1811. The center of the Five Points was the Old Brewery, the most notorious tenement in the city's history. It had once been Coulter's Brewery, where beer was produced which was known throughout the east. By the 1830s it became so run-down that it was no longer fit for use as a brewery and was turned into a dwelling where 1,000 men, women, and children lived, mostly Irish and Negro. Both the Old Brewery and the entire Five Points area were renowned for their unusual supply of pickpockets, thieves, prostitutes, beggars, and murderers. The Five Points was also the home base of several competing gangs, who prided themselves on their reputation as murderous street brawlers. The military was frequently called in by the mayor to quell rioting between gangs.

Respectable people avoided the entire district for fear of their lives. The "respectable poor," thrown in with the "vicious" by economic necessity, had no choice.

There were jobs for the Irish in New York, but they were at the lowest level of the social and economic ladder. Most had been impoverished farmers; having fled the vicissitudes of the land, they shunned agricultural work in the new country. They arrived with no job skills and had nothing to sell but their brawn and their hands. The jobs they could get were menial or manual. Men worked as rough laborers in construction and roadwork, as longshoremen, carters, teamsters, boatmen, stage drivers, and livery workers. They competed with Negroes for lowly jobs as carpet shakers, chimney sweeps, bootblacks, whitewashers, and servants. Irish women worked in homes as laundresses, cooks, cleaning ladies, and waitresses. Some found work in the factories of the clothing industry; a factory job meant long hours at low wages performing a dull routine for a faceless employer.

Living as they did in the most abject poverty, the Irish overwhelmed the city's few and inadequate public relief institutions. A majority of admissions to the almshouse were Irish; most juvenile delinquents were Irish; public hospitals and dispensaries were crowded with sick and diseased Irish; the jails were crowded with Irish lawbreakers (whose crime was usually drunk and disorderly conduct, rather than felonies). Not surprisingly, the Irish had the highest rate of insanity of any national group.

As if the lot of the Irishman were not terrible enough, the scorn of many native Americans was directed at him. The Irish were different from native Americans: they dressed differently, spoke differently, drank too much, brawled too openly, and stuck together. Furthermore, most of them were Roman Catholics, and antipathy towards Catholicism was an inheritance from English colonial days, when "Popish" priests were forbidden to enter the colony on penalty of death. There had never been many Catholics in New York; their religion was viewed suspiciously as arrogant, mysterious, and authoritarian. Then too the Irish were blamed for the very conditions of which they were the victims, just as later immigrants would be. They were accused of importing poverty, filth, and vice from Europe. American workers held them responsible for flooding the labor market and depressing wages. Many natives considered the Irish a drunken, ignorant people given to petty criminality. A typical newspaper advertisement of the period read: "Woman wanted.—To do general housework . . . English, Scotch, Welsh, German, or any country or color except Irish." [1]

However, the Irish were not entirely without friends in New York.

Tammany Hall realized in the 1830s that their numbers were too great to ignore and set about cultivating their loyalty. Tammany established bureaus to assist immigrants in obtaining naturalization papers and organized public meetings conducted by the immigrants' countrymen in their own tongue. It sought out local leaders who were influential in their ethnic community and gave them patronage to dispense. Whenever possible, party leaders supplied food, drink, clothing, shelter, and friendship.

City elections seesawed between the two major parties, the Democrats and the Whigs, for years; each had about 18,000 voters in the 1830s, and the first popular election for mayor was held in 1834. About one-third of Tammany's voters were naturalized citizens; the Whigs counted on the votes of fiscal conservatives, landowners, and nativists. Occasionally the conservative faction of the Democratic party supported the Whig candidate, which assured the election of the Whig. After Tammany adopted the Irish voters, it never espoused nativism; and the Irish were fiercely loyal to Tammany, even when Tammany took stands on economic issues which were at variance with the immigrants' interests. Both parties were responsible for massive voting frauds. Buying of votes was common. The party in power often sent inmates from the city almshouse and the prison to polling places around the city to vote several times. Out-of-town toughs were brought in to work as repeaters on election day (Philadelphia was a favorite place to recruit toughs). Gangs of ward heelers sprang up, particularly in the Irish districts where gangs already existed and had their headquarters in saloons run by local political bosses. Violence and rioting regularly occurred on election day and at party nominating conventions, as the gang of one faction sought to oust the other from control of the ballot box or the convention hall.

The traditional and comfortable way of life known to native New Yorkers appeared to be threatened by the tide of immigration, by slums and paupers, by criminals and swaggering gangs, by corrupt politics which made a mockery of democratic ideals. Out of the anger and fear of many Americans came a new political movement: nativism. In New York, it began in earnest in the 1830s with an anti-Catholic campaign on the lecture circuit, in print, and in celebrated debates between Protestants and Catholics. Several popular books of the period purported to be confessions of "escaped" nuns, revealing immoral practices in convents.

Protestant indignation coalesced into political form in 1835 with the formation of the Native American Democratic Association, which opposed permitting foreigners to hold office, opposed immigration of paupers and criminals, and opposed the Catholic Church. The Native

Americans combined with the Whigs in 1837 to win the elections for mayor and the Common Council. In the same year, a major economic depression threw one-third of the city's laborers out of work; 10,000 people were plunged into utter poverty. The city's institutions for the poor were overwhelmed, and private charity offered little more than soup lines. Because the Irish were the poorest of the poor, they were hit hardest by the depression. Though the Democrats regained control of city government in 1839, there was no real relief for the workingmen or the poor.

Tammany's interest in the immigrants was entirely self-serving. Politically, the Irish had little choice between the parties, since the Whigs espoused genteel nativism. Tammany gave small-time patronage to their Irish supporters, but did little to press for reforms that would inprove the lot of the poor. The leaders of Tammany Hall were themselves men who had become wealthy through their political connections. But the Irish gave Tammany their loyalty without making any broad political demands. Not until the 1850s did they receive a Tammany nomination to a high city office, a seat on the Common Council. Tammany's neglect of social conditions complemented the basic conservatism of most Irish immigrants, which was typified by "their dislike of utopian schemes for the reformation of society in America, their hatred of abolitionism, their contempt for feminism, and their defense of traditional morality." It was not a newer and better world that the ambitious Irishman sought, but a chance to have a share in the world as it was. In the early 1860s, the politically astute Irish took over Tammany Hall, but during the preceding decades, the party organization had only to eschew nativism and denounce temperance legislation to keep faith with the immigrant voters.

Used by his friends and abused by his enemies, mired in filthy slums and threatened by misfortune and disease, many Irish Catholics sought solace in their church. In the face of society's contempt, the Catholic Church became the unifying protector of its benighted faithful: "The Church imparted dignity. It was a symbol of strength with which the individual identified himself and the means of salvation, offering escape from the poverty and cares of the present world. . . . The harsh realities of immigrant existence in New York gave strength to the Church and made it the bulwark of the Irish community." [2]

The Catholic clergy understood that they were in a Protestant society, surrounded by hostility. Their people were poor, and the Church could offer them little more than spiritual comfort. There was nothing the clergy could do about the wretched conditions of life in the city. All they could do was to guard their flock's faith and ward off inducements to assimilate, which they saw as Protestant attempts to destroy the Church. Aside from performing their traditional religious roles, the clergy found that the only significant way they could protect their people from Protestant propaganda was to discourage their followers from using the schools of the Public School Society. Of about 12,000 Catholic children in the city in the late 1830s, only a few hundred were enrolled in the public schools.

The trustees of the Public School Society were puzzled by the Catholic clergy's attitude. They wanted all children, especially poor children, to have some education and had always assumed that the bulk of nonattendance was due to the ignorance or indifference of poor parents, who either sent their children to work or allowed them to roam the streets and wharves as vagrants and beggars. Believing that it was their solemn duty to see that as many as possible were brought into the schools, the trustees hired "visitors" to go about the community urging parents to enroll their children; they even recommended in 1832 to the Common Council that public charity be denied to parents who kept their children out of school. This harsh policy was adopted by the Common Council but never enforced.

Convinced of the superiority of their schools, the trustees could not understand the Catholics' resolute nonparticipation. The schools of the Public School Society were widely considered the best in the country. At a time when public schools everywhere were in their infancy and plagued with problems, New York's schools were open eleven months of the year and had excellent administration and carefully screened teachers. A decade of reform and expansion had tripled the enrollment to 20,000 by 1838. The system was a model of order and efficiency, embodying the virtues of good habits, cleanliness, thrift, and industry.

The trustees failed to understand the seriousness of the Catholic challenge. Their habit of castigating truants' motives and character made them slow to recognize that the Catholics rejected their schools *on principle*. Their smugness made them unable to comprehend that the size of the Catholic population would soon present a threat to the very existence of the Public School Society and to the concept of nonsectarian common school education which the trustees had promoted into official policy.

CHAPTER 4

The Catholics
Challenge the System

BETWEEN THE TRUSTEES of the Public School Society and the Catholic clergy, there was an unbridgeable cultural gulf. The clergy, champions of the most despised class in the city, knew how far apart they were; the trustees did not. The trustees, living in a world of secure comfort, extolled the virtues of democracy and exhorted the poor to share their optimism about the future. The clergy knew how remote from reality were the ideals of the trustees. The trustees' lectures on morality meant little to the hard-pressed slum dweller. The clergy comprehended the trials of the poor and believed that what they most needed was the succor of their faith. The Church and its traditions had been a part of the Irish immigrants' childhood, and in the new country it was a familiar oasis. To the Catholic poor, the Church was accepting. But the public schools were reproachful, disapproving of their habits and their morality, disparaging their family, their religion, and their culture.

The attitude of the Society towards the poorest children was typified in the 1831 *Annual Report:*

> The attention of the Board has been . . . directed to the necessity of a school near the 'Five Points', in which neighborhood, there is a large number of children, alike destitute of literary and moral instruction; and who have been so long subjected to the influence of the worst examples, that it is not thought proper to associate them with the respectable and orderly children who attend the Public Schools.

The children of the Five Points were triply disadvantaged: they were Irish, they were Catholic, and they were poor. As the number of Irish Catholics steadily increased each year, the cultural antagonism between them and the native population intensified. Because the school is an institution where culture is transmitted, both formally (in curriculum) and informally (in standards of behavior), the school was destined to be the stage on which the old culture and the new culture struggled for accommodation or dominance.

The American Catholic hierarchy decided as early as 1829 that Catholic schools ought to be established to protect the faith; in that year, the American bishops decreed:

> Since it is evident that very many of the young, the children of Catholic parents, especially the poor, have been exposed and are still exposed in many places of this Province, to great danger of the loss of faith or the corruption of morals, on account of the lack of such teachers as could safely be entrusted with so great an office, we think it absolutely necessary that schools should be established in which the young can be taught the principles of faith and morals, while being instructed in letters.[1]

However, the shortage of Catholic teachers and the impoverished condition of the Catholic community made this statement a hope and a warning, rather than a basis for action. There were then five Catholic schools in New York City, which was far from sufficient for a Catholic population of some 35,000.

In 1834, the city's leading Catholic clergyman, Bishop John Dubois, suggested changes to the trustees which might have made acceptable at least one public school, P.S.S. No. 5, on Mott Street near St. Patrick's Cathedral. The alterations he proposed would "ensure the confidence of Catholic parents, and remove the false excuses of those who cover their neglect under the false pretext of religion, which they do not practice." He assured the trustees that he had no sectarian motive, no desire to proselytize, and he was "as much averse to encroach upon the conscience of others, as to see others encroach upon" his own.

He asked that the board permit him to recommend a Catholic teacher for P.S.S. 5, subject to examination and removal by the trustees; that he be allowed the use of the school after school hours, for religious instruction of apprentices and servants; that the trustees expunge any passages in schoolbooks which inculcated Protestant sectarian principles or defamed the Catholic religion; that the untenanted upper part of the school be used for the temporary education of 200 Catholic girls until their own school, recently destroyed by fire, was rebuilt; that the bishop be permitted to visit the school occasionally and offer suggestions to the trustees for its improvement; that religious instruction be given to Catholic children between five and seven each evening by a clergyman designated by the bishop. However, if all these requests were objectionable, he then asked that the Society sell one of its discontinued buildings to St. Patrick's.[2]

Dubois' tone was courteous; he laid down no ultimatum. But the inference was clear that if the Public School Society did not adjust its schools to the special needs of the Catholic community, then the Catholics were prepared to expand their own schools. By making some accommodation, the Society might have made the Catholics feel welcome in their schools. But the trustees, insistent that their schools did not violate anyone's religious views, rejected Dubois' requests as "privileges . . . which have never been asked for, or granted to any other denomination." The Society's trustees wrote the bishop of their hope that more Roman Catholic children would attend the public schools and that more Catholics would become members of the Society and of its board. They offered to remove any material in the books "which can reasonably be objected to by any denomination." [3] Bishop Dubois never responded to their offer, and the matter was dropped, with no efforts made to eliminate the objectionable passages or to hire a Catholic teacher for P.S.S. 5.

The Public School Society failed to recognize that its version of nonsectarianism was sectless Protestantism. Its failure, too, to remove voluntarily the open slurs against Catholicism in many of its textbooks underscored the transparency of its seeming religious neutrality. The Society's inability to surmount its narrow cultural horizons made it impossible for the Society to reach a compromise with the Catholics and guaranteed that the conflict would ultimately reappear on a larger scale.

Bishop Dubois had no heart for conflict. He preferred to avoid stirring up popular antipathy to Catholics. Besides, he was aging, and the administration of his fast-growing diocese required all his energies. In 1838 he obtained the services of forty-one-year-old John Hughes as a coadjutor bishop to assist him in his duties. Two weeks after Hughes assumed his new post, the seventy-five-year-old Dubois suffered a paralytic stroke. The following year, the enfeebled Dubois was suspended from his post, and Hughes was made bishop of New York.

In contrast to Dubois, who was deferential and cautious in dealing with Protestant society, Hughes was aggressive and militant. He refused to don the meek demeanor customary for the leader of a barely tolerated minority. He was himself an Irish immigrant, son of a poor farmer; Hughes came to the United States in 1817 at the age of twenty. He worked as a gardener for Mount St. Mary's College and Theological Seminary, while preparing for admission as a student; the president of the college, who befriended and instructed Hughes, was John Dubois.

The year that Hughes was ordained to the priesthood in Philadelphia, Dubois was consecrated as bishop of the diocese of New York. Hughes was a priest in Philadelphia until he was called to New York to assist Dubois.

As an Irishman, Hughes understood the temper of his countrymen in New York. As a Catholic, he was uncompromising and fiercely loyal. Hughes thrived on conflict and controversy. He encouraged his coreligionists to fight for their political and civil rights. Forceful and energetic, he cared not a whit how the dominant Protestant society reacted when he lectured on "The Decline of Protestantism." He loved his church and was willing to enter any arena to uphold its honor. Occasionally, he sought out encounters with Protestant bigots. During 1830, while a priest in Philadelphia, he wrote a series of pseudonymous columns which were filled with slanders of the Catholic Church and submitted them to a nativist journal in New York called *The Protestant,* which unquestioningly printed them. When Hughes revealed his authorship of the diatribes, he explained that "I wanted to ascertain whether or not conscience had anything to do with the columns of *The Protestant.* I found it had not. . . . In a word, I could not find a line deep enough to fathom the editorial depravity of *The Protestant.*" [4]

During the 1840s, when violent anti-Catholic riots erupted in other cities, Hughes prepared a plan of defense for every Catholic Church in New York. In case of trouble, the churches would be surrounded by a force of 1,000 Catholic men, prepared to fight to the death. He informed the city government that "if a single Catholic church were burned in New York, the city would become a Moscow." In 1844, when religious tension was high in the city, he met with the mayor, who asked if Hughes was afraid for his churches; Hughes replied, "No, sir; but I am afraid that some of *yours* will be burned." Though Catholic buildings and homes were destroyed in Boston and Philadelphia in the same period, New York's Catholics survived the nativist hysteria without suffering major violence.[5]

Hughes devoutly believed in the Catholic mission to convert the rest of society to the true faith. It was said that he liked to quote Metternich's remark that "the Catholics and the Protestants in your country are like the iron pot and the earthen pot floating down the stream together; when they clash, the earthen pot must be broken." In 1850, while nativism was still on the rise, Hughes spoke directly to the widespread Protestant fear that the Pope had a secret plan to take control of the Mississippi Valley. Preaching in his cathedral in New York City, he said:

Protestantism pretends to have discovered a great secret. Protestantism startles our Eastern borders occasionally on the intention of the Pope with regard to the Valley of the Mississippi, and dreams he has made a wonderful discovery. Not at all. Everyone should know it. Everybody should know that we have for our mission to convert the world, including the inhabitants of the United States, the people of the cities, the people of the country, the officers of the Navy and the Marines, commanders of the Army, the Legislatures, the Senate, the Cabinet, the President and all.[6]

When Hughes became bishop of New York, there were eight Roman Catholic churches in the city, serving about 60,000 people. Each church maintained a free school; about 3,000 children attended the Catholic church schools. These schools, ineligible to receive public funds since the Bethel Baptist Church incident in 1825, were poor, overcrowded, understaffed, and badly equipped. Thousands of Catholic children attended no school at all, since the Catholic schools were too poor to expand and the clergy discouraged Catholic participation in the public schools. Only a few hundred Catholic children attended the Society's schools. These circumstances impressed on Bishop Hughes the urgency of building Catholic institutions. The financial demands on the Church in New York were so great that in the fall of 1839 Hughes sailed for Europe to obtain money for a new seminary and college (later Fordham University) and to recruit priests and teachers for the Catholic schools.

While Hughes was in Europe, an event occurred which was of signal importance to the development of public schools in New York City. It had reached the attention of New York's Governor William H. Seward that thousands of poor Irish Catholic children in New York City were not in school. Seward perceived that publicly supported schools which systematically excluded a particular group of children, even if the exclusion was unintentional, were impeding universal education. In his second annual report to the legislature in January 1840, Seward made a radical proposal:

> The children of foreigners, found in great numbers in our populous cities and towns, and in the vicinity of our public works, are too often deprived of the advantages of our system of public education, in consequence of prejudice arising from differences of language or religion. It ought never to be forgotten that the public welfare is as deeply concerned in their education as in that of our own children. I do not hesitate, therefore, to recommend the establishment of schools in which they may be instructed by teachers speaking the same language with themselves and professing the same faith.[7]

When Catholic leaders, both clergy and laymen, read Seward's address, they realized that his concern might enable church schools to obtain public funds.

Seward, a Whig, had been elected governor in the fall of 1838. In New York the Whigs swept control of the state assembly in 1837, won the governor's race in 1838, and took a majority of the state senate seats in 1839. By 1840, when Seward delivered his controversial education message to the legislature, the Whigs were a majority in both houses.

In his first year as governor, Seward toured the public schools of New York. He felt that they were excellent institutions, but he was deeply disturbed by the large numbers of Irish and German Catholic children who were not in school. He firmly believed that unequal education was the source of social inequality. He felt that the only way to end poverty and crime was to extend the blessings of education to everyone, and that the power of the few "cannot be broken while the many are uneducated." Even if wealth and other external symbols of power were taken away from the upper classes, Seward maintained, "knowledge still remains the same mighty agent to restore again the inequality you have removed." Seward was convinced that universal education was the challenge of his age, and he was determined to do whatever he could to advance that goal.[8]

Seward's convictions thrust him into the forefront of the age of reform that came to fruition in the 1840s, expressed by movements on behalf of prison reform, slaves, temperance, education, and utopianism. Many writers and public figures, aware that they were creating a new society, consciously tried to connect public policy with its social consequences. One historian, describing the period, found its essence in an offhand comment by Ralph Waldo Emerson: "We are all a little wild here with numberless projects of social reform; not a leading man but has a draft of a new community in his waist-coat pocket." [9]

Seward, born in Florida, New York, in 1801 to a well-to-do family of English-Welsh descent, was elected governor at the age of thirty-seven. He quickly became engaged in the major issues of his time, invariably articulating a humanitarian point of view. He was a shrewd politician, but was willing to risk unpopular actions rather than abandon his principles. After a trip to Ireland in 1833, Seward became an ardent advocate of Irish freedom from England; he became convinced that it was English policy to keep the Irish ignorant, and that it was ignorance that kept people poor. As early as 1839, he spoke out forcefully against slavery, long before the antislavery movement had a popular base in New

York State. As governor, he incurred the wrath of Southern states by refusing to return fugitive slaves; he promoted legal reforms which reduced lawyers' fees, a move which displeased that powerful group. His
ambitious, expensive program of building roads and canals raised the
state's indebtedness and caused the cost of borrowing to rise; the long-
range value of the improvements was great, but the charges of extravagance did Seward no good politically. He was re-elected governor by a
narrow margin and declined to run again at the end of his second term,
having moved too far in advance of public opinion for his political
good. In 1848, six years after leaving the governor's chair, he was
elected to the United States Senate by the Whig majority in the state
legislature (senators were not yet elected by popular vote). He attracted
national attention in 1850 by opposing any compromise with slave states
and asserting that "there is a higher law than the constitution." He
warned time and again that if slavery were not abolished, the country
would be plunged eventually into civil war; in 1858, he prophesied an
"irrepressible conflict" between the free states and the slave states. After
losing the Republican presidential nomination in 1860, he became Lincoln's secretary of state. In 1865, he survived an attack by a coconspira-
tor of John Wilkes Booth and continued to serve until the end of President Andrew Johnson's term in 1869; his purchase of Alaska from Russia
for $7.2 million in 1867 was known at the time as "Seward's Folly."

Some of Seward's advisors, particularly Thurlow Weed, his closest
strategist, believed that his idealism was also good politics. Weed was
editor of the *Albany Evening Journal* and a powerful figure in the Whig
party. Weed and Seward led a faction of the Whigs which was attempting to broaden the party's base. The Whigs tended to favor governmental intervention in economic and social matters, whether for regulation
of the currency, the tariff, or reform measures. The Democrats espoused
egalitarianism but, in Jeffersonian fashion, were suspicious of concentra-
tions of power and feared the intrusion of governmental action. The New
York City Whigs, who occasionally allied with the nativists in local elections, were the conservative wing of the party. The upstate Whigs, led
by Seward and Weed, felt that the extension of the vote had radically
changed the nature of politics. They believed that if the Whigs could not
identify with issues which cut into the Democrats' strength, then their
party would go the way of the aristocratic Federalists.

Seward's biographer, Glyndon Van Deusen, wrote that Seward
hoped to achieve political advantage by siding with the Catholics on
the school issue. Exploring Seward's motivations has been a favorite pas-
time of historians of these events. It may have been that his motives

were political, but the course he advocated was consistent with his deepest personal convictions. He wrote to a friend in December 1840, as the political storm he had unleashed was breaking, "Knowledge taught by any sect is better than ignorance. I desire to see the children of Catholics educated as well as those of Protestants; not because I want them Catholics, but because I want them to become good citizens. In due time these views will prevail notwithstanding the prejudices that have assailed them."[10]

The Catholic leadership of New York understood Seward's message as an invitation to apply for public funds. Soon after the governor's address, Dr. John Power, one of Bishop Hughes' vicars-general, convened a meeting of the trustees of all the city's Catholic churches, who agreed to seek public subsidy for their schools. (Hughes was still in Europe.) Power then traveled to Albany, where he was encouraged by legislators who were close to Seward. After Power's return, he and the Catholic laymen drafted a petition on behalf of the eight Roman Catholic churches, each of which supported a free school. On February 25, 1840, the Catholic plea was submitted to the city's Board of Assistant Aldermen (the Common Council was composed of a Board of Assistant Aldermen and a Board of Aldermen). Since the Bethel Baptist Church controversy in 1825, the Common Council controlled the distribution of the state common school fund.

The Catholic petition was brief and simple, avoiding allusions either to the Governor's statement or to the inadequacies of the Public School Society. The Catholics requested public aid on the basis of need. Within the next month, additional petitions were presented by the Scotch Presbyterian Church and a Hebrew congregation, letting city officials know that if the Catholics got a share of the fund, then they too wanted their portion.

The Catholics expected opposition, but they were not prepared for the nearly universal condemnation of their plea by other religious groups, by the Public School Society, by groups of public-spirited citizens, and by the press. Naturally, the leading antagonist of the Catholic petition was the Public School Society, which staunchly opposed any diversion of the common school fund to a sectarian body. The Society, as in the past, maintained that it would be a violation of the principle of separation of church and state to grant special privileges to one sect. The trustees of the Society feared that if funds were given to the Catholics, then other sects would apply until there was no money left for nonsectarian common schools. They argued that the common school fund was strictly civil and had been created to benefit schools open to all, where no one's beliefs were offended. To support a particular sect, they held, would be equivalent to taxing all for the support of the special beliefs of a small minority.

Even the Catholic press was divided on the issue. There were two Catholic papers in the city in February 1840. The *Truth Teller*, published since 1825 by laymen who supported the Jacksonian bent of the Democratic party, opposed Seward's educational proposals. The editors of the *Truth Teller* rejected any legislation which affected the religious sphere. They maintained that religious training belonged in the realm of the home and the church. Suspicious of any governmental measures which tended to foster European-style heterogeneity, they felt it was alien to a free and democratic society to prolong differences and inequalities among men. The *Truth Teller* preferred public schools which embraced all classes and religions equally.

The other Catholic newspaper was the *Register*, which had been initiated in 1839 by the Rev. Felix Varela, who was co-vicar-general of New York with Dr. Power. The *Register* attacked the *Truth Teller* for its support of the existing school system. The *Register,* which spoke for the Church hierarchy, held that Catholic schools were an integral part of the state common school system. Their reasoning was that the schools of the Public School Society were Protestant public schools, while their own schools were "Catholic public schools," equally entitled to public support. Only this system of public support, they believed, could achieve the state's object of universal education "without forcing the professors of any religion to act against their conscience." Polarization of the school issue along Catholic and anti-Catholic lines caused the *Truth Teller* to back down, rather than appear to oppose the interests of its own people.[11]

About a dozen prominent Catholic laymen organized a well-attended weekly meeting to discuss the school question. Despite their efforts to remain nonpolitical, political charges and countercharges began to fly. Three days before the municipal elections in April 1840, Whig partisans stood at the door of St. Peter's Roman Catholic Church and distributed copies of a pro-Irish speech made by Seward. The church trustees denied complicity in the incident, and the Democratic *Truth Teller* attacked the Whigs for defiling the Church in their search for votes.

Meanwhile, the Board of Assistant Aldermen assigned the controversial Catholic petition to its Committee on Arts and Sciences and Schools, which opened public hearings in March of 1840. The Catholic spokesmen stated that their schools were free and open to all, and that there was no attempt to change the religious views of non-Catholic pupils. As taxpayers, they felt that they should enjoy the benefits of the common school fund without being compelled to send their children to schools where their religion was exposed to ridicule and censure. They promised that if funds were granted to Catholic schools, religious instruction would be offered only after school with parental approval. The Public School Society denied that its books and exercises reflected adversely on the Catholic Church, but offered to expunge any offensive material. Public support of any sectarian school was contrary to the principle of separation of church and state, it maintained.

The Society was perturbed by the Catholic petition, but not unduly concerned. The trustees had every reason to believe that their schools were indeed *the* public schools of New York City. The Common Council had vindicated their position several times before, and the present Common Council worked closely with the Society. All who had seen its schools testified to their excellence. Enrollment had grown steadily, and it was the trustees' belief that in time all the children of the city who were not in private schools would be provided for by the public schools, as funds became available. The assurance of the trustees was a reflection of the general goodwill and esteem which the Society commanded in the city. The management of the schools and the trustees themselves had always been above reproach. The Society had instituted its own merit system for hiring teachers, who had to pass an examination to win their low-paying jobs. Unlike city government, there was neither nepotism nor patronage in the hiring and promotion system. The trustees were not the sort of people who had either friends or relatives who sought work in the school system.

Despite their confidence, the trustees were anxious to remove any grounds for objection to their system. They were disturbed that Catholic children, who were of the poorest and most ignorant class in the city, were not enrolled in their schools. While the Board of Assistant Aldermen was considering the question, the Society furnished Rev. Varela a set of schoolbooks at his request and promised to give his suggestions "the most serious and respectful consideration." Varela shortly afterwards responded with a cordial letter which mentioned four offensive passages in the books. The constructive tone of Varela's letter encouraged the Society to believe that the problem might be worked out amicably.[12]

These tentative negotiations were pushed into the background on April 27, 1840, when the Committee on Arts and Sciences and Schools brought in its report on the controversy, known as Document No. 80. It opposed the Catholic claim:

Religious zeal, degenerating into fanaticism and bigotry, has covered many battle-fields with its victims. . . . To prevent, in our day and country, the recurrence of scenes so abhorrent to every principle of justice, humanity, and right, the Constitutions of the United States and of the several States have declared . . . that there should be no establishment of religion by law; that the affairs of the State should be kept entirely distinct from, and unconnected with, those of the Church; that every human being should worship God according to the dictates of his own conscience; that all churches and religions should be supported by voluntary contribution; and that no tax should ever be imposed for the benefit of any denomination of religion, for any cause or under any pretence whatever.[13]

The committee said that it was true that Catholics paid taxes and were entitled to benefit by them,

but it should be borne in mind that they are taxed not as members of the Roman Catholic Church, but as citizens of the State of New York; and not for the purposes of religion, but for the support of civil government. . . . Our institutions are designed not to create or perpetuate religious distinctions, but to place all mankind upon a common footing of equality. Any legal acknowledgement of any religious denomination, as a dependent upon the public bounty for any kind of pecuniary aid or support, would be an abandonment of the great constitutional principle, that the end and aim of all just government is the equal protection of all men in the free exercise and enjoyment of the rights derived from the written Constitution of the land, or the still higher authority of nature. . . . It requires no argument to prove that taxation of all sects, for the benefit of one, is a violation of the rights of conscience.

The committee vigorously affirmed the ideal of the common school, free of any sectarianism. It rejected the alternative of dividing the school money among all sects, for this would discriminate against those who belonged to no sect and would represent the legal establishment of religion, which is expressly forbidden by the Constitution. "It is immaterial, in the eyes of the law, whether a citizen professes any or no religious faith; . . . there is no difference, in legal principles, between taxing him for the purpose of educating the young in the doctrines of many churches, to which he does not belong, and taxing the Catholic for the benefit of Protestant schools, or taxing the Protestant for the support of Catholic seminaries."

The committee directed the Public School Society to remove immediately any books which might be offensive to Roman Catholics or to any other sect.

If religious instruction is communicated, it is foreign to the intentions of the school system, and should be instantly abandoned. Religious instruction is no part of a common school education. The Church and the fireside are the proper seminaries, and parents and pastors are the proper teachers of religion. In their hands the cause of religion is safe. Let the public schoolmaster confine his attention to the moral and intellectual education of the young committed to his charge, and he fully performs the duties of his profession, discharges the trust reposed in him as a public agent, and fulfils his obligation as a citizen.

Document No. 80 was adopted by the Board of Assistant Aldermen by a vote of 16–1. When Bishop Hughes returned from Europe a few months later, he caustically summed up the clergy's reaction to the report: "We are citizens when they come to gather the taxes, but we are Roman Catholics when we look for a share of the fund just contributed." [14]

A few days after the decision, the trustees of the Public School Society appointed a committee to examine the schoolbooks for any material which was objectionable to Catholics. They obtained an interview with Dr. Power, which resulted in his requesting a set of schoolbooks. The committee waited several weeks for his suggestions, but received no word from him.

The trustees thought that the report of the Board of Assistant Aldermen had settled the dispute; the Catholics, aware that Governor Seward would give their cause a sympathetic hearing in Albany, knew that Document No. 80 was not a defeat but a necessary step on the path to Albany. To rally Catholic support for the fight ahead, the clergy launched a new Catholic newspaper, the *Freeman's Journal*, on July 4, 1840. The first issue served notice that the school question was far from closed. An editorialist wrote that a state where only one religion was practiced had no problem with religion in its schools. But in a state where religious liberty prevailed, some religious group must find their principles endangered in the schools, since it was impossible for schools to teach hostile creeds simultaneously or one creed only. Yet, the elimination of all religious instruction from the schools was also unacceptable to Catholics, "for, with a Catholic, Religion forms a vital part of education." [15]

This position defined the Catholic rejection of not just the Public School Society, but of the very idea of common schooling. By their description, the religious liberty of Catholic schoolchildren could be protected *only* in a school where the Catholic religion was taught. A school which attempted to teach all creeds or no creed at all was repugnant to them. Devout Catholics did not want their children exposed to other religions, nor did they want their children educated in a school which put error and truth on an equal footing.

In the second issue of the *Freeman's Journal*, Dr. Power responded to the Public School Society's schoolbook committee with a scathing attack on the public schools. He called them deist, sectarian, and anti-Catholic. He charged that their books contained "the most malevolent and foul attacks" on Catholicism. He strenuously objected to the daily reading of the King James version of the Bible, without note or comment, in the classrooms of the public schools. Wrote Power: "The Catholic Church tells her children that they must be taught their religion by AUTHORITY. The Sects say, read the bible, judge for yourselves. The bible is read in the public schools, the children are allowed to judge for themselves. The Protestant principle is therefore acted upon, slily inculcated, and the schools are Sectarian." [16]

The Public School Society's schoolbook committee was shocked by Power's rude and angry column. After their frank meeting with him six weeks before, they had expected his cooperation in revising the schoolbooks. The trustees naively persisted in their belief that the schoolbooks were the reason for the Catholics' discontent. They failed to understand that a mere expurgation of the books would not satisfy Catholic demands. The clergy did not cooperate in editing biased material out of the books, because they knew that to do so would remove one of the best complaints they had against the schools of the Society. It was months before the trustees understood that the Catholic clergy was battling not to change the schoolbooks, but to destroy the Society itself and the concept of common schooling which it had established as the criterion for receiving public funds.

Perhaps mindful of the unfolding controversy in New York, the American Catholic hierarchy met in May 1840 in Baltimore and directly attacked the Protestantism of the common school:

> Since it is evident that the nature of public education in many of these Provinces is so developed that it serves heresy, as well as that the minds of Catholic children are little by little imbued with the false principles of the sects, we admonish pastors that they must see to the Christian and Catholic education of Catholic children with all the zeal they have, and diligently watch that no Protestant version of the Bible be used, nor hymns of the sects be sung, nor prayers be recited. Therefore, it must be watched that no books or exercises of this kind be introduced with discrimination of faith and piety. These efforts of the sects are to be resisted everywhere, constantly and moderately, imploring the help of those who have authority to use a fitting remedy.[17]

CHAPTER 5

The Bishop
Takes Command

BISHOP HUGHES returned from his European trip on July 18, 1840, and immediately assumed command of the Catholic fight. Two days after his return, a regular meeting was held in the school house of St. Patrick's Cathedral. The meetings in recent weeks had become increasingly unruly and disorderly. Hughes realized that the intemperate nature of the meetings was imperiling the good name of the Catholics, as well as the ultimate outcome of the issue. He was distressed at the factionalism which split the Catholics along party lines; he worried that his followers might become the dupes of "political underlings who had been accustomed to traffic in their simplicity." He resented the way the Catholics blindly supported the Democratic ward bosses without regard to their own self-interest. Finding so many Catholics aroused about the school issue, he realized that this was a question which could unite Catholics *as Catholics*, without regard to party affiliation. Hughes smarted at the fact that Catholics, because of poverty and discrimination, were second-class citizens. He knew that their needs would continue to be ignored until they learned to act on their own behalf and not to be taken for granted by the ward bosses.[1]

The fight had been launched without Hughes, but it took Hughes to broaden the offensive beyond a contest for public funds. Under Hughes' direction, the Catholics fought for their self-respect and the respect of their fellow citizens. There was nothing Hughes could do about poverty and discrimination in the immediate present, but he wanted politicians and everyone else to know that Catholics intended to stand up for their rights. As militant blacks over a century later raised the banner of black power, so Hughes sought to unify and assert Catholic power.

Hughes knew that the *Truth Teller*, published by lay Catholics who were Democrats, had charged the leaders of the movement with having "political designs" since the controversy was initiated by a Whig governor. Hughes felt that the merits of the issue, which he believed favored the Catholic claim, would be forgotten in the midst of political maneuverings. He said pointedly to his followers:

> Politics must not be introduced: first, for the perhaps insignificant reason that if they be introduced I disappear from amongst you; and secondly, for the very important one that your prospects would thereby be defeated. . . . If you have any regard, then, for my feelings or your own interests, do not introduce politics. We do not meet for political purposes. I defy our enemies or our friends to show that one word of politics was ever tolerated in our meetings.[2]

Hughes asserted that Catholics were compelled to support schools which violated their rights of conscience; that these same schools, supported in part by Catholic taxes, circulated insidious attacks on Catholicism while quietly inculcating Protestant practices; and that these schools pointedly excluded Catholics by insisting on a daily reading of a version of the Bible forbidden to Catholics. The public schools, he charged, were at one and the same time infidel (because they taught no particular religion), sectarian (because of such practices as Bible-reading), and anti-Catholic.

Bishop Hughes abhorred common schools. He was no assimilationist. The arguments used in favor of common school education—its contribution to social equality and social harmony—were irrelevant to Hughes. His mission was to protect the faith of his followers, which he could do best by pursuing religious separatism. He wrote to the bishop of New Orleans on August 27, 1840, that the battle against the Public School Society "will cause an entire separation of our children from those schools and excite greater zeal on the part of the people for Catholic education." Hughes used the contest to warn his flock of the dangers of common school education and to win their support for Catholic institutions. He was able to build on the dislike of the Irish immigrant for the high-born, arrogant aristocrats who served on the Public School Society's board of trustees. Though Hughes, like the Catholic Church itself, was authoritarian and paternalistic, surely no less so than the Public School Society, he was able to use the democratic aspirations of the times to construct a case against the highly centralized private corporation which ran the public schools. Had the schools been then in the control of an elected board, or even a board appointed by an elected official, the Catholic case might have been less compelling.[3]

The Catholics met every fortnight in the basement of St. James Church. Hughes always spoke, usually at great length. He was a brilliant public speaker, alternately sarcastic, vitriolic, and humorous. His presence attracted large and enthusiastic audiences. At a meeting in July, a committee was appointed to prepare a report to the public which explained the injustice of the existing school system. Two weeks later, the committee returned its report, which was a bold plea for the support of non-Catholics. It held that Catholics represented no threat to their fellow-citizens but the denial of the rights of Catholics was a threat to the rights of all Americans:

> We are Americans and American citizens. If some of us are foreigners, it is only by the accident of birth. As citizens, our ambition is to be Americans; and if we cannot be so by birth, we are so by choice and preference. . . . We hold, therefore, the same ideas of our rights that you hold of yours. We wish not to diminish yours, but only to secure and enjoy our own. Neither have we the slightest suspicion that you would wish us to be deprived of any privilege which you claim for yourselves.[4]

The address advanced the Catholic view of the importance of religious education. "Mere secular knowledge," it held, was not enough; only the cultivation of religious and moral understanding could mold enlightened and virtuous citizens. Nonsectarianism, it was argued, necessarily banished the Christian religion from the public schools; since the negation of Christianity is infidelity, therefore "the public school system in the city of New York is entirely favorable to the sectarianism of infidelity." The Catholics feigned surprise that other Christian sects were not affronted at the use of public funds to subsidize "the sectarianism of infidelity." Their failure to protest, wrote the committee, "is not for us to say; but for ourselves we can speak: and we cannot be parties to such a system except by legal compulsion and against conscience."

The system which the Catholics believed fair was the one which had been customary in the city in the first quarter of the century, when public funds had been apportioned to church schools to educate the poor children of their own congregations. The address held that "even the least perfect religion of Christian sectarianism would be better than no religion at all." The abolition of this system was blamed squarely on the Public School Society, which relentlessly pursued aggrandizement of its power until it had achieved a monopoly of public education funds. Class resentment smouldered beneath the Catholic description of the So-

ciety's trustees, who were the embodiment of upper-class Protestant benevolence. "That corporation is in high and almost exclusive standing
with the Common Council," complained the report. The history of the
Society was recited in a fashion that pictured the trustees as ambitious
and presumptuous:

> They finally got themselves denominated "The Public School Society of
> New York," and from that time labelled *their* schools, as if they belonged
> to the community at large, "Public Schools". . . . They are merely *called*
> "public schools", but they belong to a private corporation, who have
> crept up into high favor with the powers that be, and have assumed the
> exclusive right of monopolizing the education of youth.

Nonsectarianism, the Catholics warned, was dangerous to society,
for it permits "the will of the pupil to riot in the fierceness of unrestrained lusts." While the intellect is trained, "the heart and moral character are left to their natural depravity and wildness. This is not education; and, above all, this is not *the* education calculated to make good
citizens."

The Catholics emphasized that common schooling itself was a
threat not only to them, but to all minorities, to all future dissenters,
and to the Republic itself:

> Should the professors of some weak or unpopular religion be oppressed
> today, the experiment may be repeated tomorrow on some other. Every
> successful attempt in that way will embolden the spirit of encroachment
> and diminish the power of resistance; and, in such an event, the monopo-
> lizers of education, after having discharged the office of public tutor, may
> find it convenient to assume that of public preacher. The transition will
> not be found difficult or unnatural from the idea of a common school to
> that of a common religion; from which, of course, in order to make it
> popular, all Christian sectarianism will be carefully excluded. . . . Should
> the American people ever stand by and tolerate the open and authorita-
> tive violation of their Magna Charta, then the republic will have seen the
> end of its days of glory.

The trustees of the Public School Society issued an immediate rebuttal to the Catholic address. They rejected the Catholics' assumption
that only New York City withheld money from religious societies. Public
support of religious schools had been practiced for a few years in New
York City, but nowhere else in the state; it had been discontinued in the
city because it produced rivalry among the sects and abuses of the common school fund. The trustees considered the issue of public support for

religious schools "conclusively settled by public opinion," as well as by law.[5]

The Society recounted its efforts to get the Catholic clergy's cooperation in revising the schoolbooks. The fruitless meetings, followed by the clergy's unprovoked public denunciations of the Society, led the trustees to realize that "no expurgation—nothing of a mere negative character —will satisfy the Roman Catholic clergy. If the doctrines of their Church be not taught, nothing can be which they would not pronounce heretical, and 'adverse to Christianity'. Even the Holy Scriptures are sectarian and dangerous, 'without note or comment'; and certainly no comments would be acceptable other than those of their own Church." The trustees concluded that the Catholics did not consider *themselves* to be either sectarian or a denomination, but "the Church." The Catholic position could be understood only by recognizing that, to Catholics, the teachings of Roman Catholicism were the truth, but the teachings of other religions were sectarianism.

Bishop Hughes sent a copy of the Catholic address to Governor Seward, accompanied by a warm letter, thanking him for his "high, liberal and true American views" on popular education. Hughes had an instinctive touch for politics and without self-consciousness suggested to Seward the political possibilities of befriending the downtrodden immigrants: "I have reason to believe that should our country call you yet to higher trusts in the Republic, your name will ever be cherished with a peculiar regard by the Catholics of the present generation throughout the United States." Seward responded to Hughes at once, assuring him of his support for the Catholics' claim and congratulating him for avoiding appeals to prejudice, which are "incident always to every effort to reform manners or morals." As to his political future, Seward wrote that he was quite satisfied to remain governor of New York.[6] *

DESPITE THE EXCHANGE of antagonistic statements, the Society's trustees persisted in hoping that they and the clergy might be able to iron out their differences as reasonable men. Their schoolbook committee, rebuffed weeks before by Dr. Power, arranged an interview with Bishop

* In fact, Seward's advocacy of the Catholic cause had repercussions twenty years later, when he was a leading candidate for the Republican presidential nomination. Because anti-Catholic Know-Nothings held the balance of power in two key states, Indiana and Pennsylvania, the Republican convention passed over the outspoken Seward in favor of the moderate candidate, Abraham Lincoln.

Hughes in early September 1840. Hughes met with them and requested a complete set of schoolbooks, which were promptly supplied. All that the Society had to show for its efforts at working with the Catholic Church since the spring was the delivery of a set of schoolbooks to each of the three top clergymen, Hughes, Power, and Varela. About two weeks after Hughes received his set, the committee wrote to inquire whether he was ready to suggest passages for expunging. Hughes replied that he could not understand the committee's supposition that he intended to assist them in their "laudable undertaking." Unfortunately he lacked the time to help them. No, he had requested the books in order to document the Catholics' case against the Society. "This I should perhaps have stated," wrote Hughes, "but the omission was purely accidental." [7]

Hughes had no intention of making the Society's schools more tolerable for Catholic children. He doubted the good faith of the trustees on the book question, in any case. He said often that while the trustees might delete obnoxious material for the present, there was no assurance that those passages would not appear in future editions or under future boards of trustees. To the regular Catholic meeting, Hughes derided the trustees and likened them to an officer in England designated "The Keeper of the King's Conscience." He said, "The trustees of the Public School Society are become the guardians of the consciences of both the Catholics and the Protestants. . . . They stand as umpires between the churches, and they profess to regret that the Catholic clergy have not met them to obtain their confidence, and to have a joint examination and expurgation of the public school books." Hughes believed that it was important not to collaborate with Protestants in such matters, because while some objectionable material might be removed, whatever remained would appear to have Church sanction.[8]

The Society's schoolbook committee, intent on eliminating Catholic objections, doggedly proceeded with its task of revision, but without Catholic assistance. Some passages were stamped out with ink; some pages were pasted together; some volumes were discontinued. In textbooks and library books, the committee found much that offended Catholics. It was not that the Society was bigoted, but *society* as a whole was. The books available to the schools reflected popular anti-Catholicism. Even some Catholic schools used the offending books for lack of alternative selections. One library book, *An Irish Heart,* warned that if immigration from Ireland continued, "our country might be appropriately styled the common sewer of Ireland." A textbook, the *New York Reader,* contained "A Dialogue between Fernando Cortez and William

Penn," in which Penn denounces Cortez, the "Papist," for the murders of the Inquisition and of innocent Indians. Another reader gave biographical data on "Huss, John, a zealous reformer from Popery. . . . He was bold and persevering; but, at length, trusting himself to the deceitful Catholics, he was by them brought to trial, condemned as a heretic and burnt at the stake." In another brief biography, Martin Luther was described as "the great reformer. . . . The cause of learning, of religion, and of civil liberty, is indebted to him, more than to any other man since the Apostles." The most generally used English reader of the day, by Lindley Murray, pictured Martin Luther, the Protestant reformer, rousing people from the Dark Ages. The Catholics hated the implication that all learning was stultified until the Protestant Reformation.[9]

One story drew Catholic ire because of its espousal of the equality of all religious beliefs. The story, which was supposed to exemplify religious tolerance, showed a father taking his son out on a Sunday morning to see the many different kinds of churches that exist in a free country. The son asked, "Why . . . do not all people agree to go to the same place and to worship God in the same way?"

> "And why should they agree?" replied his father. "Do you not see that people differ in a hundred other things? Do they all dress alike and eat and drink alike, and keep the same hours, and use the same diversion?"
>
> "In those things they have a right to do as they please," said Edwin.
>
> "They have a right, too," answered his father, "to worship God as they please. It is their own business and concerns none but themselves."
>
> "But has not God ordered particular ways of worshipping him?"
>
> "He has directed the mind and spirit with which he is to be worshipped, but not the manner. That is left for every one to choose. All these people like their own way best."

Bishop Hughes thought this story equated religious beliefs with trivial choices like ways of eating and dressing. He quipped that the boy showed more sense than his father. As the defender of a church which believed in infallible authority, he could not condone teaching children to judge for themselves in the realm of religion. This passage typified for him the teaching of infidelity.[10]

DURING LATE SEPTEMBER 1840, while the Public School Society was busily expurgating the textbooks, the Catholics readied a new appeal for public funds, this time directed to the city's Board of Aldermen. If the Catholics won, the negative decision of the Board of Assistant Aldermen would be reversed; if they lost, they could then carry their plea to the state legislature in Albany.

Where their earlier petition had asked for funds on the basis of need, the new petition made the major issue the Catholics' rights of conscience. Once again, the Catholics denounced the Society's biased schoolbooks and claimed that a public school education was deleterious to children's morality. Once again, they characterized the Society as an arrogant and wealthy monopoly which callously disregarded the rights of Catholics. In a new twist, the Catholics complained that the Society lacked the confidence of the poor, since it was necessary for the Society to hire attendance officers to coerce vagrant children to attend school. As a gesture of compromise, the Catholics offered to place their schools under the supervision of the Public School Society, in order to allay fears that public funds would be used for sectarian purposes. If they were given public funds, they promised all religious instruction would be offered after school hours. Their petition held that parents had the right to control the education of their children and to have teachers whom they trusted.

In answer to the new Catholic petition to the Board of Aldermen, the Public School Society submitted a brief remonstrance, referring to the lengthy arguments advanced previously in the Board of Assistant Aldermen. The Society answered the Catholics' new charge succinctly: "The records of the schools will demonstrate that the industrious and respectable portions of the laboring classes repose entire confidence in the public school system and its managers." Again, the Society steadfastly opposed any grant of public funds to sectarian institutions, under any conditions; there could be no "apology for breaking down one of the most important bulwarks of the civil and religious liberties of the American people." [11]

The Board of Aldermen invited both sides to speak before them in a great debate on October 29 and 30. It was highly unusual for petitioners to appear personally before the board, but an exception was made because of the interest and importance of the question. After months of agitation and meetings, the issue had become the city's most heated controversy, the subject of angry editorial denunciations and pronouncements from the pulpit. The Jews were the only denomination which did not join in the general attack on the Catholics; the *New York Herald* noted with amusement that the Jews and Catholics were united for the first time in 1,840 years.[12]

Hughes determined that he, rather than a Catholic lawyer, would represent the Catholic cause at the great debate. Some of his followers expressed concern that he might sustain insults to his dignity in the highly charged situation. He replied to them that "so vital and important do I consider the question, that I conceive that I cannot be anywhere more in keeping with my character as a bishop than when I stand before you pleading the cause of the poor and the oppressed. And so near is the question to my heart, that I can bear insult from morning till night." [13]

The debate was held at City Hall, before the full Common Council; the Board of Aldermen invited the Board of Assistant Aldermen to hear the debate with them. The council chamber was filled with a vast crowd, and the corridors were so packed that the speakers had difficulty pushing their way through. Bishop Hughes arrived accompanied by several Catholic priests and prominent laymen.

The Public School Society was represented by two trustees, both prominent attorneys, Hiram Ketchum and Theodore Sedgwick. In addition, several Protestant ministers appeared to speak in opposition to the Catholic petition.

Speaking first, Hughes delivered a three-hour attack on the Society's schools and a defense of the Catholic request. He dissected all the previous statements of the Society, sentence by sentence, to demonstrate that Catholic motives had been misinterpreted. He insisted that they did not seek funds for sectarian purposes, as the Society had charged, and would charge again that day. The Catholics would not expose their children to the ingrained bias of the public schools. Hughes mocked the Society's efforts to expurgate its books of bigotry. He wondered why Protestants should require Catholic assistance to detect slander:

> Are we to take the odium of erasing passages which they hold to be true? Have they the right to make such an offer? . . . Have they given us a pledge that they will do it, or that they will not even then keep them in? . . . And, then, after all the loss of time which it would require to review these books, they can either remove the objectionable passages, or preserve them as they see fit. An individual cannot answer for a whole body. . . . We may correct one passage to-day, and another next week; and then another body may come into power, and we may have to petition again and again. [14]

Hughes saw nothing in the Constitution or the laws of the state which prohibited public aid to religious societies. Their exclusion from the common school fund in 1825 had occurred because of the abuses of one of them. He bitterly condemned the interpretation of the laws which required that common school education be confined to a secular education:

To make an infidel what is it necessary to do? Cage him up in a room, give him a secular education from the age of five years to twenty-one, and I ask you what he will come out, if not an infidel? . . . Now I ask you whether it was the intention of the Legislature of New York, or of the people of the State, that the public schools should be made precisely such as the infidels want? . . . They say their instruction is not sectarianism; but it is; and of what kind? The sectarianism of infidelity in its every feature.

The Society was characterized by Hughes as an organization of wealthy men who had "step by step" enlarged their sphere until they monopolized public education funds. He expressed astonishment that men of good intentions "would sooner see tens of thousands of poor children contending with ignorance, and the companions of vice, than concede one iota of their monopoly, in order that others may enjoy their rights." Hughes derided the Society's fear that other sects would demand a share of the school fund if the Catholics won theirs; since so many other sects had joined in opposition to the Catholics, he believed that they must be well satisfied with the schools of the Society. But if any should want public funds to educate their poor, then Hughes thought they too should have it.

Speaking on behalf of the Public School Society, Theodore Sedgwick responded that there were three possible ways in which education and religion might be related. First, there was a completely secular education, such as a master gives an apprentice, and in which religion plays no part. Then there was moral education, which instructs children in the fundamental tenets of duty which are common to all religions, such as admonitions against lying and stealing; this was the education given by the Public School Society. And there was religious education, which inculcates specific sectarian dogma. Sedgwick emphasized that "no man can undervalue" the importance of a religious education, "but the State does not intend to give it." He saw no point to the Catholic petition other than to establish sectarian schools for religious purposes, for their major objection to the existing schools was that they conveyed no religious instruction.

Hiram Ketchum, the Public School Society's other attorney, personally attacked Hughes for descending "into the arena." He charged that the Catholics' regular meetings were like political rallies and that Hughes was trying to bring political pressure on the Common Council. He referred several times to Hughes as "the mitred gentleman," which struck Hughes as an appeal to bigotry.

Hughes replied that the speakers for the Society had posed two alternatives: "Either the consciences of Catholics must be crushed and their objections resisted, or the Public School System must be destroyed." He complained that the two lawyers had not spoken to Catholic grievances at all, but had merely urged the defeat of their petition. He spoke sarcastically of Ketchum's criticism of him:

> This gentleman . . . really seems to me to confirm all I say on the ground we have taken. I know he lectured me pretty roundly on the subject of attending the meetings held under St. James Church. I know he did more for me than the Pope: the Pope "mitred" me but once, but he did so three or four times during the course of his address. He read me a homily on the duties of station; and he so far forgot his country and her principles, as to call it a "descent" on my part, when I mingled in a popular meeting of freemen. But it was no descent; and I hope the time will never come when it will be deemed a descent for a man in office to mingle with his fellow-citizens when convened for legitimate and honorable purposes.

Hughes admitted that Catholic priests had discouraged their followers from patronizing the public schools. "So long as they are attached to our religion, it is my duty, as their pastor, as the faithful guardian of their principles and morals, to warn them when there is danger of imbibing poison instead of wholesome food."

Because the fight was led by a Catholic bishop, the issues were cast in religious terms. There were, however, other tensions which transcended the overt debate. Even had there been no religious difference between the immigrants and the natives, there were sufficient grounds for cultural conflict. The newcomers were discriminated against not only because they were Catholics, but because they were immigrants, they were of a minority ethnic group, and they were poor. Had their cause been led by an Irish immigrant who was not a clergyman, it is possible that the goal might have been more Irish teachers, more Irish trustees, sympathetic textbooks, and so on. In this era, however, religion was of great importance to Catholics and Protestants alike, and religion became the language in which the cultural clash was expressed. Stimulated in part by its own traditions of exclusivity and in part by the pervasive aura of Protestant bigotry, the Catholic Church resolved to have its own schools rather than to use its grievances to restructure the Public School Society. The Church did not want public schools which were truly common schools, where Irish children and other Catholic children might go without fear of prejudice. Because of the Church's view of the inseparability of religion and education, the kind of nonsectarian common school which had developed in New York and in other parts of the country was wholly inimical.

AFTER THE GREAT DEBATE ended, the Board of Aldermen referred the question to a committee, where it remained for the next ten weeks. In the intervening period, the aldermanic committee attempted to reconcile the Society and the Catholics. Each side submitted compromise proposals. The Society offered again to submit its books to the Catholics. The Catholics offered to place their schools under the Society's supervision, to submit their teachers to the Society's examination, and to open their schools to government inspection. Neither proposal satisfied both groups. The Catholics were unwilling to agree to any plan that did not recognize the distinctive character of Catholic schools, and the Society would not agree to the public funding of any distinctly sectarian school, no matter who its teachers or how its funds were spent.

The committee, at the Society's request, inspected the schools of both contestants. The committee reported that the public schools were "admirably adapted to afford precisely the kind of instruction for which they were instituted." The public school classes

> exhibited an astonishing progress in geography, astronomy, arithmetic, reading, writing, &c, and indicated a capacity in the system for imparting instruction far beyond our expectation; and though the order and arrangement of each school would challenge comparison with a camp under a rigid disciplinarian, yet the accustomed buoyancy and cheerfulness of youth and childhood did not appear to be destroyed in any one of them.

The Catholic schools that the same committee visited were

> lamentably deficient in accommodations and supplies of books and teachers. The rooms were all excessively crowded and poorly ventilated, the books much worn as well as deficient in numbers, and the teachers not sufficiently numerous. Yet, with all these disadvantages, though not able to compete successfully with the public schools, they exhibited a progress which was truly creditable; and, with the same means at their disposal, they would doubtless soon be able, under suitable direction, greatly to improve their condition.[15]

On January 11, 1841, the special committee urged the Board of Aldermen to reject the Catholic petition. Its recommendation was adopted by a vote of 15–1. The lone dissenting alderman was Daniel Pentz, who afterwards was a great favorite of the Irish Catholics for his stand on their behalf.

Now the Catholics, having exhausted their appeals at the local level, could direct their claim to Albany, where they knew they could count on the support of Governor William Seward.

CHAPTER 6

First Round in Albany

IN THE FALL OF 1840, Governor William Seward was re-elected, carried to victory on the coattails of General William Henry Harrison, the successful Whig presidential candidate. In 1838, Seward had won a majority of over 10,000 votes; in 1840, his margin of victory was halved. He was unquestionably hurt by the school issue. The expected defection of Irish Catholics from the Democratic ticket did not materialize; at the same time, some nativist Whigs refused to vote for Seward. Seward was accused of scheming to put Protestant children into Catholic schools, of trying to undermine the Protestant religion, of making a deal with the Pope, of subverting republican institutions, of being a secret Jesuit. His son later wrote that Seward was urged to "explain away, withdraw, or recant the unpopular doctrine by some public avowal, in order to save his election. To all such his answer was firm and decided. He believed the principle to be right; and not less so because it was unpopular for the moment. He should adhere to it, let the election go which way it might." [1]

Bishop Hughes was disappointed that his countrymen had not supported Seward. He wrote to Seward after the election, "We are almost ruined by the havoc which politics have made among our ignorant and misled people." He knew that most Irish gave unquestioning allegiance to Tammany Hall. He considered the pro-Democratic *Truth Teller* to be a "vile print," whose editors were "miserable traffickers" in the credulity of the Irish Catholics. [2]

Six days before the Common Council rendered its negative judgment on the Catholic petition, Seward delivered his third annual message to the legislature. He strongly reiterated his dissatisfaction with New York's public schools, because of their failure to educate large numbers of foreign children, and served notice that the Catholics' fight would now be carried to Albany:

Although the excellent public schools in the city of New York are open to all, and have long afforded gratuitous instruction to all who seek it, nevertheless the evil there exists in its greatest magnitude. Obviously, therefore, something more is necessary to remove it than has yet been done. . . . No system is perfect which does not accomplish what it proposes; our system, therefore, is deficient in comprehensiveness in the exact proportion of the children that it leaves uneducated. . . . Knowledge, however acquired, is better than ignorance. . . . Neither error, accident, nor prejudice, ought to be permitted to deprive the State of the education of her citizens. . . . I solicit their education less from sympathy, than because the welfare of the state demands it, and cannot dispense with it. . . . I desire the education of the entire rising generation in all the elements of knowledge we possess, and in that tongue which is the common language of our countrymen.[3]

Though Seward withdrew his original controversial proposal to teach foreign children in their native tongue, he could not resist declaring that he had no fear of "the influence of any language or creed among an enlightened people." His message left no doubt that he would support any reasonable plan to advance universal education in New York City, whether secular or sectarian.

Nativist members of his own party in New York City saw no reason why a Whig governor should court the Roman Catholics. But Seward's re-election, narrow though it was, left him free to pursue his convictions without regard to the nativists in his party. He was prepared to take the political consequences; he was not a candidate for re-election in the following election. When he first made his educational recommendations in January 1840, he obviously did not anticipate the passions it would arouse. He wrote to an elder of the Methodist Episcopal Church in November 1840 that he "had never heard of the New York Public School Society its foundation its powers or its funds nor of the disputes which had existed concerning the apportionment of the public school monies." In another letter, he expressed his inability to understand why Americans, all descended from immigrants, should hate foreigners or wish to deny them the rights and privileges of citizenship:

I ventured to promise myself that one of the chief benefits I might render the State was, to turn the footsteps of the children of the poor foreigners from the way that led to the House of Refuge and the State-Prison, into the same path of moral and intellectual cultivation made so smooth and plain for our own children. . . . If there was one policy in which I supposed all republican and Christian citizens would concur it was this. I found, however, to my surprise, that the proposition encountered unkind reception. . . . Nevertheless, I am not discouraged by all this, I am only determined the more conclusively to discharge the responsibility resting upon me, . . . to provide for the support of the glorious superstructure of universal suffrage—the basis of universal education.[4]

Seward's public commitment to assist the Catholic cause guaranteed that the next round in the school fight would occur in Albany. Seward and Hughes felt that the Catholics had a good chance of winning, since the Whigs controlled both houses of the legislature.

To organize for the battle at Albany, Bishop Hughes summoned a meeting at Washington Hall, corner of Broadway and Reade Street, on February 11, 1841. Hughes derided the Common Council's rejection as "the defeat of justice by authority." Catholics were not friendless, he promised: "This is a question of right, and though a whole Board should be found to bend the knee to the Baal of bigotry, men will be found who can stand unawed in its presence, and do right." The Catholics that night formed a Central Executive Committee on Common Schools, which called protest meetings in wards all over the city, organized local committees, and gathered signatures on petitions to the legislature.[5]

About 7,000 signatures and a Catholic statement were brought to Albany by Joseph O'Connor, a prominent Catholic lawyer. O'Connor presented them to Gulian C. Verplanck, a Whig state senator from New York County who was known to be friendly to the Catholic cause (none of the legislators at the city or state level was Catholic). O'Connor conferred with the secretary of state, John Spencer, who was also ex officio superintendent of common schools for the state. At the next regular meeting of the Catholics on April 3, O'Connor reported back that he had found at Albany "honest, liberal, honorable men—men of both political parties who publicly and fearlessly acknowledged that there was something wrong in the city of New York—that there was here an unnatural corporation, such as one as should not exist under a republican government." [6]

The statement delivered by O'Connor was presented to the state senate by Verplanck on March 29, 1841. It was titled a "Memorial of Citizens of New York for an amendment of the common school laws." The petitioners were advised by Verplanck and others to call themselves citizens, rather than Catholics. After the memorial was read in the senate, it was at once referred to John Spencer. Spencer, though he was the governor's chief educational agent, was not an educator; he was a lawyer and a politician who had held numerous elective and appointive offices. Before the school controversy came to an end, Spencer was appointed secretary of war in President Tyler's cabinet, and later secretary of the treasury.

While Spencer was studying the memorial, the city of New York held its local elections for mayor and the Common Council. Among those seeking re-election was Daniel Pentz, the alderman who had cast the dissenting vote on the school issue. Large handbills appeared all over the city, urging Catholics to turn out and support their candidates. One poster read, "Catholics Arouse! . . . combine—agitate! agitate! agitate! . . . *the balance of power is in your hands*—Use it!" Another handbill urged Catholics to vote for Pentz, "the friend of the Catholics." The pro-Democratic *New York Evening Post* called these handbills a transparent Whig deception intended to split the Democrats. On April 13, a Democratic mayor and Common Council were elected, but Pentz, also a Democrat, was defeated. The Catholic *Freeman's Journal* expressed indignation that Pentz was singled out for defeat, obviously for his independent stand on the school question. Catholic editorialists warned their opponents not to complain if the Catholics should follow their example in the future and vote their self-interest.[7]

On April 26, John Spencer presented his recommendations to the state legislature, and at once transformed the politics of the New York schools. After Spencer's report, the Public School Society found itself on the defensive, struggling to survive. Spencer sided completely with the "citizens" who petitioned the legislature for a change in the school system; he never once referred to them as Roman Catholics.

Spencer changed the nature of the controversy by redefining it. He held that the entire dispute was befogged by the religious issue, when in fact there were broad educational questions at stake. Disagreeing with the Public School Society's concept of the common school, he contended that the school should be whatever the community around it wanted it to be. The issue, then, was larger than whether or not there would be religious instruction in the public schools; it was, instead, a question of who would control the schools and whether there would be any constraints on their power.

Should there be limits on the power of the majority? This dilemma would reappear time and again in the American public schools, cast in two versions, either as a conflict between the will of the majority and the rights of the minority, or between the will of the people and the judgment of the expert. The American concept of lay control of public education implies the persistence of these issues as inherent tensions.

If the public school belongs to the public, then the majority of voters or school board members has the power to determine the purpose and direction of the school. Under this definition, if the community elects a school board that wants all children to learn Roman Catholicism or Protestant fundamentalism, then the state, as well as dissenting teachers and parents, must bow to the decision of the majority. This interpretation of untrammeled local control has given rise to the Scopes trial in 1925, and to countless other school battles where local boards have attempted to purge books and teachers in order to uphold traditional religious, political, or social dogma.

However, there is another definition of the public school which accepts public control but limits the community's power to inculcate its views. The public school, it holds, is an instrument of the state, and the state should neither suppress opinions nor hold any of its own. The role of the school is to encourage inquiry, not to impose interpretations. In this definition, the school must eschew any form of sectarianism, whether religious or political; must respect the academic freedom of its teachers and students; and must teach only those values which are commonly held. This was the vision of the common school that Horace Mann was developing in Massachusetts in the 1840s, firmly disassociating the school from any partisanship.

The tension between these two definitions of the common school can never be altogether resolved, because each, in its pure form, is unsatisfactory. The school that is strictly controlled by its community poses a threat to minority rights and academic freedom and acts as a spur to factionalism within the community, as competing groups seek control; the school which embodies only one doctrine is unlikely to hold the allegiance of those who disagree, unlikely to retain public confidence, and unlikely to remain a public school. Yet no school can wholly eliminate the teaching of values and beliefs, for to do so would make it impossible to distinguish between right and wrong, in human relations as well as in history. An uneasy compromise is struck by accepting a notion of "common values," but what constitutes common values at one point in time may not a decade later. The need to adjust to dissident views guarantees a constant potential for conflict between the school and the community.

John Spencer concluded that the "common" values of the Public School Society were not common enough and that the remedy was to place control of the public schools in the hands of the public. Spencer's analysis was by no means impartial. He had received his appointment from Governor Seward, and the governor had strong convictions on the subject. Spencer had the delicate political task of satisfying the demands of the Catholics without alarming non-Catholics. An outright recommendation of public assistance to parochial schools would have been politically unacceptable. Realizing this, Spencer attacked the Public School Society as a private, closed corporation which determined its own policies, selected its own officers, handled large amounts of public funds, and was not subject to state supervision. The power of this organization, he charged, was "hostile to the whole spirit of our institutions." He painted a picture of "many thousands" of citizens demanding the right to control their children's education through elected officials.

The autocratic nature of the Society, Spencer claimed, was responsible for its low enrollment. Using very rough estimates, he held that fewer than half the children in New York between the ages of four and sixteen were in school. He described an attitude "prevalent among the people, that an attempt is made to *coerce* them, directly or indirectly, to do something which others take a great interest in having done. They are not left, or called on, to act *spontaneously*—to *originate* anything, or take any part in matters which they are told most deeply concern themselves." The Society was not equipped to govern the public schools because it operated

> without allowing to the parents of the pupil the direction of the course of studies, the management of the schools, or any voice in the selection of teachers; it calls for no action or cooperation on the part of those parents, other than the entire submission of their children to the government and guidance of others, probably strangers, and who are in no way accountable to these parents.[8]

To right this wrong, Spencer proposed that all schools in the city which educated poor children receive a share of public funds. He held that it was better to have a large number of schools, whatever their quality, than to have a limited number of excellent schools. He accepted the Catholic view that nonsectarianism was a form of sectarianism. Nonsectarianism was approved by men who had no creed, but was likely to offend those who wanted the schools to transmit their creed to their children.

Spencer argued that it was wrong to exclude sectarianism from the schools. If it was the will of the people to have religious training in the public schools, then so be it. In this Christian country, he wrote, "public sentiment would be shocked by the attempt to exclude all instruction of a religious nature from the public schools."

Spencer proposed that the state adopt a policy of "absolute nonintervention" in religious practices in the schools. The state should not legislate on the matter at all; it should be left to the discretion of the people in each district. In other parts of the state, he wrote, the nature and extent of religious instruction

> is left to the free and unrestricted action of the people themselves. . . . The law provides for the organization of districts, the election of officers, and the literary and moral qualification of teachers, and leaves all else to the regulation of those for whose benefit the system is devised. The practical consequence is, that each district suits itself, by having such religious instruction in its schools as is congenial to the opinion of its inhabitants.

The difficulty with this proposal was the great dissimilarity between the heterogeneous, shifting population of New York City and the fixed, homogeneous populations of the small towns and villages in the rest of the state. Within any section of New York City could be found numerous religious and ethnic groups, which were unlikely to agree on sectarian practices of one kind. To surmount this objection, Spencer suggested that schools in New York City be controlled by whoever established them. In this way, Catholic schools would be designated district schools and remain under Catholic control.

The Public School Society failed, wrote Spencer, because it was centralized; no single religious policy could suffice for all the diverse elements in the population. Decentralizing the schools would permit every group to have its own religious training. He predicted that a healthy rivalry among the school districts would improve the quality of instruction, since schools would receive funds on the basis of attendance rolls. Further improvements would result from the "care and protection" which "the people themselves" would provide for the schools.

To implement his proposals, Spencer recommended that each ward in the city be considered a separate school district, as distinct as any town school system; that each ward elect trustees to a district board; and that each ward elect a commissioner to represent them on a city-wide board. The schools of the Public School Society would continue to operate with their own board of trustees, but under the general supervision of a city-wide Board of Commissioners of Common Schools.

BISHOP JOHN HUGHES was delighted with Spencer's report; he saw it as "a blow . . . from which the Public School Society will never recover." He had the good political sense not to show his pleasure publicly. The usual Catholic meetings were temporarily suspended to avoid inflaming Protestant opinion. He did not want to give the impression that Spencer had placated or championed the Catholics. Hughes wrote to Spencer at this time and revealed his intense hatred for his adversaries: "I think that the blow which has been given to this complex monopoly, of mind, and money, and influence, in the city of New York, must be followed up, until they have virtually 'ceased to reign.' That point once gained, their powers of mischief will have lost their impulse to activity against you or his Excellency the Governor." [9]

A bill embodying Spencer's recommendations was prepared, and the State Senate Committee on Literature held hearings. Hiram Ketchum appeared for the Society. He pointed out to the committee that the petitioners were not simply a group of discontented citizens, but were specifically the Roman Catholics, whose similar request had been twice denied by the city's Common Council, a popularly elected body. Their request was rejected, said Ketchum, not because they were Catholics (Protestants' requests had also been turned down), but because they wanted funds for sectarian schools. He appealed to the committee not to pass legislation which failed to respect the wishes of the majority of New York City. He attacked the facts in Spencer's report, especially his figures on the number of children not in school. The figure Spencer used was based on children between four and sixteen; Ketchum noted that most children entered school at six and left school at age twelve to work. He denied that the Society was not accountable to the public; the regulation of its funds by the Common Council was a direct form of democratic control. He defended the private, corporate structure of the Society by likening it to the boards of the almshouses, hospitals, and other charities which expended public funds for benevolent purposes. Ketchum strenuously objected to Spencer's idea about religious instruction in the schools, calling it "the tyranny of the majority." Catholics should not be taxed to support the inculcation of the Protestant religion, he argued, nor should Protestants be taxed to support the Catholic religion. Under Spencer's plan, he charged, the minority would be compelled "to worship according to the opinions of the majority. . . . If there were five hundred in one district, and but one man in that district that protested, he would have a clear right to do so." [10]

The Society opposed the introduction of the district system on three grounds: first, that it tended to become associated with partisan politics; second, that a city with a constantly shifting population required uniformity in the schools, not competing systems within the city; and third, that the district system would not be able to appease conflicting religious views except by driving out the minority view or by adopting the nonsectarianism of the Public School Society.

As debate began in the senate, the atmosphere became heated. An anti-Catholic diatribe was placed on the desk of each senator. The city's Common Council asked the senate to defer action in order to give the aldermen time to study the proposal. The animated debate was brought to an unexpected halt by a motion to postpone consideration of the issue until the beginning of the next session of the legislature, in January 1842. The friends of the bill were taken by surprise, and the motion passed 11–10.

Supporters of the bill were outraged. First to react was Governor Seward. Previously, in the spring of 1841, he had nominated Hiram Ketchum to fill a vacancy on the state circuit court ("to placate rabid Protestant opinion in New York," wrote Glyndon Van Deusen, Seward's biographer). A week after Seward's school bill was shelved, Seward withdrew Ketchum's nomination. Ketchum helped to defeat the bill, at least temporarily, and lost the judgeship.[11]

The Catholics resumed their meetings. Bishop Hughes devoted three evenings in June to attacks on Ketchum's speech to the senate committee. After Hughes reviewed Ketchum, a prominent Presbyterian minister, Reverend William Brownlee, announced his intention to review Hughes. Brownlee was well known for his nativism and anti-Catholicism, and his discourse on July 8, 1841, set off a small riot between Protestants and Catholics in his audience. The next night Catholics barred him from entering the North Dutch Church, where he planned to lecture again, and another riot ensued. Catholics and Protestants were polarized by the issue, and the breach between them widened as the fall elections approached.

CHAPTER 7

"No Flinching!"

THE NOVEMBER 1841 ELECTION was crucial to the settlement of the school question, which was the dominant issue in the campaign. Thirteen assembly seats and two senate seats were at stake in the city. Late in May, a nativist group headed by Samuel F. B. Morse formed the American Protestant Union, which planned to endorse only those candidates who were "safe" on the school issue. Protestant and Whig newspapers urged voters to support only candidates pledged to the Public School Society. The Whigs decided to abandon their own governor by running a ticket pledged to oppose any change in the existing system. Gulian Verplanck was the only Whig favorable to the Catholics who was offered his party's nomination; when he declined to run, the ticket was unanimous against change.

The Democrats' dilemma was even more ticklish than that of the Whigs. If they alienated either their Protestant or their Catholic voters, defeat was certain. Tammany Hall wriggled out of the predicament in customary style, by trying to please all factions simultaneously. When the list of nominations was made public, the Democratic nominees for the assembly included Daniel Pentz. The *Tribune* commented that "Mr. Pentz is put on to catch the Catholics." The position of the other nominees on the school issue was not known. The Democratic *Evening Post* assured its readers that the Democratic candidates "are not for allowing any religious denomination whatever to put in a claim for any portion of the school money." On the surface it appeared that the entire city was arrayed against the Catholics.[1]

With the election only a few days away, Bishop Hughes decided that it was time to act. On the night of October 29, "the friends of civil and religious freedom" met at Carroll Hall, on the corner of City Hall Place and Duane Street. Hughes announced to a crowded hall that since so many of the candidates on both tickets were pledged against the

Catholics, it had become necessary to run an independent ticket friendly to the Catholic cause. He revealed a slate which endorsed ten of the thirteen Democratic nominees for the assembly and which proposed independent Catholic nominees for the three remaining assembly seats and the two state senate seats.

The reading of the names was received with loud cheering and applause. The Bishop's emotional appeal for unity was reported the next day in an extra edition of the *Freeman's Journal:*

> You have now, gentlemen, heard the names of men who are willing to risk themselves in support of your cause. Put these names out of view, and you cannot, in the lists of our political candidates, find that of one solitary public man who is not understood to be pledged against us. What, then, is your course? You now, for the first time, find yourselves in the position to vote at least for yourselves. You have often voted for others, and they did not vote for you; but now you are determined to uphold, with your own votes, your own rights. (Thunders of applause, which lasted several minutes.) Will you, then, stand by the rights of your offspring, who have for so long a period, and from generation to generation, suffered under the operation of this injurious system? (Renewed cheering.) . . . Will you let all men see that you are worthy sons of that nation to which you belong? (Cries of "Never fear—we will! We will, till death!" and terrific cheering.) Will you prove yourselves worthy of friends? (Tremendous cheering.) Will none of you flinch? (The scene that followed this emphatic query is indescribable, and exceeded all the enthusiastic and almost frenzied displays of passionate feeling we have sometimes witnessed at Irish meetings. The cheering, the shouting, the stamping of feet, waving of hats and handkerchiefs, beggared all powers of description.) Very well, then; the tickets will be prepared and distributed amongst you, and, on the day of election, go, like free men, with dignity and calmness, entertaining due respect for your fellow-citizens and their opinions, and deposit your votes.[2]

The day after the ten Democratic assembly nominees received the Catholics' endorsement, a concerted effort was made to put them solidly on both sides of the issue. The *Evening Post* published a card to the public signed by ten Democratic candidates, stating their unequivocal opposition to any appropriation of the school fund for sectarian purposes and disowning the Carroll Hall endorsement. Of the ten who signed the card, seven had been endorsed by the Catholics. The three who did not sign the statement in the *Evening Post*, Daniel Pentz, William B. Maclay, and John L. O'Sullivan, were "out of town." Before the election, Pentz and O'Sullivan also issued statements of opposition to funding sectarian schools. Maclay was the only Democrat who did not announce his opposition to the Catholic claim, but he was so widely thought to be on the side of the Public School Society that the nativist American Protestant Union endorsed him.

The press reacted excitedly to the bold Catholic ploy of endorsing a separate slate. All of the papers condemned Bishop Hughes, except for the Catholic press and Horace Greeley's *Tribune*. Greeley, an intimate of Governor Seward, urged voters to support the Whig ticket no matter how they felt on the school issue, and, simultaneously, accused the Democratic candidates of "cowardice, insult and treachery" for trying to hold both sides of the question. The rest of the Whig press denounced the Democratic disclaimers as frauds and charged that there was collusion between the Catholics and Tammany Hall. The *Post* countered by charging that the Carroll Hall meeting was "a plot concocted at Albany, to bring odium upon the democratic ticket," since all the Democratic candidates had clearly stated their opposition to the Whig governor's scheme for allocating money to the Catholic Church. The nativist press crowed delightedly that Hughes' entry into politics proved all their fears of a Roman Catholic plot to gain control of the country.

The angriest personal denunciation of Hughes came from James Gordon Bennett, editor of the *Herald*. He thought Hughes' actions were part of a plot to discredit the Democratic ticket and called Hughes "the mere tool of Governor Seward." He charged him with "a bold, daring, reckless, unexpected attempt to control politics by the force of religious sentiment," thereby "sullying his garment with political dirt. . . . This intrigue has been negotiating between Governor Seward and Bishop Hughes for nearly two years past." Bennett recommended that clergymen "must be confined to their own proper business—saying masses for the dead—forgiving the sins of the living—giving the sacrament—marrying young couples at $5 a head—eating a good dinner and drinking good wine at any generous table—or toasting their shins on a cold night at their own fire-sides." He called on Catholics to "abandon the impudent priest and his separate ticket." He warned the mayor to have sufficient military force on hand during the election, because Hughes' movement might lead to religious violence "of the darkest Popish ages of Roman religious tyranny." He felt that Hughes was trying to mold the Catholics into an independent political force which could become the balance of power in city elections. For months after the election he kept up an intense personal barrage against Hughes. Only the *New York Times* defended Hughes' right "to mingle in politics, express an opinion, deposite his vote, etc. as anyone else."

Two days before the election, the American Protestant Union and the nativist *Journal of Commerce* announced a bipartisan Union ticket to rebuke "priestly dictation." So ambiguous were the pronouncements of some Democratic candidates that three of them, including William B. Maclay, received the endorsements of both the Catholics and the American Protestant Union.[3]

The election was a Democratic victory throughout the state. The Democrats regained control of both houses of the legislature from the Whigs. The balance in the state senate changed from a 21–11 Whig majority, to a 17–15 Democratic majority. In the assembly, the 66–62 edge held by the Whigs was resoundingly reversed to 33 Whigs and 95 Democrats. In New York City the Democrats managed to win ten of thirteen assembly seats, and one of two senate seats. The ten winning Democrats were those who had received the Catholic endorsement. The winning candidates received from 16,000 to 18,000 votes each. So close was the election that only a few hundred votes separated the losing candidates from the winners. The five independent Catholic nominees for assembly and senate received little more than 2,000 votes each. Given the closeness of the election, these votes were sufficient to deny victory to the Democrats who failed to receive the Catholic endorsement.

The press looked at the 2,000-odd votes received by the independent Catholics and thought that the election was a decisive rejection of the Catholic position. But Hughes knew otherwise. He knew that his gamble had paid off. He had conclusively demonstrated to Tammany Hall that it needed Catholic support to win. In a city where Whigs and Democrats were almost evenly divided, the 2,000 votes which Hughes influenced were crucial. The Democrats with the Carroll Hall designation had won, and those who did not have it had lost. Hughes had achieved what James Gordon Bennett feared: on this issue and at this time, a small but effective band of Catholic voters held the balance of power in city elections.

A week after the election, a large Catholic meeting convened and publicly warned the Democrats that unless the Democratic-controlled legislature acted on their demands, another Catholic challenge would be injected into the city's spring elections for mayor and the Common Council.

When the state legislature convened, the Democrats were in a quandary. Something had to be done on the school question in order to forestall Catholic political action in the spring elections; Tammany Hall did not wish to risk the loss of valuable patronage controlled by municipal officials. And they knew that whatever they did, they would have to bear the responsibility for it with Protestant voters. Worst of all, they were saddled with this burden because of a Whig governor. Nonetheless, they determined to pass a bill which would placate the Catholics, regardless of campaign pledges to the contrary. The committee in each house which would consider the school issue was carefully chosen to expedite favorable action. In the assembly, William B. Maclay, the New York City Democrat who had been unpledged on the question and endorsed by both extremes, was selected as chairman of the Committee on Colleges, Academies, and Common Schools. In the senate, the Democratic leadership did not put either New York City senator on the crucial Committee on Literature, knowing that both were pledged to preserve the Public School Society.

Governor Seward's annual report to the legislature at the beginning of the 1842 session again called for restructuring of the New York City school system. He stated that the Public School Society, "after a fair and sufficient trial," had failed to gain the public's confidence and was unable to provide universal education. It was not for society to conform itself to the public schools; rather, the public schools must adapt themselves to the demands of society. He promised to approve any adequate remedy devised by the legislature which vested control of the schools in elected commissioners. Since parents and taxpayers have a right to participate in the control of public education, "what has hitherto been discussed as a question of benevolence and of universal education, has become one of equal civil rights, religious tolerance, and liberty of conscience." [4]

A month later, the Maclay committee in the assembly reported back a bill entitled "An Act to Extend and Improve the Benefits of Common School Instruction in the City of New York." The Maclay report easily translated the school question into terms which Democrats could accept and defend to their constituents. Opposition to monopolies was a prime Democratic stance, and Maclay cast the present issue in the Democratic tradition of antimonopolism. He castigated the Public School Society as an "irresponsible private chartered company. Such a concentration of power into mammoth machinery of any description is odious to the feelings, and sometimes dangerous to the rights of, freemen." He asserted that the Society had defended its monopoly for years "against the struggles of discontented masses of the population."

Not only was the Society a powerful and privileged monopoly which violated popular rights by its very existence, held Maclay, but it signally failed to provide universal education. He claimed that 96 percent of the children in the rest of the state were in school, while fewer than 60 percent of the city's children were in school. This was to him "incontrovertible proof . . . that nearly one half of the citizens of the metropolis protest against the system, and demand its modification." The Public School Society complained that the figures used by Spencer and now Maclay were distorted to damage the Society; they claimed that the estimates did not include the thousands of children in pay schools. They also insisted that their own attendance figures represented an average quarterly attendance, while the rural figures counted any student who attended school for any length of time, whether a month, a week, or a day.[5]

Maclay stated that popular dislike for the existing school system ran so deep that it could only be remedied by "bringing home the education of the young of the city to the business and bosoms of their parents. The common school system of the State successfully and admirably accomplishes that object; and the committee therefore recommend that the system shall, as far as it is practicable, be extended to the city and county of New York." The bill proposed that the city's wards be treated as separate towns for school purposes, each electing its own trustees, inspectors, and commissioners of school monies. The Maclay bill made no mention of religious instruction, implying that it accepted the Spencer doctrine of "absolute non-intervention" into each district's sectarian practices.

One assemblyman suggested that the bill be submitted to a referendum in the city; he felt that the matter was a local issue, and that there was as much propriety in obtaining a majority vote as in the case of the Croton Water Works bill, which had been submitted to the city voters a few years before. Maclay strongly opposed this suggestion. He said that referenda should be called only on very important and rare occasions, and he did not consider this to be one of those occasions. Other New York City Democrats spoke against the proposed referendum, and it was defeated.

While the Maclay bill was pending, the friends of the Public School Society rallied to its defense. A report appeared in the *New York Evangelist* which drew on official records of the city and state to discredit some of Maclay's charges. What appeared to be universal education in the country districts was in fact faulty record-keeping; many of the country districts recorded more children in school than actually resided in their district. The report concluded that no more than 6,000 children (or 10 percent) in the city were without education. This proportion was low compared to that in the neighboring city of Brooklyn, which was organized under the district system, where 56 percent of the school-age children received no schooling of any kind.

The Commissioners of Common Schools of Brooklyn commended the public schools of New York and criticized their own district system:

> In our public school system we are far, very far behind (those of New York). . . . We regard them as admirable models of imitation. . . . That the defective condition of our schools results mainly from their being conducted on the isolated district system, we entertain no doubt; nor is this a hasty conclusion. This conviction, long since entertained, has been deepened by time and investigation; for, however well adapted that system may be to a country or village population, our own experience, and that of other cities in our State, have fully evinced that it is not adapted to the exigencies of a city population.[6]

The commissioners hoped that a central board of education would be created in Brooklyn to replace the nine distinct corporations then responsible for public education in that city. They felt that the high mobility of a city population made a uniform system necessary.

Such arguments were well and good, but they had now become irrelevant. The future of the school system of New York was to be settled on a political basis by the state legislature. Each side argued that its conclusions were most in keeping with inviolate American principles; the rhetoric on both sides was high-flown and self-righteous. But it wasn't rhetoric which passed bills, it was political pressure.

The Whig press in the city, trying to halt the Maclay bill's momentum, got behind a campaign to collect signatures on behalf of the Public School Society. About 20,000 names were gathered, and a mass rally was called for March 16 in City Hall Park. The sponsoring papers drummed up as much attention as they could and predicted a crowd of 20,000. Large placards went up about the city announcing the rally. Unfortunately for the Public School Society, the widely heralded rally was an unmitigated disaster. About 5,000 persons attended, and almost 1,000 of them were hecklers. The speakers' platform was surrounded by a solid ring of Irish Catholics. Amid jeers and catcalls, the first speaker, with trembling hands, read a series of resolutions. Such cries as "cock-a-doo-dle-doo" and "sit down, old square-toes," overwhelmed the speakers. Within half an hour the meeting was hastily adjourned, with the speakers in full flight and the platform in the possession of the rowdies. No one heard the Society's resolution not to oppose the election of school commissioners and their willingness to accept any other educational agencies which had no religious ties. The fiasco in the park seemed to demonstrate to the deliberating assembly that support for the Society had been exaggerated.[7]

As the bill moved closer to passage in the assembly, a Whig assemblyman proposed an amendment which required that "no religious doctrine or tenet shall in any manner be taught, inculcated, or practised in any of the common or district schools in the city of New York." Maclay argued against this amendment on the ground that it was not contained in the general state school law, and he saw no reason to stigmatize the people of the city of New York with a special prohibition, not trusting their own good judgment on the matter. This amendment was defeated, and the bill passed the assembly on March 21 by a vote of 65–16. Fifty-six Democrats and nine Whigs voted for the measure; nine Democrats and seven Whigs opposed it. In the New York City delegation, the Whig members opposed it, and all but one Democrat supported it (the dissenting Democrat abstained).[8]

The next day, March 22, Bishop Hughes wrote to Governor Seward, congratulating him on the assembly's passage of this "Bill which adopts your views and vindicates them under slightly disguised language!" Hughes remarked on the irony of having the bill enacted by Seward's opponents, when his own party had been too blind or too obstinate to favor it. He promised that while the bill was pending in the state senate, its friends would stay quiet. He hoped that Seward and Weed could get senate Whigs to support the measure, since the Democrats favor it "from necessity, and against their will." The controversy over the schools "has taught our people a lesson which their former leaders are sorry they ever learnt. And if the other party should not shut the door against them, they will remain as they now call themselves—'independent'— even after the bill shall have been carried. I told you long ago, that the magic wand was broken, and the Whigs *alone* can restore its power in the hands of their opponents." The "magic wand" referred to Tammany Hall's power over the Irish voters. Hughes warned that if the Whigs opposed the Maclay bill in the senate, "it would throw our people en masse back into the faithless arms that have embraced and wheedled them so long." Hughes added that while the bill was not exactly what the Catholics wanted, "it goes so near it that we shall be willing to give it a fair trial. . . . May I ask of you the favor to destroy this as soon as read." [9]

The Maclay bill faced certain trouble in the conservative senate. The Democrats held only a two-vote edge there, and the New York City senators were opposed to it. The Senate Committee on Literature substantially amended the legislation to bolster its chances for passage. The amended bill provided for a central board of education which would insure a degree of uniformity and harmony among the districts. And even more important, the principle of absolute nonintervention was replaced by an explicit proscription of sectarian instruction in the public schools. Without banning religious training in the new system, the bill had no possibility of success in the senate. Despite these changes, the four-man delegation from New York County remained opposed. Maclay conferred with the two Democratic senators and reminded them that the defeat of the bill would help their political opponents in the imminent charter elections in the city. Maclay and the chairman of the Committee on Literature at last got the support of one of them by agreeing to require that school elections take place apart from general elections.

New York City's election day, April 12, drew closer, and the amended Maclay plan was still under consideration in the senate. The Catholics feared that the bill would be defeated if it did not pass before the election. It was decided to enter an independent slate of candidates at the last minute. On April 7 a nominating session was held at Carroll Hall, and Catholic candidates for mayor and the Common Council were announced. The Catholics resolved "to maintain no connection with men tainted with aristocratic impurities of this Society, or willing to sustain the anti-Republican principles on which it is based." [10]

As the Catholics had hoped, New York City Democrats pressured their representatives in Albany to bring the bill to a successful conclusion in order to head off the rebellion of the Catholic voters.

The Democratic leadership in the city and in the legislature decided that the bill had to pass before the city election, and it did. On the evening of April 8, the senate passed the amended bill by a vote of 13–12. A tie vote would have defeated the bill. One New York City Democratic opponent of the bill had conveniently retired for the night when the vote was taken; he later insisted that he had been paired with a supporter of the bill, who in fact had already been paired with another opponent.

The next day the assembly overwhelmingly approved the amended version of the bill, and Governor Seward at once signed it. On April 11, the Catholic ticket withdrew from the municipal election and threw its support to the Democrats. On election day, street fights raged between mobs of nativists and Catholics. A mob marched on Bishop Hughes' residence and stoned it. They knocked down the doors, smashed the windows, and broke up the furniture. Hughes and his staff were away at the time. The police and the mayor rushed to the area to try to disperse the mob; Mayor Robert Morris called out the militia to protect St. Patrick's Cathedral on Barclay Street. That day, Mayor Morris, a Democrat, was re-elected, while control of the Common Council went to the Whigs by a margin of 9–8.

No matter which way the election went, the issue was settled. For the first time in its history, New York City would have a school system that was directly controlled by the people and entirely financed from the public treasury. Almost inadvertently, the people of New York had a public school system.

BETWEEN
THE WARS

1842–1888

CHAPTER 8

The Ward School System
Takes Root

THE FIRST GREAT SCHOOL WAR was over. The new school law passed, but with a proviso denying funds to religious schools. The Catholics did not win, but the Public School Society lost. Governor Seward wrote to his friend Bishop Hughes that

> notwithstanding the ill grace with which that measure was adopted, and the offensive and grudging spirit which marks it, the proceeding is an acknowledgement of the vices of the old system and its unequal operation. . . . Every stage in the history of this strange controversy increases the strength of those who demand reform and the weakness of those who cling to error.[1]

Hughes was skeptical about the new law, for he did not believe that complete religious neutrality in the schools was either attainable or desirable. The *Freeman's Journal* gave the Catholic reaction to the bill:

> We have to consider more the *evils* from which it relieves us, than the positive benefits which it confers. . . . Heretofore, the will of a hundred trustees was the rule to which all were required to submit; now, every citizen, Catholic as well as Protestant, has as much a voice in determining whatever appertains to public education as any of the oligarchy.[2]

Hughes could take comfort in the humbling of the high and mighty Public School Society. The Society had been benevolent, patronizing, condescending, cheerful in travail, tolerant, pious, self-congratulatory—so good, in fact, that the Catholic leadership of the movement had grown positively to detest them.

79

Throughout the controversy, to those who listened carefully, Hughes made it clear that he never intended to have anything less than Catholic schools, either with or without public funds. In April 1841, before Spencer's report appeared, Hughes told a meeting of Catholics, "We must then look forward to the organization of schools . . . so that if we are ultimately obliged to educate Catholic children at the expense of second taxation . . . then, indeed, we shall study that ours will be a thorough education, and a thorough Catholic education." Through the pressure of his movement, he drove Protestant sectarianism from the schools, but he doubted that it would be a lasting result.[3]

Hughes' fears about sectarianism in the new public schools were immediately justified. The first Board of Education, elected in June 1842, was dominated by opponents of the Catholics. The first county superintendent of education was Colonel William L. Stone, a prominent newspaper editor and nativist. Daily Bible reading was official school policy. The Catholics protested the use of the Protestant Bible, but, in 1844, the Board of Education ruled that Bible reading without note or comment did not constitute sectarianism.

Bishop Hughes (he became Archbishop Hughes in 1850) was the leader of the Catholics in New York until his death in 1864. After the conclusion of the school controversy, he made the establishment of a Catholic school system the first priority of the Church. The struggle evoked broad Catholic support for a separate school system and for strong Catholic institutions. The fight unified Catholics and sharpened their group consciousness, as Hughes hoped it would. But there was a price to be paid. The conflict fanned the flames of nativism, which became a national force.

While Catholic historians have usually written of Bishop Hughes as a heroic, pugnacious defender of the faith who dared to champion the rights of his downtrodden followers, some modern Catholic historians have taken a different view. David O'Brien in 1966 and Andrew M. Greeley in 1967 asserted that Hughes' militancy contributed to the growth of nativism and that separate schools became necessary to shield immigrants from the consequent hostility. O'Brien held that because of Hughes the Catholic population was turned away from assimilation and became ghettoized, believing firmly in the superiority of parochial education and in the enmity of Protestant society. O'Brien believed that Hughes "did little to assist his immigrant followers to understand their surroundings and to live in peace with their neighbors. He taught them that a strong, militant, and politically united Catholic bloc could defend its interests, but he neglected to instruct them in the requirements of the common good." Greeley called this aspect of Hughes' influence "a major disaster," and said that while Hughes did not create the Catholics' "ghetto" mentality, "he symbolized it, reinforced it, and worst of all, failed to provide a working alternative in precisely the time and the place where such an alternative could have been of critical importance."[4]

As for the Public School Society, its days were numbered. The new law permitted it to continue operating its schools, but it had to request funds from the new Board of Education. The outcome of the controversy proved that the Society was not the powerful corporation that its enemies described. True, its trustees were wealthy, upper-class men. But in the America of 1842, the political power of the aristocracy was not what it had been. The democratization of American life had rendered the social position of the trustees insignificant in comparison to the voting power of the masses. And in the final analysis, with a Whig governor and a Democratic legislature anxious to diminish its authority, the Society was politically impotent. In its annual report for 1842, the trustees stated:

> After a successful career of 37 years . . . it has pleased the Legislature of our State to enact a statute which the Trustees fear will result in subjecting their noble Institution to the blighting influence of party strife and sectarian animosity. The glory of their system, its uniformity, its equality of privilege and action, its freedom from all that could justly offend, its peculiar adaptation to a floating population embracing an immense operative mass . . . is dimmed, they fear, forever.

Many of the harsh charges against the Society's reputation were repudiated. The new superintendent, William L. Stone, reported to the Board of Education in 1842 that he had completed a tour of the public schools of the Society and found them to be "in excellent condition. . . . By far the greater number exceed any other common schools, in their neatness, the perfection of their order, the soundness of their instruction, and the general efficiency of their discipline." He discounted the figures which John Spencer used to demonstrate the Society's inability to educate the children of the city. Stone said that fewer than 7,000 children were out of school, not 32,000, as Spencer had reported.[5]

Over the ensuing decade, the elected Board of Education gradually undermined the Public School Society's viability, first by denying them the right to open new schools, then by cutting back on the funds needed to repair and maintain existing schools. In 1853, the Public School Society, though its enrollment over the decade remained stable at about

25,000, decided it was useless to continue competing with the new ward schools, which parsimoniously controlled the purse strings of the Society. In that year the Public School Society merged with the Board of Education and ceased to exist. The Board of Education's annual report for 1853 commented on Seward's educational proposals which precipitated the original controversy:

> This recommendation of the Governor was extremely unacceptable to a large portion of the people of the city, and had it not proposed to preserve the schools of the Public School Society which had, deservedly, the confidence and affection of so large a number of the citizens, it is doubtful whether the popular will would have allowed the recommendation of the Governor to go into useful effect.

The Society, maligned in the heat of battle, had served a valuable purpose. Thomas Boese recalled that "in 1835, in a ninth class of thirty-two boys, there were two future judges of the Supreme Court, at least one member of the Legislature, a City Register, several principals and assistants, and one Assistant City Superintendent of Schools, one clergyman, and three or four highly successful merchants. These were nearly all sons of men who earned their bread by daily toil." On the negative side of the ledger, the schools of the Society were never able to surmount the stigma of pauper education, a stigma which the trustees did little to dispel.

Most of the Society's teachers became teachers in the new ward schools, bringing with them rigid and mechanical methods, even though the Lancasterian system was not employed. Under Lancasterian influence, memorization of facts became the standard of success (see Chapter 1). As one historian of education wrote, "The spell of Lancaster has hindered all reform movements." The legacy of Lancaster was evident in another historian's description of "the school machine" in early American public schools.

> From the lowest grade to the highest, the pupils followed an endless succession of book assignments which they learned out of hand to reproduce on call. The chief end of pupils was to master skills and facts as directed by a teacher who in turn was under the automatic control of a printed course of study, a set of textbooks, and the necessity of preparing her class to pass certain examinations on the content of a specific number of printed pages. . . . The business of a school being what it was, any movement, any conversation, any communication, were out of order. The spirit of control was military and repressive. . . . (Children had to be) submissive to the rule of the drill-sergeant in skirts who unflinchingly governed her little kingdom of learn-by-hear-and-recite-by-rote.[6]

THE FIRST CENTRAL BOARD OF EDUCATION was composed of two elected commissioners of common schools from each of seventeen wards; each ward also elected five trustees and two inspectors. Each ward controlled its own schools. Ward trustees hired and fired teachers, subject to the inspectors' approval. Each local board determined its course of study, selected its own books, purchased its own supplies at its own price, and sent the bill to the Board of Education.

Even before the merger of the two systems, many teachers left the Society to teach in the new ward schools, where wages were higher. Thomas Boese, author of a history of the New York schools in 1869, wrote that an "escape from the cold, unsympathetic, almost military rule of the Society was equivalent to an emancipation." Still, life for the ward teachers was not easy. Classes were usually as large as eighty or more children. There was no job security or tenure; retention of teachers' jobs frequently hinged on the re-election of their political patron as a ward trustee.[7]

After the absorption of the Society schools by the ward schools, the Board of Education renumbered all the schools, permitting the original public schools to retain their numbers. Boese says this was done to "preserve a certain historic identity and continuity, and is one of many indications that the Board of Education, the legal successors of the Society, was to identify its history with their history, and to recognize its origin as in one sense the primal form of their own organization."[8]

Over the next half century, until the next massive reorganization of the city school system in 1896, the central board and the local ward boards engaged in jealous battles over their respective powers. The central board was empowered to make general rules and regulations, but was not able to compel compliance by the ward boards. Time and again, the central board and local boards fought over the power to appoint and dismiss teachers, to procure supplies, to determine salaries, to establish holidays, to control expenditures for repairs, cleaning, building, fuel, and every other kind of outlay. According to Boese, these disagreements

> led to much confusion and irregularity, and to great inequalities in the distribution of advantages to different sections of the city, though all were sustained by the common treasure. Some boards were nearly, or quite indifferent as to the interests committed to their charge, while others, by various means, obtained undue advantages. The members of the Board of Education itself being ex officio members of the boards in their several wards, naturally looked more to the local interests which they specially represented than to those more general ones for which they were nominally chosen.[9]

Sectarian antagonisms were driven from the public schools, though only because the Catholics retired to build their own school system. But just as the trustees of the Public School Society warned, politicians became attracted to the schools because of the opportunities they offered as a source of patronage.

The involvement in partisan politics, the lack of uniformity from district to district, the narrow perspective of local boards, the constant jostling between local boards and the central board—all of these were portents of the conflict, ineffectiveness, and inefficiency which would call forth a new generation of school reformers in years to come.

WHEN THE PUBLIC SCHOOL SOCIETY was absorbed by the Board of Eduiation in 1853, its passing was not mourned. Though it had been held in high regard, it was an anachronism in a democratic nation. The first historian of the New York schools, Thomas Boese, wrote that the ultimate victory of the ward schools

> was based on a DIRECT and IMMEDIATE APPEAL TO THE PEO-PLE. No body of men, no matter what their character or social standing, were placed between them and their children. If they have one interest which, in this land of *self-government*, they should jealously guard, and keep as closely as possible under their control, surely it is the selection of those into whose hands is committed that most sacred and responsible trust, the education of their offspring.[10]

The most distinctive feature of the new ward school system was the direct, annual election of all school officers, both to the local boards and the central board. During the course of the next half century, this feature was amended, eroded, and finally abandoned altogether.

The first alteration of the school elections occurred after six elections. It became apparent, in the words of school officials, that voter apathy "rendered the election no election at all." In 1848 the state legislature merged the school elections with the general elections. In an age when massive election frauds were routinely committed by both major parties, separate elections were supposed to keep the schools free of party politics. Instead the separation merely guaranteed low participation at election time.[11]

After the elections were merged, the office of school trustee at once became enmeshed in ward politics; each party entered competing slates in school elections. So long as the ward schools competed with the Public School Society, each system maintained rigid economy of operation, in order not to compare unfavorably with the other. After the Society was dissolved in 1853, financial as well as political restraints were removed from the ward school system. School construction costs rose sharply, and overall school expenditures increased at a much faster rate than attendance.

Since each ward was treated like an independent district, each board of trustees had a large number of jobs and contracts to dispense. While the county superintendent had the responsibility for granting certificates to prospective teachers, the teachers could be hired only by the local boards. The superintendent's examination was no great hurdle, as there were teachers in the schools as young as sixteen. With no established procedures or standards for hiring, the chief requirement for one to obtain a teacher's job was to know a trustee. Trustees often hired teachers as favors for friends or political allies. From time to time, charges were brought against trustees accused of selling appointments or promotions. The most infamous such scandal occurred in 1864, when the entire board of trustees of one ward as well as some principals were suspended for forming a "ring" to extort payments from teachers and contractors.[12]

The natural outcome of the ward system of schools, in which each ward was independent of the others, was expensive parochialism. The districts competed with each other for pupils, for money, for teachers, for whatever advantages might be gained. Schools were built intentionally close to ward lines in order to draw students away from other districts; as a result, some schools were overcrowded, while others were half-full. Since commissioners—that is, members of the central Board of Education—were elected from each ward and sat in as members of the local ward boards, their primary loyalty was to the ward. Consequently, the central board was weak, and devoted most of its time to "log-rolling" on behalf of narrow local interests. There was no agency or individual in the city who made educational policy decisions in the interest of the city as a whole.

Because of the near-autonomous status of the local boards, the central board, when it recognized abuses, was powerless to correct them. In 1851, the president of the Board of Education expressed his regret to his colleagues that teachers had no established tenure of office, but were "subject to the caprice of those who have power to remove them." He pointed out that "on a change of ward officers, all the teachers in a school have been dismissed together, with certificates of good character and conduct!" He felt that the absence of "reasonable security" was demoralizing to the teachers.[13]

In that same year, 1851, the various laws affecting New York's schools were consolidated into a single law, which contained three new features: First, the Board of Education was authorized to hire a city superintendent (until then, the county superintendent had been hired by another municipal agency). This represented a first step towards unified educational administration of the city, though at the time the office had little more than visitation powers. Second, the central board was authorized to centralize purchasing of supplies. And third, the school law of 1851 stated bluntly that "No school officer shall be interested in any contract" paid for with school funds. This prohibition was strengthened by another law, passed in 1854, which required the central board to remove any commissioner, inspector, or trustee "who shall be or become directly or indirectly interested by way of commission or otherwise, in any contract or undertaking for the furnishing of any supplies of books or materials, or for the performing of any labor or work for any of the schools or buildings." Corruption in public life was so commonplace in the 1850s that this admonition had little effect other than as an admission that corrupt behavior existed. The same act in 1854 increased the size of the local boards from five to eight members.[14]

Early in 1857, a letter to the *New York Times* summarized the weaknesses of the school system. The writer urged the centralization of the schools under an elected board of education, "composed only of our most respectable and intelligent citizens." With the schools under the regulation "of a single responsible authority . . . any neglects (could) be traced directly to their source, and the people would be saved from the humiliation of seeing one of their most important and sacred interests confided to ignorant, inexperienced and unscrupulous hands, as is very often the case under the present arrangement." The letter-writer wanted the hiring of teachers taken away from the local trustees and put in the hands of the city superintendent. He charged the local trustees with hiring many "young and totally incompetent teachers" and using political influence to override decisions of the central board. The correspondent scored the practice of giving the youngest classes to the least competent teachers and deplored the sterile, disjointed curriculum which required memorization of isolated facts year after year.[15]

Discontent with the schools was broad enough in 1857 to cause the state legislature to appoint a commission to study the New York school system, but sentiment for change was diffuse and lacked the spur of a dramatic issue. On the other hand, there were 264 elected school officials allied to maintain the status quo. In the same year that the investigation was conducted, William Marcy Tweed ("Boss Tweed") was a member of the central board and served as a member of the important Committee on School Furniture, which assigned large contracts.

Abuses in the ward school system were widespread, and dissatisfaction grew. In 1864, the legislature passed a bill which reorganized the ward system into a district system. It divided the city into seven school districts. Under the old ward system, there were forty-four members of the central board, two elected from each ward. Under the new district system, each district elected one school commissioner each year for a three-year term, making a central board of twenty-one members. While the central board was separated from ward politics, the ward was still the basic political unit of the system. The trustees continued to be directly elected in each ward. The number of trustees on each board was reduced from eight to five.

Under the 1864 law, the commissioners were no longer ex officio members of local boards, which freed them from narrow regional interests. The trustees retained full power of appointment over teachers and janitors, but were required to submit nominations for principal and vice-principal for the approval of the central board. For the first time, teachers were guaranteed by law they could be dismissed only for cause related to their morality or competency; they were guaranteed ten days' notice of dismissal and received the right of appeal to the state superintendent of public instruction. Furthermore, the law made financial peculation by school officers a misdemeanor punishable by a fine and a prison sentence.

The 1864 bill was vigorously attacked by the *New York Sun* as an embodiment of "the vicious centralizing principle . . . the great evil of our time." The *Sun* suggested that "a few simple reforms, such as the selection of teachers by competitive examination, and the removal of favoritism, party or otherwise, in their appointment, and above all the election of competent and efficient school officers, will effect a more healthy beneficial change in our school system than the narrow centralizing process." But the *New York Herald* disagreed. It hoped that the new law would lead to an elimination of "the disgraceful abuses which have so often been exposed in the management of our common schools. . . . Our public schools, in the hands of a lot of wretched ward politicians, instead of being a great public benefit, are very rapidly being converted into an irredeemable public nuisance. . . . The men who, through political influence, succeed in obtaining control over the schools are, morally and intellectually, unfit." [16]

The caliber of the central board tended to be higher than that of local boards, and demands for reform invariably originated in the central board. Though spoilsmen like Tweed were elected to the Board, the president of the Board was generally a man commanding respect. During Tweed's tenure, for instance, the Board president was a prominent anti-Tammany Democrat who later served as chief justice of the Superior Court of New York.

Civic leaders who expected the district system to cleanse the schools of petty corruption and inefficiency were quickly disappointed. Appointments, promotions, and contracts were still considered patronage, since there was neither a merit system for jobs nor competitive bidding for contracts. A history of the city which was published in 1869 noted that "when the Trustees and Commissioners happen to be educated men of character, the schools they have charge of are well managed; but when as very often happens, they are ignorant and unprincipled, the schools suffer thereby. There is, therefore, a marked difference in the schools. Some of them are excellent, and others the opposite." Among the elected school trustees were "keepers of groggeries, who can hardly write their own name. . . ." [17]

Yet school affairs seemed honest and pure in comparison to city politics. The Tweed Ring, under the direction of William Marcy Tweed, grew more powerful during each year in the 1860s. Tweed, born in New York in 1823, was the son of an Irish-born chair maker. Tweed tried his hand in his father's business, but failed. His talent was for politics. As a young man, he was foreman of the city's best-known volunteer fire company, known as the Americus or Big Six. The emblem of the company was the head of a tiger, which caricaturist Thomas Nast later used to symbolize Tammany Hall. In 1851, Tweed was elected an alderman and served when the Board of Aldermen was so corrupt that it came to be known as "the Forty Thieves." In 1853, Tweed won election to Congress. After losing his race for re-election, he was elected a member of the Board of Education in 1855. Then, in 1857, he was elected a member of the Board of Supervisors, a post he held until 1870. The last position was the keystone from which Tweed developed his power, for the Board of Supervisors controlled public improvements, taxation, and the finances of various city departments. As a member of the Board of Supervisors, Tweed was able to hand-pick election inspectors, a process which facilitated voting frauds. In 1863, Tweed picked up two additional positions of great importance. His appointment as deputy street commissioner gave him patronage of thousands of laborers' jobs. In the same year he was chosen Grand Sachem of the Tammany Society and took control of the party organization. In 1864, he added yet another post to his résumé when he was chosen by the seventh ward school board to fill a vacancy as a ward trustee.

Tammany's base of power was the huge immigrant vote, which was half of all the city's voters. About half of the city's population of one million were foreign-born; more than half the immigrants were Irish, and a third were Germans. With the help of Tammany judges, immigrants were naturalized with astonishing speed, sometimes as many as 1,000 a day. Tammany workers helped the immigrants fill out the necessary forms and then provided witnesses who would swear that the immigrants had resided in the country long enough to become citizens. Thousands of illegal naturalization papers were sold to prospective voters.

Tammany handed out jobs to the faithful; district leaders personally distributed coal and clothing to the needy. But the organization did nothing to alter the squalor in which many immigrants lived. In 1864 almost half a million people lived in crowded tenement houses with inadequate ventilation, poor light, little or no yard space, and foul privies. The city inspector attributed the high death rate among the poor to "the wretched habitation in which parents and children are forced to take up their abode." Thousands of families lived in pest-ridden, dank cellars; another 20,000 lived in flimsy squatters' shacks. The poor lived in worse conditions in New York than in any other American city, with higher rates of crowding and higher rentals. When the state legislature finally passed a tenement-house law in 1867, setting minimum requirements for light, ventilation, and distance between houses, the law was effectively nullified by a provision which permitted the city's Board of Health to grant variances, which it did freely. The Board of Health, like other city agencies, was controlled by Tammany, the "champion of the immigrants," which missed no opportunity to cash in on loopholes.[18]

The lot of poor immigrant children in the city was particularly hard. In the late 1860s, the director of a private welfare agency in the center of the worst Irish slum wrote in the *Evening Post* that between 10,000 and 30,000 vagrant children lived on the city streets. At nights they were to be found "sleeping in carts, cellars and hallways, on steps, in boxes; in short, anywhere they can find a spot not too likely to be stumbled over in the night." During the day, these children, some as young as six years old, blacked boots and sold newspapers, matches, and other small articles; many were beggars and thieves. That writer, like many others, called for the enactment of a compulsory education law. But the fact was that the schools were already desperately overcrowded. The state superintendent of public instruction reported that many New York classrooms, intended to hold from 50 to 70 pupils, had between 100 and 150 pupils under one teacher.[19]

The school system's involvement with ward politics seemed to some anti-Tammany citizens to be the cause of its inefficiency and extravagance. A drive was mounted in early 1867 to abolish the entire system of governance and replace it with a paid commission, appointed by the governor. Removing municipal functions from the control of the Tammany-run city government and placing them under the jurisdiction of a state commission was a device frequently employed by an anti-Tammany coalition of Republicans and "good government" forces in the legislature. Legislation introduced in 1867 would have transferred the entire control of the New York and Westchester schools to a "Metropolitan Board of Education," just as New York's police force had been converted into the state-controlled Metropolitan Police Force in 1857.

Reaction in the city's newspapers to the proposed school commission was sharply divided. The Democratic *New York Evening Post* opposed "placing the proper functions of municipal government in the hands of independent boards, having no connection with the city government proper, and no real responsibility to the people whose valuable interests they have the disposal of." The proposed bill "would remove even the modicum of interest now felt by the people in regard to the education of their children, by placing the control where the people have nothing at all to do with the schools." The *Post* proposed that the school system become one of the departments of the city government, run by a commissioner appointed by the mayor. However, opponents of Tammany had nothing but scorn for a proposal to deliver the entire system over to the unrestrained wishes of the mayor.

The Republican *New York Tribune* attacked the existing system: "The wild and pernicious spirit of rivalry between the local boards, and the self-interest greedily at work within them, only tend to keep up a system of ignorant, willful wastefulness most damaging to education and the city." To those who protested the relinquishing of city responsibility to a state commission, the *Tribune* responded: "Look at the roll of the Boards of Aldermen and Councilmen; inquire into their intelligence, their morals, their pursuits in life, etc. and even supposing them to be right-minded men, who would think of selecting a committee from the members to guard the health and lives of their millions of constituents?" The *Tribune*, like the bill's sponsors in the assembly, complained that many school trustees were semiliterate or liquor dealers.

E. L. Godkin's *Nation* attributed overcrowding and other school problems to ward rivalries and the desire for patronage, which caused schools to be built in the wrong places. "The school officers, like all other municipal officers elected by the people, have been constantly descending in the scale. Some of the trustees can hardly write their name in the visitor's book, some murder the People's English and have not even a conception of grammar, while some have even been seen drunk in the schools. One principal stated he was more sure of his position by patronizing the groggeries of his officers (all of whom but one were in the liquor business) than by attending to his school." The *Nation* charged that the Board was riddled with favoritism and nepotism.[20]

The Board of Education appealed to the legislature not to pass the bill, contending that it would be "a grievous wrong to deprive the citizens of the City of New York of the right, conferred upon those of every other county of the state, of selecting the officers who control the education of their children." Lacking any strong popular support or organized backing, the bill to establish a metropolitan commission failed. Tammany was pleased; on the eve of its complete takeover of city politics, the Ring did not want any further municipal powers turned over to the state.[21]

Tweed developed a monumental scheme for plundering the city. He realized that the state legislature held the all-important power to reorganize city government, and so, in 1868, had himself elected to the state senate, where he took personal charge of vital Tammany bills. In the same year, he consolidated his power base in the city and state by electing a Tammany man, John T. Hoffmann, as governor; a Ring insider, A. Oakey Hall, as mayor; and another intimate, Richard Connolly, as city comptroller. Further evidence of Tweed's power was demonstrated when the Democratic National Convention, held in New York at Tammany's invitation, selected Horatio Seymour, Tammany's candidate, as the party's standard bearer.

In its efforts to gain control of city agencies, the Tweed Ring would not ignore the public school system.

CHAPTER 9

The Tweed Ring in Charge

THE NEW YORK CITY public school system was bound to interest the Tweed Ring. The schools spent almost $3 million each year and employed 2,400 teachers, as well as janitors and clerical employees. Public dissatisfaction with the school system's "extravagance" created an atmosphere in which the Ring's bold appropriation of the system was initially construed as a "reform" for the sake of efficiency and economy.

The first step in the Ring's takeover of the Board of Education was the attempted ouster of Thomas Boese, the loyal clerk of the Board since 1858 (and later the school system's first historian). Boese had blocked the Ring's efforts to control the Board's disbursements and to levy political assessments on teachers. In January 1869, a Tammany member of the Board moved to replace Boese with a Tammany assemblyman, but the Board sustained Boese. The *New York Times* reported that: "The only opposition to (Boese) comes from the political 'Ring' which has control of the Democratic party in this city." [1]

In February 1869 the leading good-government organization in the city, the Citizens Association, blasted the school system. The Association complained that the system of independent wards fostered local jealousy and led to unnecessary extravagance. It further complained that there was no check on the expenditures of the schools by any elected official chosen from the city as a whole. The Association's leading member was Peter Cooper, industrialist and philanthropist, the most venerable and respected citizen of New York. Its chief agent, Nathaniel Sands, wrote its attacks on the schools; he soon surfaced in another role.

In March, soon after the attack of the Citizens Association, Tweed introduced into the legislature an act to dismiss the elected Board of Education. Under the proposed act, the mayor would appoint an interim board of twelve to serve for eighteen months, when an election would be held for a new board. The act required the Board of Education to submit its annual budget to a special city Board of Audit consisting of the mayor, the comptroller, and the head of the Board of Education.°

The *Times* immediately assailed Tweed's bill to reorganize the Board of Education, saying, "It originated with SWEENEY & CO." (Peter B. Sweeney, one of Tweed's cronies, was considered the "brains" of the Ring.) The *Times* editorially warned that the bill would put the schools directly under the domination of the Tammany organization; that the mayor would be sure to put submissive men on the board and that "the employees and teachers of the public schools can then be compelled to pay their share toward the support of the party the same as other city departments." The *Times* maintained that the legislation was revenge on the present Board for refusing to replace Boese.

At a tumultuous public meeting of the Board of Education on April 8, Thomas Boese submitted his resignation. He said it was well known that he had incurred the "displeasure of those who are the reported controllers of the political affairs of this City. This led, as you are aware, to a demand on their part that I should be removed from my official position. Your refusal to accede to this demand has, as alleged, resulted in an attack upon the reputation of the Board of Education. . . . It was sought to punish me for refusing mandates that would degrade our schools." He did not wish the members of the Board to suffer "political ostracism" for their defense of him. One Board member resigned on the spot, complaining that he had been the victim of slander and branded a thief and scoundrel for serving on the Board. Others insisted that there had never been a dishonest transaction on the part of the Board and vowed to hold their ground until legislated out of office. Another member charged that other Board members had induced Boese to resign, in hopes of persuading the critics of the Board to withdraw their reorganization bill in Albany. If such a quid pro quo were true, he said, he

---

° In the same year, Tweed introduced another piece of school legislation which was intended as patronage for his Catholic supporters. Though the bill did not mention Catholic schools, it authorized the city and county of New York to give aid to nonpublic schools where more than 200 students were taught free. This virtually defined Catholic schools, since the city's few Protestant and Hebrew schools were not large enough to qualify. When Tweed's Committee on Charitable and Religious Societies reported the bill favorably, the Republican press howled about "Popery" and the Republican majority in the senate killed it. Tweed then tucked it away in a complex budget bill prepared by his Committee on Municipal Affairs. In this form, it passed the legislature. The protests from the city were so vociferous that the Democratic party feared reprisals at the polls. The following year, aid to parochial schools was repealed, but the repeal did not take effect until the schools received two years' subsidy.[2]

"would not remain another day in the Board after it had been prosti-
tuted to the service of the Sweeney Ring, as other departments of the
City Government had been." He too resigned on the spot.[3]

The Ring had no intention of renouncing the reorganization bill,
even after Boese resigned. Forcing his resignation may have been a ploy
to establish domination over the Board—to find out who would go along
with dictation and who was stubborn—or it may have been simply an
exercise in power, a flexing of the muscles; for in any case Boese would
have been the first employee fired once the new law passed and a new
Board was appointed by Mayor A. Oakey Hall.

Almost all of New York's daily newspapers opposed the new bill,
and most thought that the Roman Catholic clergy stood to benefit by
the proposal. This was a widespread reaction, both because of Tammany's debt to the Catholic vote and because of the habitual Protestant
fear of a Catholic plot to take over the public schools. Only the *Sun*, a
pro-Tweed paper, looked favorably on the legislation. The *Tribune*,
which had roundly condemned the Board of Education two years before, now came to its defense, calling the bill "a conspiracy to destroy
the Board of Education," planned by Tammany Hall for the purpose of using the schools for patronage. "The public schools of this city would not
only be creditable to any community, but are an honor to the metropolis
itself. Of all the elective offices none can compare with this in . . . honesty and ability." [4]

Undoubtedly the public was thoroughly confused. The bill's supporters claimed that the present Board was corrupt and extravagant; its
opponents called it a Tammany power-grab. Since the bill guaranteed a
new election after a year and a half, there did not seem to be any irre-
mediable abuse underway. Except for newpaper editorials, there was no
strong expression of opposition to the legislation. The Republican General Committee in the city, which passed resolutions against the bill, did
not even bother to send them to Albany. The legislature passed the bill
on April 30, 1869.

Within a few days, Mayor Hall announced his appointments to the
new central board. Among his appointments were a few previous members, a "representative of the Israelites," a "representative of the Germans," a reformer, and one who "has had much experience as a student
of human nature." At the time, Hall appeared to have appointed a respectable board to replace one whose integrity and efficiency had been
questioned. It seemed especially remarkable that Hall had named Nathaniel Sands, chief agent of the reform Citizens Association, to the new
Board. However, the Board's independence was at once in doubt when

one of its first acts was to install William Hitchman, Tammany speaker of the state assembly, as clerk of the Board. The *Times* commented that this plum was Hitchman's reward for steering the school reorganization bill through the legislature.[5]

At its first meeting, the Board took an appropriately conservative stance, as became a group chosen to eliminate extravagance. William Wood, temporary chairman of the Board, denounced ornamental education. Subjects like music, drawing, and languages were simply robbing the taxpayers, he held. All that children needed to learn was how to read and write and a familiarity with the first four rules of arithmetic; that "was all that was required in this free country . . . to succeed in life, and nothing more should be taught in the public schools." He believed that the schools could save money by eliminating unnecessary subjects and at the same time make room for the 40,000 "street arabs" who were not in school. Such policies, he thought, would cause the rich to withdraw their children from the public schools and put them in private schools, which would spring up in abundance, thereby saving the taxpayers even more money.[6]

One of the first reports from Mayor Hall's new Board was a prediction that close to a million dollars could be saved in the school budget. The committee measured classrooms and decided that twice as many students could be fitted into the available school space; it also recommended using assembly rooms for classes. The only economy move to have lasting and beneficial effect (though its success was not for reasons of economy) was the Board's decision to establish a normal school in place of some 400 "supplementary" classes where female students were prepared as teachers; this proposal was the basis of the Normal College, later called Hunter College.

The Board appointed Nathaniel Jarvis, a Tammany politician, as its agent to distribute the excise funds which the legislature had set aside for parochial schools. Jarvis had charge of some $220,000 in the first year. At the time he had a political appointment as a court clerk; the following year he was chosen a member of the Tammany Society's inner council.

In 1870 Tweed achieved his greatest triumph: the legislature passed a new charter for the city which concentrated virtually unlimited power in the Ring. Under the guise of winning home rule for the city, the Tweed charter garnered the praise of the leading newspapers (although not Horace Greeley's *Tribune*), prominent citizens, and the reform-minded Citizens Association, because it made the mayor "accountable" for the conduct of his administration and established home rule. In addition to replacing metropolitan commissions with municipal departments whose heads were appointed by the mayor and responsible only to him, the new charter created a Board of Audit which controlled all city expenses. The members of the Board of Audit were the key members of the Ring: the mayor (A. Oakey Hall), the comptroller (Richard B. Connolly), the commissioner of public works (William M. Tweed), and the president of the board of parks (Peter B. Sweeney).

The new charter gave the Ring complete control, both political and financial, of the city government. No time was wasted in utilizing this new grant of power. There were several new departments to staff and millions of dollars in city expenditures to "audit." In its first year, the Board of Audit paid over $12 million to claimants; Tweed later admitted that two-thirds of that amount was divided among himself, Sweeney, Connolly, and Hall.

Tweed muted criticism of the Ring by spreading complicity as broadly as possible. Some of the city's leading merchants and businessmen became accomplices by winning lucrative, padded contracts; others found they had to inflate their bills to the city or fail to be paid at all. Most of the press was bought off by city subsidies, such as advertising and expensive gifts to reporters. Even the *New York Times* was held in check because one of its three directors was a stockholder in Tweed's New York Printing Company. Shortly after his death in the summer of 1870, the *Times* threw off all restraint in pursuing official corruption and private connivance with the Ring.

In searching out the tentacles of Ring power, the *Times* discovered that three agents of the reform Citizens Association, including Nathaniel Sands, had accepted city jobs. Sands, while a member of the Board of Education, had accepted a $10,000-a-year job in the tax commissioner's office. The *Times* realized that Sands, who had written the Citizen Association's attacks on the previous Board, had led Peter Cooper to endorse the Tweed charter.

The *Times* began to scrutinize carefully the personnel and decisions of the Board of Education and found much to criticize. The "respectable" twelve men appointed the year before by Mayor Hall included no fewer than seven officeholders. An eighth member had city contracts to supply furniture. A ninth member, the president of the Board, was elected a judge of the Court of Common Pleas on the Democratic ticket in 1870 at a time when the Ring strictly controlled and sold nominations. When the president of the Board was elected to the bench in 1870, the vacancy on the Board was filled by the city's receiver of taxes.

It was never charged that members of the Board personally profitted through their service. But the Board compliantly awarded contracts to Ring-connected firms. A Tweed-owned stationery firm supplied the schools with blank books, paper, pens, and ink. The ward trustees, elected on Tammany tickets, made personal and political appointments in their schools without fear of restraint. One letter to the *Times* charged that unfit teachers were being hired over the objections of the principals and the city superintendent.[7]

Before the expiration of the term of the appointed Board, Tweed unexpectedly introduced an amendment to the City Charter to abolish the Board of Education and convert it into a department of the city government. Tweed promised that the present Board would be reappointed, that there was no political motive; the change was intended solely to promote greater efficiency in the administration of the schools. Almost alone, the *Times* saw through Tweed's plan. The real purpose of the act, the *Times* predicted, was not to put the Board out, but to keep it in. The appointed board was due to expire at the end of 1871 and be replaced by an elected board. Tweed's amendment proposed to make *all* school officials serve by appointment, rather than direct election. Other papers saw the move as a commendable attempt to bring the schools into a closer relationship to the city government and to streamline the schools' administration. A few weeks after it was introduced in the spring of 1871, the amendment was pushed through the legislature without debate.[8]

For the first time since the ward schools were established in 1842, popular election of ward trustees was abolished. Under the new act, trustees were to be appointed by the mayor for a five-year term. The low repute into which the elected ward trustees had fallen, particularly in the past two years, caused few people to speak out in defense of them. Public control of the schools was at last centralized in only one elected official, the mayor, who held the power to appoint all school-board officers, local and central.

A few days before the legislature passed the Tweed school bill, the *Times* reported on the Board of Education's proposed budget for 1871. The Board requested $3,950,000; the *Times* pointed out that the last budget of the so-called extravagant board had been $3,150,000. Even more surprising were the attendance figures. Since 1842, average attendance had grown steadily; in 1868 it had been 86,000. In the first year of the Tammany Board of Education, though population had continued to increase, the average school attendance was down by 10,000. Thus, despite loud declarations from the Board about the importance of educating the city's thousands of vagrant children, attendance declined while expenses rose sharply.

As the *Times* predicted, Mayor Hall reappointed the entire board to serve as the unpaid commissioners of the new Department of Public Instruction. At its first meeting, one of the commissioners delivered a valedictory to the old, independent Board of Education, which, he said, had done its work and now belonged to the past. Establishing the new Department of Public Instruction, he stated, did not represent a change in the system or the administration or the course of studies. "The recent act established a connection between the administration of public instruction and the Municipal Government. The Department of Public Instruction is in name and in fact a branch and department of the City Government. If instruction is the business of the state, this is as it should be." Another commissioner hailed the change from elected to appointed trustees: "The trustees of the schools, with some noble exceptions, are the plague-spot of the system because they have the power of putting in the teachers. And we know very well that they are influenced by nepotism with regard to these teachers in an extraordinary manner, so that, however stupid a teacher may be, if she can say, in Tony Lumpkin's words, that 'her Grandmother is an Alderman and her Aunt a justice of the peace', she is very sure of receiving a nomination." [9]

Conditions in all branches of the city government were so abysmally corrupt that, in June of 1871, the *Times* had to admit that the Education Department had done good work that year, "almost alone among the public departments of this city." [10] The days of the Tweed regime were numbered, however. For more than a year the *Times* had punched and gibed at the Ring without any hard evidence of wrongdoing. In July 1871, the *Times* gained access to incriminating documents, figures secretly copied from the auditor's books by a dissident Democrat. What the editors had heard as rumors and could only allege, they were at last able to prove.

During July and August 1871, the *Times* demonstrated to New York that the Tweed Ring had stolen millions of dollars from the city. A public mass meeting was held at Cooper Union on September 4, 1871, which led to the formation of a Committee of Seventy. Public indignation was aroused to a high pitch. Comptroller Connolly agreed to cooperate fully with the reformers. The reformers swept the fall elections for the legislature and minor city offices. Even so, Tweed was re-elected to his state senate seat. After the election, Peter B. Sweeney left the country. Mayor Hall refused to resign and served out his full term. He was succeeded in 1873 by a reform mayor, William Havemeyer. Tweed was indicted, arrested, tried, and jailed. After a dramatic escape and a year in hiding, Tweed was recaptured and brought back to the Ludlow Street jail. He testified at length before a committee of the Board of Aldermen in exchange for a promise of freedom. The promise was never honored. He died in jail in 1878.

Tweed was the only member of the Ring to spend time in jail. He could have named names; the list would have included prominent men in both parties. Both parties, for obvious reasons, preferred to forget the Tweed era.

The school board had been at least a passive participant in the general corruption. It complied willingly with political dictation of appointments and contracts. For example, while Tweed was powerful, the Board of Public Instruction eliminated all Harper Brothers' textbooks, because of attacks on Tweed and Mayor Hall in *Harper's Weekly* by the caricaturist Thomas Nast. After the Ring lost the 1871 elections, the Board of Education promptly restored the Harper Brothers' books to its lists.[11]

The reform Committee of Seventy's new City Charter re-established an independent Board of Education, in a form similar to the 1864 district system. There was one major difference: no school official would be chosen by election. Furthermore, school officials were not permitted to hold any other public office. The reconstituted Board of Education contained twenty-one members, appointed by the mayor. The Board was responsible for appointing five trustees in each ward; the mayor appointed three inspectors in each of the seven districts. The trustees regained full power over the appointment of teachers and janitors; their nominations for principal and vice-principal were made subject to the approval of the central board. Not a single member of the original board appointed by Mayor Hall was reappointed by reform Mayor Havemeyer.

The school law of 1873 restored to New York its previous system of strong community control. Although the local trustees were selected by the central board, they were by no means subservient to the body that appointed them. In fact, the ward trustees had far greater powers than the central board: hiring teachers, selecting sites for schools, awarding contracts for fuel, books, and other supplies. The local district inspectors had the duty of overseeing the work of the ward trustees; they countersigned all bills and payrolls, and their approval was necessary to remove a teacher from her job. The central Board of Education was the legislative body of the system; its single greatest power was its responsibility for selecting local trustees.

The legacy of the Tweed era, so far as the Board of Education was concerned, was a strong conviction that the schools ought to be detached from elective politics. Almost a full century would pass before local boards were again elected by the voters.

CHAPTER 10

Inside the Classroom

THE DECADE from 1864 to 1873 had been marked by major reorganizations and political involvement in the schools. What was most surprising, in light of the extent of political change, was the relative absence of educational change. Descriptions of the schools in the 1850s were echoed in the 1860s, 1870s, 1880s, and even the 1890s.

Change came to the schoolroom very, very slowly. When the ward schools were first established, they followed the precedent of the Public School Society in rigidly separating boys and girls at the grammar level. A Board of Education document in 1853 commented: "The considerations of propriety are so obvious that they need not be enumerated, much less dwelt upon." Parents of daughters objected even to the period of mingling between the sexes which occurred at times of arrival and departure from school. In 1870, only five of ninety grammar schools were coeducational. This pattern prevailed even in 1890, with only fourteen mixed grammar schools of a total of one hundred and eight.[1]

In the early years of the ward system, the city superintendent and his assistants made a practice of visiting every class in every school at least once a year, to examine the pupils publicly. The teachers and principals were graded on the basis of this examination, and the scores were turned over to the local board of trustees. The trustees used these marks as a basis for promotion; low marks sometimes led to a cut in salary for the teacher. The teachers were not permitted to know what marks they received or in what subject their class did well or poorly.

This practice, which developed rapidly into a rigid system, became the organizing principle of the public schools. Teachers spent most of the academic year preparing for the examination. Students memorized their lessons so that they could recite them word-for-word when called upon. Classes of 60 to 100 children studied subjects like a precision drill team, in lock-step fashion, none moving faster or slower than the others.

All these preparations were aimed at attaining a perfect performance for the supervisor's examination.

In 1867, a delegation from the Baltimore schools came to visit the New York public schools. At each school they toured, mass demonstrations were performed for their benefit. The pupils, assembled together in a large room created by pushing back sliding partitions, displayed their prowess in calisthenics, reading, singing, and simultaneous recitation, regulated by the music of a piano. The achievements of the New York schools filled the Baltimoreans with admiration:

> In the Primary Department the precision with which the pupils performed was remarkable. Every eye appeared to be riveted upon the teacher during the exercises. The motions of the pupils were simultaneous with hers, and executed with great regularity. . . . In passing through the classrooms of the Girl's Grammar School, frequent attractions were presented in the promptness with which the pupils returned answers to the questions propounded by the teachers. . . . In the Primary Department, the recitations seemed to be regulated by the calisthenic movements, which are very pleasing to the children.

At another school, the committee witnessed two assemblies, one of boys, the other of girls. Each assembly of 400 pupils read and spelled in concert. A piano's cadence regulated the movement of the children from classrooms to assembly hall.

> All the changes were performed in marches, some in the usual step, others in the double-quick time of the military development. At the same school, 1200 primary students performed together, singing and acting out the songs in unity. The teachers of the primary department of No. 14 are devoted to their school. They are as proud as the children, of their ability to work through their varied exercises, and enjoy great pleasure in drilling them in their performances.

The committee also visited one of the city's five colored schools, which had teachers of both races, a colored male principal for the boys' department and a white female principal for the girls' department.[2]

Teachers were assigned to classes on the assumption that the youngest pupils, those in the primary department, required the least attention. The highest salaries were paid to the teachers of the most advanced grammar grade; primary teachers earned the least. New, inexperienced teachers began teaching primary children; a promotion meant teaching a higher grade. The primary departments were also the most crowded. It was not unusual for a young female teacher, just out of grammar school herself, to be placed in charge of a primary class of almost one hundred six- or seven-year-olds.

Ordinarily, promotions occurred when a vacancy appeared at the top of the hierarchy within a single school. If a principal died or retired, every teacher in the school expected to move up one slot. Principals and teachers usually worked until they were incapacitated, since there was no retirement fund until the 1890s. Whenever the trustees tried to promote a teacher out of turn, the other teachers complained bitterly to the central board. The rigid seniority system which informally evolved bore no relation to merit, but it did assure the teachers a degree of job security.

Children attended their neighborhood schools, and it was accepted as an obvious fact that children in poor districts were less able as students than children in better districts. A description written in the late 1860s held that "the pupils are as different as they conveniently can be, and vary with the district. In the lower wards, and on the east side of the town, they are mostly of foreign parentage, and very inferior to those of the schools in Twelfth, Thirteenth, and Twenty-eighth streets. Nor are all the pupils, as is often supposed, the children of poor people, though the majority are. In some of the districts the scholars belong to the best families in the City, their parents sending them to prove their democratic principles." [3]

The political power of the Irish was reflected in the numbers of Irish teachers employed by the ward trustees. After 1870, about 20 percent of the teaching staff was Irish. In wards which were heavily Irish, it was not unusual to find public schools staffed almost entirely by Irish teachers. Though the Catholic clergy denounced the public schools, there were large numbers of Irish children who attended them; in fact, many Catholic spokesmen conceded that children with ambition were better off in the public schools than in the parochial schools. Still, while local boards in homogeneous wards sometimes preferred teachers of their own background, community control was not used within the school itself to further the values or ideology of any particular group, even if it was the local majority. In these, as in ethnically mixed wards, it was understood that the public schools were to provide a *common* education and not to advance the special interests of any one group.[4]

Except for the authority to examine prospective teachers and to visit classes, the city superintendent had little power. In the 1860s, the city superintendent complained in his annual report about the unequal classification of primary pupils; he described a school of 736 pupils and 10 teachers: 8 of the teachers taught a total of 305 pupils, while 1 teacher had a class of 269, and another a class of 162. He did not think it was necessary to have a class of under 60 children, but he did not believe that any class should exceed 100.

Until the organization of the Normal College in 1870, and for many years after, teacher training was either superficial or nonexistent. The Board sporadically maintained a Saturday normal school for teachers, but it was not really sufficient to upgrade the level of preparation. There were 2,206 teachers in 1868, of whom 2,030 were females. Of this total, 241 had teachers' certificates from the state superintendent of public instruction; twenty-six were graduates of the State Normal School at Albany; the remaining teachers held certificates from the local ward trustees. The inadequate preparation of most teachers contributed to their dependence on mechanical methods of teaching.

School officials preferred having female teachers, because they could save money by paying women less than men. Women were glad to have teaching jobs, because it was one of the few respectable occupations open to them. And women were blatantly discriminated against by the school system. They were paid less then half of what men received for the same job. A bylaw required the immediate dismissal of any female teacher who married (until 1904, when the state Court of Appeals ruled the bylaw invalid). Although a high school for boys, the Free Academy (later the City College of New York), was organized in 1849, only the need for better-trained teachers, and the hope of economy, led to the establishment in 1870 of the Normal College, which accepted women. There were no female school officials until 1886, when two women were appointed to the Board of Education, setting off a great controversy.

For most of their first fifty years, the ward schools struggled to survive recurrent economy drives, often mounted by leading citizens. Conservative organizations held fast to the view that the duty of the state was to provide only as much education as was needed to eliminate illiteracy and to protect itself from the dangers of mass ignorance. Higher education was considered the prerogative of private enterprise. Typical of this sentiment was the report of a committee of the New York Municipal Society in 1878, which held that taxpayers resented both the City College and the Normal College for providing higher education to public school students at public expense. The committee held that "the proper limits of public education" had been crossed and complained that the present system was "a departure from a system good enough for our fathers fifty years ago." But not all opposition came from the upper class. Some Catholic spokesmen expressed distaste for the egalitarianism of the public school and complained that children were being educated above their station, which made them insolent and intractable.[5]

Even the Democratic governor, Lucius Robinson, attacked the provision of higher education at public expense as "legalized robbery," in his inaugural address in 1877. The following year, the president of the New York City Board of Education spoke out forcefully in defense of the public schools. He complained that in each of the past five years, the roll of pupils had grown while the money allotted the schools had actually diminished. While the Board of Education's power to provide accommodations was curtailed, the demand increased with each year. The Board president was particularly alarmed by the governor's criticism: "In a Republic I think it is the part of wise statesmanship to give the people that secondary or higher education, at the expense of the State or municipality, which they themselves demand." Only a monarchy, he said, limits public education to the basics of reading, writing, and arithmetic. New York must offer public schools good enough to attract the children of the wealthier classes to them. "Good schools, both for primary and secondary education . . . would, with *rapid transit,* help to bring back to the city that great middle class which, during the last ten years, has been absolutely squeezed out of it, leaving here only the very rich and the very poor." [6]

The Board president who issued this liberal statement of policy was William Wood, the same William Wood who a decade earlier had opened the first meeting of the Tweed Board of Education with an attack on "unnecessary" subjects like foreign languages, music, and drawing, and anticipated saving money by driving the children of the well-to-do out of the public schools. By 1878, he had come to understand the larger role of the public school and to hope that the school could be the instrument of a better, more democratic society. The radically changed views of this doughty, conservative Scotsman were but a reflection of the ferment which was stirring the educational world. New ideas were struggling to be born, and New York was destined to be one of the prime battlefields for the new education.

SECOND
SCHOOL
WAR

The Rise
of the Expert

CHAPTER 11

The Birth of
a Reform Movement

NEW YORK CITY has always had its poor and has always had an insufficiency of either will or resources to solve their problems. In the nineteenth and early twentieth centuries, governmental efforts to assist the poor were minimal. While government was grudging, the Democratic organization was not. Tammany Hall had its own system of poor relief, in exchange for party loyalty. The high-minded citizens who occasionally combined to turn the Tammany rascals out showed no interest in the poor. The reformers were concerned about honest government, because waste and theft led to higher taxes and affronted their sense of morality as well. The indifference of the well-born reformers to the daily misery of the masses served to reinforce the fealty of the immigrant poor to Tammany Hall.

Until the 1880s and 1890s, the prosperous classes rationalized the contrast between their own comfort and the distress of the poor with an implicit belief in social Darwinism, the application of Darwin's determinism to social conditions. "Survival of the fittest" justified the gross inequalities in American life without taxing the conscience of the rich. By this theory, the poor were poor because they were too lazy or too stupid or too irresponsible to take advantage of the opportunities about them. Those who held power believed they had achieved it through their industry and intelligence. It was the "natural" workings of "immutable" laws which brought the strong to the top, while the weak remained mired in poverty.

These attitudes about the causes of poverty determined the efforts to ameliorate it. Since poverty was thought to be a personal problem, poor relief necessarily took the form of charity. Under the direction of a commissioner of public charities and correction (the juxtaposition of "charities" and "correction" in one agency was itself revealing of public attitudes), the city maintained insane asylums, a poorhouse, a workhouse, and a few hospitals. Public institutions were chronically overcrowded. The largest of the public hospitals was Bellevue, which had only 768 beds. The city's limited resources were supplemented by numerous private organizations. Foundling homes, street baths, day-care centers for small children, orphan asylums, homes for destitute children, dispensaries, clinics, hospitals, homes for the aged, lodging houses for vagrants, and numerous other services were provided by hundreds of private charitable and benevolent organizations.

Yet despite what must have then seemed like a mammoth effort to relieve poverty, the slums of New York grew in population and intensified in misery during the last decades of the nineteenth century. Unprecedented numbers of immigrants poured into the port of New York; many moved on, but thousands remained to crowd into the already pestilent slums. By 1890, there were 1.5 million people in New York; almost 43 percent were foreign-born, and 80 percent were either foreign-born or of foreign parentage. New York was accustomed to a large foreign contingent in its population, for even as early as 1860 the majority of New Yorkers were foreign-born.

But after 1870, the source of immigration began to shift. Previously most immigrants came from England, Ireland, and Germany. In reaction to economic and political conditions in Europe, the numbers of immigrants from these countries began to decrease, while those from southern and eastern Europe began to rise. The "new immigration" was composed predominantly of Italians and Eastern European Jews. By the last decade of the nineteenth century, the "new immigration" accounted for more than half the number of all immigrants into the United States.

The Jews and Italians moved into slums previously occupied by Irish and Germans. The tenements into which they crowded were built to get the greatest number of humans into the smallest space possible, with no thought for conditions of sanitation, light, or air. Garbage collection and sewage disposal were haphazard, particularly in the poorest districts. In the densely populated lower part of the city, playgrounds and open spaces were nonexistent.

The human costs of industrialization were exacted most heavily in the slum, where men, women, and children toiled long hours in miserable conditions for subsistence wages. Industrial accidents, economic recession, a closed factory, the replacement of hands by machines—any one of these eventualities could suddenly throw a family or a hundred families beyond the brink of poverty into pauperism. With the new economic conditions brought about by industrialization, it became more difficult to sustain the belief that poverty was entirely the consequence of personal unworthiness.

During the last quarter of the nineteenth century, labor unrest and increasing immigration combined to shake many comfortable New Yorkers out of their accustomed complacency and to make many feel that their way of life was threatened. Infant labor organizations, in New York as well as elsewhere, engaged in bold strikes, directly confronting the established order. And the steady growth of the Lower East Side slum could scarcely be ignored. Teeming with hundreds of thousands of Russian and Polish Jews, it was a hotbed of socialism, anarchism, militant unionism—"radical" philosophies that suggested discontented and aroused masses.

The slum was the symbol of everything that educated and sensitive New Yorkers despised. It was poverty, illiteracy, ignorance, squalor, disease, filth, crime, vice, and degradation. It was the bastion of the corrupt political machine.

To conquer, subdue, and transform the worst that was in America, a reform movement was born in New York City in the 1880s. For some reformers, the motive was fear, fear of the spread of anarchism and discontent, fear for their own comfortable way of life. For many others, the motive for reform was compassion, the compassion born of a deep commitment to religious principles and ideals of social justice. Some reformers agitated for improved tenement-house laws, others for laws to eliminate sweatshop labor and child labor. The first settlement house in New York was opened in 1886. The Neighborhood Guild, later called University Settlement, was soon joined by College Settlement, East Side House, and the Hebrew Alliance. The settlements launched numerous programs of social service in the slums. Perhaps even more significant than their work in the slums, however, was the advocacy of social reform which settlement workers brought to the middle-class and upper-class world. Settlement workers, themselves largely from the privileged classes, informed and interested their peers in the problems of the poor. They became the most articulate and persistent lobbyists for social change.

Reform moved into the political arena in the 1880s, but with no program other than that of throwing out Tammany Hall. The political reformers had yet to absorb the programs of the social reformers. Political reformers won occasional victories, but with no permanent impact on conditions in the city. The sterility of political reform during this period was demonstrated in 1886, when the reform organization combined forces with Tammany Hall to prevent Henry George, who had widespread labor support, from winning election as mayor. The candidate who repulsed George's strong appeal to the masses was Abram S. Hewitt, a successful industrialist–Congressman and one of New York's "best men." Hewitt's willingness to cooperate with Richard Croker, boss of Tammany, made it clear that the reform leadership of the city— composed of Protestant clergymen, merchants, civic leaders, prominent lawyers, and philanthropists—was interested in good government, but not in the alteration of basic economic arrangements. Two years after Hewitt's victory, the labor party was too factionalized to enter a candidate, and Tammany Hall's candidate easily defeated Hewitt, completely eliminating reform influence from city government. Not until six years later, in 1894, were the reformers strong enough to mount a challenge to Tammany.

The ineffectiveness of reform on the political front did not deter the social reformers, who were inspired by a sense of mission. Though they worked on a broad range of interrelated problems, the common denominator for almost all the social reformers was their belief in the importance of education. Education was the key to the future for individuals and for society as a whole. The reformers' belief in the power of education to right the wrongs of society echoed the views of Governor Seward. If anything, faith in education as the great agent of democracy had become more widespread in the intervening years. What had changed was the understanding of what was meant by "education." When Seward and his idealistic contemporaries advocated universal education, they had little more in mind than the elimination of illiteracy; in the 1840s, that alone represented a monumental step forward for society.

But by the late 1880s, farsighted educators began to develop larger goals for education. New ideas and philosophies about learning and child development set the stage for the blossoming of a broad-based movement for educational reform. Innovative educational ideas were brought to America by educated immigrants and by American educators who studied abroad. The traditional system of teaching through memorization came under attack from followers of the Swiss reformer Johann Heinrich Pestalozzi and his German admirer, Johann Friedrich Herbart. Pestalozzi's object-teaching method, which he devised and practiced in the late eighteenth and early nineteenth centuries, relied on the use of concrete objects to encourage powers of observation and reasoning. Herbart studied Pestalozzi's work and evolved a philosophy of education which stressed the importance of psychology and ethics in education;

his revolutionary contribution was the proposal that education be formulated with an understanding of the human mind and of the social ends which it accomplished. Another German, Friedrich Froebel, pioneered in the study of preschool children, opening the first kindergarten in 1837. His followers, many of whom made a cult of the kindergarten, opened the first American kindergarten in 1857 and organized Kindergarten Associations in many cities. A display of kindergarten materials at the Philadelphia Centennial Exposition of 1876, attended by nearly 10 million people, helped broaden awareness of and demand for kindergartens.

The Centennial Exposition provided would-be educational reformers with another innovation which showed how education could be made relevant to an industrial society. The Russians exhibited drawings, models, and tools of a new curriculum which taught industrial and mechanical arts. The manual-training approach seemed to offer a practical alternative to the dry, unimaginative curriculum that dominated the public schools. It attracted admirers who believed it would be more attractive to poor children who avoided school than the present curriculum and would train them for useful work.

These reforms were related in that they rejected the basis of the old education, in which the teacher had a specific unit of information to drill willy-nilly into the submissive child. The new education sought to make school more like life, as in the kindergarten, where the inquisitive nature of the child is engaged in learning and playing. And it sought to make schools conscious of the social and practical consequences of their curricula.

The new education drew heavily on the work of psychologists G. Stanley Hall and William James, who maintained that curriculum had to proceed from a knowledge of children and their development. Hall, a pioneer in the field of child study, held that the school must be fitted to the child, not vice versa. James' pragmatism and behaviorism encouraged the reformers' belief that through education it was possible to mold the character of young children in ways that would ultimately result in a better society.

Many educational reformers drew strength from their conviction that the school was the key to the battle with the slum. Other reforms would ameliorate poverty. School reform, they thought, would break its grip.

EDUCATIONAL REFORM moved sporadically in New York in the 1880s. At the beginning of the decade, the Ethical Culture Society, led by Felix

Adler, established the Workingman's School for the children of the poor. It incorporated the kindergarten and manual training. In 1881, Grace Dodge and other socially prominent philanthropists organized the Kitchen Garden Association to help slumdwellers by teaching them such practical subjects as household management. The Kitchen Garden Association was reorganized in 1884 as the Industrial Education Association to agitate for the inclusion of industrial education and manual training in the public schools. The Association took a strong interest in the need for trained teachers, and in 1887, under the leadership of Professor Nicholas Murray Butler, helped to establish a new teacher-training college. Nicholas Murray Butler was then an associate professor of philosophy at Columbia College and the president of the Industrial Education Association. By 1892, the new college was chartered as Teachers College and a year later became associated with Columbia.

The newly opened settlement houses were directly involved in educational programs whose range and diversity reflected the paucity of the public schools' offerings. The settlements had recreational and athletic programs, arts and crafts courses, first-aid classes, trade education, adult literacy classes, kindergartens, drama groups, English classes for foreigners, playgrounds, and every other kind of class, club, or program that seemed to fill an unmet need. The immensity of the undertaking, the sheer volume of unmet needs, made the settlement workers keenly aware of how little the public schools accomplished in the slums. Thousands of children were turned away from sorely overcrowded schools to join the small army of juvenile vagrants on the city's streets; thousands of others worked in sweatshops twelve hours a day, in clear violation of the compulsory education law, adopted in 1874 but rarely enforced.

Many slum schools were in decrepit condition. And what went on in the classrooms bore no relation to the pupils' lives outside. School was a place to memorize facts, rules, and definitions and to be tested on a carefully predetermined body of knowledge by persons in authority. Each grade, from lowest to highest, followed a minutely detailed course of study prepared by the city superintendent. The lowest grade studied numbers: "*Counting* by ones to 100, by twos and threes to 50; also, counting backward by ones from 10; *adding*—by ones, twos and threes, to 20; *figures*—to be read to 100 and written to 30." Pupils were directed to study form and drawing by learning the following: "sphere, cube, square, oblong; position of straight lines—vertical, horizontal, oblique; angles—right, acute, obtuse; surface, face, edge." The exactness of these course descriptions, which defined the course of study for every subject in every grade, was both a crutch and a harness for the teachers. It gave direction to the numerous teachers who had little or no training, but it positively discouraged creativity or imagination. Since teachers had such large classes, there was neither time nor incentive for a teacher to find new ways to teach; and since ultimately she would be judged by her pupils' memory-work, it would have been foolish to try to teach them anything different.[1]

The New York Board of Education in the 1880s was composed for the most part of successful, conservative businessmen. They were not deaf to the complaints of the reformers, but they preferred to make changes in a cautious manner. They were well aware that the schools were not providing seats for all the school-age children in the city, but their budget requests to the city administration were regularly pared down, even for ordinary operating expenses. In the late 1880s, the Board established three new evening high schools for working children and set up a highly popular evening lecture series for working men and women. In response to reform critics, the Board experimentally introduced manual-training courses like crafts and woodworking for boys and sewing and cooking for girls. By 1890, almost 20,000 children were pursuing a manual-training course of study.

In another move to placate critics, Mayor William R. Grace appointed two women to the Board of Education in 1886; one was reformer Grace Dodge. The outgoing president of the Board sharply criticized these appointments. Six years later, when the term of the only remaining female member expired, he urged the mayor not to appoint another woman to the Board. The pressure, he said, came from "a great multitude who can hold meetings on the slightest provocation, appoint committees and circulate petitions whenever a chance is offered to put a woman where a man was before." He contended that the presence of women made vigorous debate impossible, because of the necessity of being courteous to them. "I pray you to withstand the petitions of the enthusiastic and ill-informed, and await with patience the passing of the skirt brigade." [2]

The issue which proved most troublesome to the Board of Education was the lack of accommodations. It was common knowledge that thousands of children lived on the streets and never attended school, despite the employment of a dozen agents to enforce the compulsory education law. There simply were not enough classrooms for the children of New York. The situation distressed everyone: school officials, reformers, parents, and the press. School officials were hard pressed to keep up with the fast-growing population and failed to anticipate the population shifts which left downtown schools empty, while uptown wards were overcrowded. The central board was handicapped in its planning because the local boards had the initiative for requesting new schools but sometimes failed to exercise it promptly. The central board blamed the local boards, the local boards blamed the central board, and the problem worsened each year.

So persistent were complaints from parents and the press that, in 1884, Mayor Franklin Edson asked the president of the Board of Education whether it were true that there were "100,000 of Arabs in the City of New York, or children who live on the street and do not attend school." The Board president called the charges false, and said that there were about 5,000 truants below 14th Street (downtown), but about 15,000 children out of school above 14th Street, "and their non-attendance is almost entirely due to a lack of school accommodation." [3]

In 1888 the reformers' persistent charges about the sterile curriculum and mechanical methods of the schools prompted the Board of Education to appoint a special committee to evaluate their criticism. After several months of visiting schools, the committee became convinced that most complaints stemmed from the superintendents' ritualistic examinations of teachers and pupils. This method, with no basis in statute, had grown from a practice to a system. "To characterize it in moderation warrants the statement that it is bad, detrimental to the best interests of education, and is in fact the cause of much that is now injuring and impairing the usefulness of our public schools."

The committee described how the city superintendent or one of his assistants examined the children of every class at least once during the year. The superintendent gave the principal of each school one to six days' advance notice of impending examinations. The examining superintendent spent an average of a half-hour with each class. The marks assigned by the examiner went onto the teacher's permanent record and determined her pay and her prospects for promotion.

The result of this "barbarous" system was that teachers tended to conceal all weaknesses, rather than ask a superintendent for help. Students spent all year preparing for the next examination. "Accuracy of statement and correctness of answer are counted far beyond their true value, and even among the youngest children the memory is used as an educational tool to a degree that should not be tolerated even in the education of adults." [4]

The committee recommended changing the emphasis of the supervisors from examination to assistance. It urged that the best teachers be excused from regular inspection and that others receive regular guidance from the superintendents.

A few weeks before the Board committee issued its self-critical report on the examination system, a new reform organization announced its existence with a well-articulated blast at the public school system. This organization, called the Public Education Society of the City of New York, was headed by Nicholas Murray Butler, a twenty-seven-year-old professor who was destined to be the city's leading spokesman for school reform. Butler was convinced that a thorough overhaul of the school system was imperative and pinpointed three areas in which the schools were defective: in failing to provide sufficient accommodations; in their outdated methods and curriculum; and in the system of school administration by local boards.

Butler charged that the primary classes were dangerously over-crowded, that the lowest grade had an average of eighty-seven pupils per teacher, and that physical conditions in antiquated schools endangered the pupils' health. What was worse, 3,000 children had been refused admission to school in 1887, and almost 4,000 in 1888. One immediate improvement, he suggested, would be to copy London's example and build roof-top playgrounds on schools.

The curriculum, he complained, was "incomplete and disconnected" and failed to offer training in any one subject. Teaching was absolutely uniform, based entirely on a teaching manual which did not incorporate any of the educational advances of the times. He decried the emphasis on memory work, which made "no reference to the general growth of the pupil's mind from year to year, nor to their increasing power of observation and reasoning. . . . This evil is greatest in the primary grades. Children of five and six years of age are subjected to an examination in thirteen subjects and marked upon them, when many of the little ones have not been in school three months." [5]

Going beyond these issues, which had already attracted the attention of the Board, Butler attacked the ward-trustee system itself. He believed that educational decisions should be left to professional educators and was appalled at the authority of the ward trustees. He proposed that all educational affairs be put in the hands of a council of professional educators, while all business matters be reserved for a lay Board of Education. The primary weakness of the existing system, he held, was divided responsibility, which led to inefficiency. Confusion among trustees, inspectors, and the central board led to situations where some buildings were in disrepair and remained unsafe over protracted periods. Some wards were so large that five trustees were not capable of supervising the schools of the ward. And the process of removing a teacher from her job was so complicated, involving trustees, inspectors and the central board, that it rarely ever occurred. Butler urged the appointment of a commission to study and revise the school law.

The Board of Education was not prepared to take seriously so revolutionary a proposal as Butler's, particularly his implied elimination of the trustees. The Board knew there were problems, but believed they could solve them within the existing framework. The Board's first stab at reform was to adopt most of the recommendations of its special committee which had studied the examination system. It set up two categories of teachers, exempt (from supervisors' visits) and nonexempt. Only the teachers with the best records, based on past examination marks, were placed on the exempt list. All other teachers, as well as anyone who had taught less than five years, were nonexempt and would continue to receive supervising visitations.

The change was well intentioned, but it had little effect on the problems it sought to cure. The city superintendent complained that the change multiplied his paperwork and caused dissatisfaction among teachers who thought they belonged on the exempt list. In the classroom, nothing changed. Less than two years later, the classification of teachers was abolished because of its demoralizing effect on teachers who felt they were branded as inferior and because of the city superintendent's continuing opposition.

In 1890, the Board of Education's efforts to protect teachers from unfair practices led to a head-on dispute with the trustees, which had far-reaching effects. The Board asked the state legislature to require trustees to fill vacancies by hiring from a pool of teachers who had lost their jobs when attendance decreased or a school was closed. These teachers had to be paid, since they had not been discharged for cause. The ward trustees ignored the Board's urgings to hire these teachers before taking on new teachers; they felt that the central board was trying to usurp their full autonomy over the hiring of teachers and to establish the precedent of a "list" of eligible teachers. Some ward trustees feared that the central board's next step would be to propose the application of civil service principles to the hiring of teachers.

The trustees formed a city-wide organization to fight the proposed legislation in Albany, and defeated it. Resentment against their strong, unified political front set off a reaction in the press and on the Board. The anti-Tammany newspapers were convinced that the trustees proved that they, like every other arm of city government, were Tammany-controlled and politically motivated. The *Times* began to talk loosely of a "ring" and a "machine" in the schools and asserted editorially that the 4,000 teachers in the city were hired on the "spoils plan, promotion being made according to the political pull of the aspirant." These charges reflected the anger within the central board at finding its reform efforts stymied by the political power of the trustees. One commissioner described the conflict in an open Board meeting: "Last year we had a bill providing that teachers thrown out of employment by dwindling schools downtown, must be employed before new teachers were appointed to vacancies in other parts of the city. A gang of trustees went up to Albany and defeated this most proper measure, because they wanted patronage for their friends living out in Connecticut and New Jersey." [6]

From this time on, the ward trustees' public image began to erode steadily. Since the school system had been reconstituted in 1873, after the Tweed Board, the schools had stayed out of municipal politics. Though occasionally a school officer was tried for malfeasance, no one raised the charge of corruption against the school system as a whole or any of its parts. But after the trustees defeated the Board of Education legislation in 1890, insinuations and charges of petty corruption, patronage, and political influence became more and more frequent. Institutional friction grew out of the underground and above-ground warfare between the local boards and the central board. The local boards accused the central board of overstepping its legal authority and trying to absorb purely local responsibilities; the central board, or at least some of its members, claimed that the local boards pursued their narrow ward interests without regard to the needs of the city.

In 1891 a minority of the Board suggested that the legislature establish a commission to revise the school laws; they were sure that an impartial panel would recommend diminishing the power of the ward trustees. The majority of the Board was not willing to turn the power of rewriting the school law over to an outside commission and rejected the proposal. Instead, the Board appointed a committee of seven of its members to report on school reform, a compromise suggestion made by Commissioner Adolph Sanger. The Committee of Seven, chaired by Commissioner Charles Bulkley Hubbell, promptly urged the Board of Education to seek legislative authority to redistrict the schools. One uptown ward had more than 50,000 pupils, while a downtown ward had not a single school. The uptown trustees, who controlled huge budgets and hundreds of jobs, had more real power than the Board of Education. Some uptown trustees had turned down appointment to the Board of Education. Because of their power, these trustees were able to defeat the Board of Education's legislation in Albany. The committee presented redistricting as a plan to equalize the burdens of the ward trustees, but the trustees rightly saw it as a plan to reduce their political power.

The Board of Education approved a redistricting bill, but the trustees went to Albany and defeated it in 1891 and again in 1892.

The full report of the Special Committee of Seven, ready late in

1891, was a painfully honest document which called for sweeping changes. It recommended numerous reforms that outsiders had been pressing for years, such as kindergartens, high schools, and manual training. While New York still thought of such programs as experimental, Boston already had twenty-nine public school kindergartens and eight public high schools. While visiting other school systems, committee members saw "teachers who after teaching for a score of years are still studying to be better teachers . . . superintendents who are acknowledged to be moral and social forces in their communities and widely recognized authorities in matters pedagogic." To avoid sounding unduly critical, they quickly added: "We are glad to find that most of these features are familiar in our own system." In fact, they were quietly criticizing stolid City Superintendent John Jasper, who had held office since 1879 with no more vision than that of an efficient bookkeeper.

The most controversial section of the report dealt directly with the trustees' power to hire and remove teachers. The report stated that though the majority of the city's 4,000 teachers were "faithful, well-educated and conscientious," there were "too many in this great corps who are incompetent and unfitted for their work." The fault lay with the method of appointment. "Everyone familiar with the working of our system knows the pressure that is sometimes brought to bear on the Trustees for the appointment of teachers by the latter's ofttimes influential friends. . . . Given a license to teach and influential friends, the possessor is likely to secure a position for life in our public school system under the present methods. The system of appointment is in our opinion radically wrong." To remedy the situation, they proposed three changes: first, a three-year probationary period for new teachers; second, a requirement that all teachers present a certificate of good health and retire automatically at age seventy (some teachers were deaf, disabled, or infirm, but held onto their positions through the protection of compassionate trustees, who knew that retirement often meant poverty); and third, a competitive examination for teachers who completed the probationary period. "We favor the application of civil service rules in a modified form . . . realizing however that it is impossible to have reflected upon an examination paper some of the essential qualities for successful teaching." They proposed that the superintendent establish an eligible list of teachers who had passed the examination and that every appointment be made from this list.[7]

The report of the Special Committee of Seven was a political bombshell. The trustees and organized teachers' groups (there were eleven different teachers' associations) immediately opposed the report.

The educational reformers applauded it, but they were as yet too few to muster political support for drastic changes. The general public was apathetic about the public schools, and there was no single issue which roused public attention. The bill reducing the powers and duties of trustees died in committee at Albany; Superintendent Jasper explained drily, "There was a strong outside opposition to it." When a bill to facilitate the removal of incompetent teachers also failed, Jasper commented: "There is little probability of its ever passing, on account of the opposition of interested parties." [8]

Political support for school reform might have been roused by the mayor, but none of the city's mayors during this period showed any interest in the schools. Hugh Grant, a Tammany stalwart who defeated Abram Hewitt in 1888, was re-elected in 1890, despite revelations of widespread municipal corruption by an investigative committee from the state legislature. In 1892 another Tammany man, Thomas Gilroy, was elected mayor.

The mayors of New York in the late 1880s and early 1890s received two kinds of political pressure concerning the schools. On the one hand, they received appeals for support from organized trustees and teachers who felt threatened by the proposals of the central board. Since demands for reform usually emanated from Tammany's enemies, Tammany mayors had no conflict in supporting the trustees and teachers; school reform bills failed to get Tammany support in the legislature, which doomed them. On the other hand, the mayors were beseiged by complaints from parents whose children could not get into school, a mounting problem since the early eighties. No mayor had an answer for this problem, other than to deny that it was his fault. The city schools were simply overwhelmed by the unpredictable population upsurge, as well as by the unexpected shift in residential neighborhoods. Just as the Board of Education began concentrating its limited funds in the uptown wards, the depleted downtown wards filled up with immigrant Jews and Italians, who crowded into dilapidated school buildings and needed more schools and more teachers.

Many letters came to the mayor's desk each year, complaining of the lack of accommodations. One mother wrote to Hewitt that her seven-year-old son had been out of school for two weeks because of illness; when he tried to return, he was refused readmission due to overcrowding. Hewitt sent her letter to the president of the Board of Education, who replied that the boy should apply to school at the opening of the next term, where he would have to "take his chances with the other children." Mayor Grant, in his four years in office, received a steady flow of petitions, letters, and notices of mass meetings to protest the lack of school accommodations. One school principal wrote to Mayor Gilroy to plead with him to prod the commissioners on their promise to build a new school in place of his present, cramped building: "I have refused admission to nearly three hundred children since September 11. These are all the children of poor parents who, at the most, cannot afford to keep them in school but a short time." [9]

Everyone recognized the problem, but no one seemed to have the power to solve it. The schools were trapped in an antiquated organization, with separate divisions for primary, girls' grammar, and boys' grammar. The primary schools, which covered the first three years of schooling, were severely overcrowded, while there were vacancies in many grammar schools because so few children (fewer than 25 percent) finished the seven-year course. Strict adherence to this classification made it impossible to shift pupils and make more seats available. It was a time-honored, *traditional* form of organization, defended by trustees, teachers, and parents, as well as principals, who foresaw that consolidation would cause a reduction in the number of principalships.

The Board of Education was further handicapped by the failure of the municipal government to give it sufficient money. Since the reorganization of the schools in 1873, final approval of the education budget was left to the Board of Estimate and Apportionment. In effect, this meant that the mayor determined the education budget, since he and his appointees controlled the Board of Estimate. The budget requests of the Board of Education were cut every year, despite the growing numbers of pupils. There was hardly enough money available to maintain ongoing commitments, yet pressures mounted for new schools, school repairs, kindergartens, salary increases, manual training, and other new programs. Between 1890 and 1895, school expenditures per capita actually *decreased*.

The deteriorated state of most school buildings was a constant source of embarrassment to the Board of Education. Settlement-house workers complained loudly to newspaper reporters of the unsanitary conditions in the slum schools and of dangerous overcrowding. A typical inspector's report, reprinted in the papers, described a grammar school: "We cannot speak very cheerfully about the condition of this school. It is not fit to be called a school-house. . . . It seems to have been neglected by everybody connected with the school." The sanitary inspector for the Board of Health informed the Board of Education that heating and ventilating changes had to be made in the majority of schools and that in many of these schools the health of the school children was endangered. The Board urgently requested funds to make the necessary repairs; the mayor granted them half what they believed absolutely vital to make the buildings minimally safe.[10]

In March 1892, the Board of Education initiated an investigation into charges of corrupt practices in the construction of school buildings, a function of the local ward trustees. It was discovered that in some instances inferior materials had been used, appurtenances had been omitted, cheap construction and poor performance had been overlooked, and contractors had been paid for work which had not been completed. The Board of Estimate promptly cut off the investigation's funds. A few months later, Mayor Grant launched his own investigation of the Board of Education by the Commissioners of Accounts, with the intention of shifting the blame for the schools' problems from the mayor's office to the school officers.

The Board of Education had always decried the Board of Estimate's parsimonious policy towards the schools. Now the mayor's Commissioners of Accounts struck back with a report which criticized the Board for sloppy accounting practices, for inefficiency and extravagance. It was not, said the report, the fault of any individual school officials: It was "the system." The report seemed to suggest that the combination of powerful local boards and a weak central board was antiquated and inappropriate to the needs of a modern city:

> The Educational Department of our City is a beautiful instance of cumbersome machinery and interminable circumlocution. Responsibility for appointments of teachers and principals, excessive expenditures, useless repairs, etc., cannot be charged to the commissioners, any more than it can be laid at the doors of the various Boards of Trustees and inspectors. The Commissioners of Education are hampered by the present system. It is a relic that has come down to us from other days, when doubtless it worked to satisfaction, but our modern education and the spreading of our population makes changes imperative.[11]

After its sweeping indictment of the school system, the report offered few substantive proposals for change, other than to suggest that all repairs be done in a central workshop. That the report contributed to undermining public confidence in the schools was of little moment to the mayor, who was interested only in defending his administration against the Board of Education's complaints. Ironically, the Tammany mayor's report indirectly aided the cause of those who wanted to overthrow the system of local control of schools and who identified local control with Tammany politics and corruption.

CHAPTER 12

"Save, Save the Minutes"

WITH GROVER CLEVELAND at the top of the Democratic ticket in 1892, Tammany Hall had no difficulty electing Thomas Gilroy to the mayoralty. Although the reformers did not enter a challenger to Gilroy, they were by no means inactive. Even in early 1892, momentum was gathering which would force a direct confrontation between the reformers and Tammany Hall in 1894.

The seeds of a movement were sown in early 1892 when a Presbyterian minister, the Reverend Charles H. Parkhurst, launched a moral crusade against the police and the city administration. Dr. Parkhurst charged that the city and the police permitted vice to flourish. He shocked New York by making his accusations from the pulpit of his church, the fashionable Madison Square Presbyterian Church. He later substantiated them by personally touring the haunts of the underworld in the company of a private detective. He claimed that the city allowed illegal saloons and brothels to operate without hindrance, and that either the police were totally incompetent or were paid to ignore these wide-open dens of vice.

For at least a decade, reformers had talked to themselves, unable to find a wide audience for their pet issues. Reverend Parkhurst's sensational charges of crime, corruption, and official collusion aroused the interest and indignation of merchants, lawyers, businessmen—solid citizens who were shocked and embarrassed to discover that their city tolerated immorality on a vast scale. Dr. Parkhurst's followers were convinced that the city government had to be "purified" and that this could not happen until Tammany was removed from power.

As Dr. Parkhurst swelled the ranks of reform with the morally indignant, the writings of Jacob Riis motivated still others to action. Riis' moving accounts of life in the slums reached a large readership and stirred both compassion and fear:

As we mold the children of the toiling masses in our cities, so we shape the destiny of the State which they will rule in their turn, taking the reins from our hands. In proportion as we neglect or pass them by, the blame for bad government to come rests upon us. . . . Clearly, there is reason for the sharp attention given at last to the life and the doings of the other half, too long unconsidered. Philanthropy we call it sometimes with patronizing airs. Better call it self-defense.[1]

The Children of the Poor (1892) delineated specific issues for which reformers could press. Riis urged the extension of kindergartens ("one of the longest steps forward that has yet been taken in the race with poverty"), because "the more kindergartens, the fewer prisons." He led the fight to establish open-air playgrounds in the great slum below 14th Street, where indoor school playrooms were typically dark, musty, and rat-ridden. He pleaded for the establishment of truant schools and truant homes, instead of the jailing of children alongside hardened criminals.

Riis believed that children were creatures of environment and opportunity, and that the brutal influence of the slum directed many children to a life of crime and pauperdom. His books and articles vividly portrayed the hand-to-mouth existence of young vagrants and the joyless drudgery of child laborers. He claimed that as many as 50,000 children were not registered in school; even had they all applied, the schools were too crowded to accept them. If the school was the chief weapon in the war against the slum, as Riis believed it was, then these large numbers of truants were portents of social disaster. He was imbued with faith that the school was the key to the city's future and that it should be actively and consciously used as a tool of reform, bringing cleanliness, order, the English language, enlightenment, and even "the love of the beautiful" into the dirt and disorder of the slum.

For all his passion and concern, Riis did not romanticize the immigrants; though he was an immigrant himself, having come from Denmark in 1870, he shared the prejudices of his time. He frequently referred to the immigrants as the "refuse" of Europe and advocated immigration restriction to protect the United States from a continued invasion of dirty, ignorant paupers. Had he appealed only to compassion, he might not have mobilized support for the battle against the slum. On the contrary, his concern was thoroughly pragmatic: "The immediate duty which the community has to perform for its own protection is to school the children first of all into good Americans, and next into useful citizens."[2]

As 1892 drew to a close, the public schools were under a cloud. The public complained loudly about the lack of accommodations, the politicians berated them for extravagance, the reformers belittled their failure to exert educational leadership. Since the blame could not be attached to any individual or group, it became conventional to say that the fault lay in "the system." Problems seemed to be intensified by the highly decentralized nature of the system. Local control appeared to frustrate the public interest rather than to advance it.

Even the customary defenders of the school officers admitted the need for rearranging the relationship between the central board and local boards. *School,* a publication which was started in 1889 and was oriented towards the interests of the teachers, wrote in 1892:

> As a body the school officers of New York are as fair-minded men as any who administer public interests anywhere. There are indolent and indifferent members among them who take little interest and care less for the requirements of their office . . . but their influence has always been small. . . . The evils that often arise through trustee or commissioner come from the extended and complicate character of the whole school system, which not infrequently causes measures that are of importance in one locality to be hurtful to other parts of the city.[3]

As the demands for change multiplied, the central board grew frustrated at its inability to implement beneficial reforms. City Superintendent John Jasper recommended consolidating underutilized schools; the local trustees, not wanting to diminish their domain, blocked it. The central board urged local boards to rent temporary space to relieve congested schools; the suggestion was frequently ignored. No one seemed to have the power to solve any of the schools' problems. One ex-president of the Board of Education was quoted in an interview as saying that "the chief defect" of the school system "is the difficulty of fixing a responsibility. If any blame is incurred in a ward, the trustees can throw it on the inspectors, and the inspectors may shift it on the commissioners, and the commissioners may throw it back on both inspectors and trustees." No one was accountable.[4]

Within the Board itself, pressure for change was building. In the fall of 1892, the Board was sharply divided over the question of a proposed budget item of $26,000 for kindergartens. Some Board members maintained that kindergartens were a fad, that the present primary teachers could give whatever instruction of the kind that was needed, and that the public schools were not intended to be nurseries; the leading critic especially objected to the importation of the "barbarous foreign name." The leading proponent of kindergartens was Adolph Sanger. Sanger, a leader in the Jewish community, had held many appointive and elective offices by grace of Tammany Hall and had never been considered a reformer on the Board. But his service on the Special Committee of Seven in 1891, which studied the schools of other cities, turned him into an advocate of school reform. He insisted that kindergartens were no longer an experiment, but were a widely recognized and approved method of instructing the young. He wanted the Board to commit at least $100,000 to them. After intense debate, the Board approved the figure of $26,000.[5]

In December 1892, a letter in the *Tribune* implored the Board of Estimate not to eliminate the $26,000 for kindergartens: "An appropriation now of $26,000 for kindergartens, it is urged, may do away with the necessity, twenty years hence, of ten times that sum for homes for paupers, asylums for the insane and jails for the criminals, because the kindergartens teach children to work as well as to study. Therefore, say the petitioners, the kindergarten for the children of workingmen is the first great secret of guarding against strikes." [6]

The Board of Estimate cut out almost 10 percent of the education budget, despite the steadily increasing enrollment. Funds for kindergartens were reduced to $5,000, enough to open half a dozen kindergartens. Reform achieved its first success, limited though it was.

An event of much greater significance for reform was the election in January 1893 of Adolph Sanger as president of the Board of Education. The Board president had great power to establish Board policy by designating committee chairmen, by setting up special committees and by his own public statements. Sanger was strongly identified with Tammany Hall; he had been the first popularly elected president of the Board of Aldermen. Because of his personal stature and political acumen, he was ideally situated to advance school reform.

Sanger at once injected a new spirit into the Board. In an interview shortly after his election, he said that kindergartens would be opened wherever space was available in primary schools. Sanger promised that there would be "more teaching and less reciting. The mechanical methods hitherto pursued must be dropped and more rational ones adopted." He wanted the Board to add another year at the end of grammar school to give commercial and manual training for children who were not going to college. He pledged to press for adoption of a merit system for appointing and promoting teachers, "to do away with the appointment of teachers at random, or from motives of political preference or personal favor." [7]

Sanger moved quickly to create a vehicle for educational reform by setting up a new Committee on School System and directing it to come up with suggestions for reforming the New York schools. He asked the Board to authorize an eligible list of teachers, from which trustees would make their selections, based solely on considerations of merit. He suggested training classes for new teachers and something resembling in-service training for regular teachers. He wanted teachers to be not only better trained, but better paid. He urged a salary scale for teachers, which would depend on fidelity and ability. Sanger hoped to abolish the archaic method of basing pay on the teacher's grade level, which unfairly stigmatized the lowest grades; he believed that the best and strongest teachers should be assigned to the youngest children. He urged Board members not to permit the "mere spirit of conservatism to interfere with the accomplishment of educational reforms. . . . We should not complain if the vigilance of our citizens or the spirit of the public press manifests by occasional criticism their interest in our schools. . . . If these complaints are founded on fact, it becomes our duty to investigate the shortcomings, whatever they may be, and to apply the remedy as speedily as possible."

Sanger's burst of dynamism met little opposition within the Board, but encountered resistance among trustees and teachers. The trustees correctly saw a direct challenge to their most important power, the selection of teachers. And teachers were not at all pleased by talk of a merit system. For one thing, many of them had genuine cause to fear an examination of their competence, because of their lack of training or their lack of physical fitness (some were deaf or senile). Others sensed an insult to their dignity in the proposal to examine them for their competence. But most teachers objected because they had a stake in the existing arrangements. When a principal was brought in to one school from another, all the teachers in the school felt passed over; their attitude was that the outsider should stay in her own school and wait her turn for advancement. One school had the same principal for fifty-five years; no one had received a promotion in that school for fifteen years. Teachers did not realize that competitive examination would open up job opportunities in all schools for all teachers.

IN THE SAME MONTH that Adolph Sanger was elected president of the Board, January 1893, an article appeared in *Forum,* a New York monthly with a national circulation, that broadened the ranks of educational reformers. *Forum* had been publishing a muckraking series on the public school systems of various cities for several months, written by Joseph Mayer Rice, a pediatrician and journalist. Rice, imbued with the ideals of the "new education," judged each city's schools by whether they had incorporated the scientific humanism that was the hallmark of progressive educational philosophy.

His critique of the New York schools was devastating. He described one primary school in great detail. The school's principal of twenty-five years had been "warmly endorsed" to him by City Superintendent Jasper. The principal's philosophy, wrote Rice, was that when a child enters school his vocabulary is so small as to be worthless, and his power to think so feeble that his thoughts are worthless. The pupil "should not be allowed to waste time, either in thinking or in finding his own words in which to express his thoughts, but . . . he should be supplied with ready-made thoughts as given in a ready-made vocabulary." The curriculum of this three-year primary school consisted of sets of questions and answers, which the pupils were expected to memorize *verbatim*. The principal's ideal was to have the children answer the prescribed questions without hesitation.

The school's maxim, said Rice, was "Save the minutes." To accomplish this end, "the school has been converted into the most dehumanizing institution that I have ever laid eyes upon, each child being treated as if he possessed a memory and the faculty of speech, but no individualities, no sensibilities, no soul." Children were not permitted to do anything which was "of no measurable advantage," such as an unrequested movement of the head or a limb. Seeing the children staring rigidly forward, Rice asked the principal whether they were allowed to move their heads. She answered, "Why should they look behind when the teacher is in front of them?" The unspoken spirit of the school, wrote Rice, was, "Do what you like with the child, immobilize him, automatize him, dehumanize him, but save, save the minutes."

Rice observed a typical recitation. To save time, the children recited their lesson in absolute order, one after the other. As the first child neared the end of the answer, he began to descend to his seat, while the second child began to rise to continue the recitation. Watching the clockwork ascent and descent of the children, Rice imagined that children in adjoining seats were on opposite ends of "an invisible seesaw." The children stared fixedly ahead, even when passing materials along, looking neither to the materials nor to their classmates.

At the "sense training" lesson, each child received a flag of a different shape and color. The teacher gave a signal, and the first child leapt up and loudly and quickly described his flag: "A square; a square has four equal sides and four corners; green." As he fell back in his seat, the next child on the row sprang up. The children repeated these definitions week after week, year after year, until they completed primary school.

In another class, the teacher drew a line on the blackboard and asked, "What is this?"

> The first pupil replied, "It is a line."
> The teacher: "What kind of a line?"
> The second pupil: "It is a straight line."
> The teacher then drew a crooked line and asked, "What is this?"
> The third pupil: "It is a crooked line."
> The teacher: "Wrong."
> At once the fourth pupil said: "It is a line."
> And the teacher: "What kind of a line?"
> Fifth pupil: "It is a crooked line."

The teacher explained to Rice that the student who made a wrong answer was new to the class and had not yet learned the set routine of two questions and two answers.

In reading class the children were taught to read the established number of words for that grade and no more. The children were not encouraged to learn words which they had not yet been taught. Reading was taught by recognizing and memorizing individual words. The teacher said to her class: "The other day I went down to the river and saw something with a whole lot of people on it floating on the water." She then wrote the word "boat" on the blackboard and asked the class what they thought she saw on the river. One called out "ship," another "steamer," a third "boat." In that way the children learned to recognize the word "boat."

Rice reported that children learned arithmetic by reciting "one plus one are two, two and one are three, two and two are four," etc. Later, the lesson was made more "concrete," in the words of the teacher, by saying, "one orange and one orange are two oranges."

The typical New York primary school, wrote Rice, was a "hard, unsympathetic, mechanical drudgery school." Its most salient characteristic was its severe discipline, which produced "enforced silence, immobility and mental passivity." Rice condemned the absence of professionalism or educational leadership in the New York schools. Since mechanical teaching was deemed satisfactory, there was "absolutely no incentive to teach well." And teachers were not imperiled by poor work, since teachers were rarely ever discharged; their job was protected by a cumbrous machinery which assured teachers and principals a guarantee of life tenure so long as they eschewed immoral behavior. Rice complained that the system's 4,000 teachers were supervised by only eight assistant superintendents, too few to be of any help to struggling teachers. Even if a superintendent found that a teacher was incompetent, nothing was likely to happen to her. The city superintendent, Rice wrote, exerted no pedagogical influence over his subordinates. Not even the principals were in a position to aid and direct teachers, since the younger teachers generally had more training than their principals.

Rice blamed the sorry state of the schools on the complexity and inefficiency of a decentralized system. The central problem, in his view, was that no one in authority was accountable for errors. He listed the school officials: twenty-one commissioners of the Board of Education; twenty-four inspectors; twenty-four boards of ward trustees, five per board. "When anything goes amiss, it is impossible to discover which one of these 165 persons is responsible. 'No one is responsible for anything' has become a by-word among those who in any way seek to fix responsibility." No one will accept responsibility for the unsanitary condition of the school buildings or for the irrational course of studies, he asserted, but "when appointments are to be made, every one is on the alert. . . . Everything appears to be involved in a most intricate muddle."

Rice's solution to the "intricate muddle" was to propose a radical reorganization of the duties of the Board of Education. He suggested that the Board divide its responsibilities, keeping to itself those that were strictly financial, turning over to educational experts all matters pertaining to education. He felt that the level of teaching could be raised if there were closer supervision of the teachers and if the supervisors emphasized assistance. He suggested dividing the city into at least twenty districts, each one under the watchful eye of a superintendent responsible for his teachers and schools. The city superintendent would be responsible for general conditions throughout the city and would devote all his time to the teaching of teachers. Each district could hold conferences to upgrade the teaching throughout the district and to promote unity, though not uniformity. He predicted that a "healthy, competitive spirit" would develop among the districts.

The educational work of the districts would be determined by a Board of Superintendents, made up of all the district superintendents and the city superintendent. This board would have the power to appoint principals and teachers on the nominations of district superintendents, as well as the power to discharge them.

The Board of Education would be limited strictly to the business management of the schools. Rice had no objection to the continued existence of local boards, so long as they had no control over the educational work of the schools.[8]

The Rice article created a sensation. Though Board President Sanger advised schoolmen to hear criticism with an open mind, Rice was roundly condemned by teachers, supervisors, trustees, and commissioners. He was accused of generalizing about all schools on the basis of only one school. No one denied what he had written; instead they attacked his credentials for writing about the schools.

The immediate effect of the article was to make educational reform a major issue in the city. Reformers had been frustrated because the Board of Education responded to their demands by adopting changes in a cautious, piecemeal fashion. But now school reform had a battle plan. Riis had delineated specific programs to fight for; Rice provided the means to achieve the end: a complete reorganization of the system. Rice even included a specific blueprint. The work of education would have to be put in the hands of professional educators; only then would politics be removed from educational decisions. The school system should be organized as efficiently as a great business. No sensible businessman would turn decision-making over to a board of laymen who knew nothing of the business. For the same reason, education should be left to professional experts. Professionals should determine the appointment and promotion of teachers according to civil service principles; professionals should prepare a course of study which would take into consideration modern methods and modern curriculum research. The reformers were convinced that so long as power was divided among a multiplicity of boards, so long as unqualified laymen were empowered to make educational policy, then the schools of New York would continue to flounder and fail.

What impeded professionalization? Local control of the schools. The reformers saw the greatest single barrier to change in the person of the ward trustee. He represented all that was anathema to the reformer. He received his place through political appointment, without any professional qualification, yet had full power of appointing and promoting teachers, of selecting textbooks and of supervising the condition of school buildings. The caliber of the central board seemed to be of the same respectability whether appointed by a Tammany mayor or a politically independent mayor; the reformers, acknowledging the honorable character of the central board, directed their scorn at the trustees. The fact that the trustees were unknown outside their own wards made them easy targets. After all, no individual trustee would rise to defend himself against a general charge of malfeasance, and no one would be so foolish as to testify on behalf of some 120 diverse men. Just the fact that they were attacked tended to besmirch them. It was a principle with the reformers that where there was politics, there was corruption, thus these men must have accepted their unsalaried jobs only for the patronage involved. And where there was patronage, there was Tammany. It took no great feat of imagination for reformers to conclude that their fight against the trustees was but another facet of the long fight against Tammany Hall.

The growing din of criticism finally led the Board of Education to take action. The Committee on School System framed a bill asking the legislature to authorize a commission to revise the school laws. The committee recognized that any measure of permanent reform would be impossible to accomplish without legislation from Albany, because of the power of the ward trustees to block any reforms initiated by the board itself.

The chairman of the Committee on School System, Charles Strauss, explained to the *New York Times* why the new commission was necessary. The schools were governed by laws adopted some forty years before, when the schools had only 30,000 pupils. In 1893, the pupil population was ten times that number, and control by local boards simply was no longer appropriate, as it had been when the city was a collection of localities. "There is a lack of responsibility in every department of the school system. There is too much red tape. If a teacher wanted to have a nail put into the wall, he could not do it without asking the janitor; the janitor would have no authority to do it, unless he obtained permission from the trustees of the ward, and they would have to ask the Board of Education to authorize the spending of money to buy the nail and the hammer to drive it." Strauss, a member of Tammany Hall like Sanger, gave another example:

> A janitor in an uptown school died some time ago at the advanced age of 105. He was a veteran of the Mexican War. He was utterly unable to attend to his duties, and his son, an active young man of seventy-five, did his father's work. This thing continued for twenty-five years, and yet the Trustees would not remove the old man and the Commissioners had not the power.
>
> And yet, when we attempted to remedy these things through the Legislature in acts that referred to specific abuses, the persons entrenched in their positions by the defects of the system would go up to Albany with a powerful lobby and defeat remedial measures. When the bill was aimed at the commissioners, they would defeat it; when aimed at trustees, they would defeat it; when aimed at the teachers, their association would lobby, and again would the measure be lost.[9]

Strauss did not want to do away with the trustees. Instead he proposed to redistrict the city, increasing the number of school districts from twenty-four to forty or fifty of equal size. He wanted to leave the trustees their power to appoint teachers, but subject to the absolute veto of the Board of Education.

The Board of Education bill passed without difficulty in Albany. It empowered Mayor Gilroy to appoint a commission of five members. The reformers, knowing that the bill had been prepared by Tammany-man Strauss and that the commission would be appointed by Gilroy, expected little to come of it. They feared that the commission was a device to avoid action, that the problems would be studied, a report would be issued, and nothing further would happen.

But Gilroy surprised everyone by appointing a strong and independent commission. He rebuffed the teachers' demand for a representative, refusing also to appoint a member of the Board of Education. The commission members were David McClure, a respected Catholic layman; Oscar Straus, a leading member of the Jewish community; Thomas Hunter, president of the Normal College; Stephen H. Olin, an outspoken reformer; and E. Ellery Anderson, an uptown ward trustee. Olin was the only member identified with school reform.

The 1893 commission held public hearings; it was the first time since the 1850s that a state commission had publicly examined alternatives for improving the school system. Rice's ideas were reflected in testimony to the commission. The theme of centralizing the business management of the schools and strengthening the role of professional supervisors was popular among speakers. Even Anderson, chairman of the commission, referred to the Rice article as proof of the need for change.

Late in 1893, the commission released its report, strongly urging professionalization and centralization. It proposed strengthening the central board, reducing the power of the local boards, abolishing the inspectors and creating a new, powerful Board of Superintendents. The professional staff would absorb most of the trustees' powers; the trustees would become nothing more than "frequent visitors who are called upon for advice and suggestion." The commission felt that the trouble with the existing system was that every decision required the participation of two or more bodies of officials: "A system so complicated of necessity results in inefficient action, protracted delays, always vexatious and, at times, almost disastrous, and renders it almost impossible to fix responsibility where it justly belongs." [10]

The radical changes proposed by the 1893 commission contributed to a sense of crisis about the schools. No one was pleased with the schools. Everyone had a favorite scheme for reorganizing them. Commissioners and ex-commissioners pontificated to the press on this or that pet proposal. The trustees were angry at the Board, the teachers were distraught about a threat of salary cuts, the Board was angry at the trustees, the reformers wanted to overthrow the entire system, and, all the while, the real difficulties were mounting. In the fall, at least 10,000 children were denied admission to school because of lack of space. Sanger insisted that the Board of Education was unable to solve its problems until the law was changed to give it full power over site acquisition and construction of new buildings. He said: "I have in my desk an appeal signed by 951 children in a certain uptown section of the city asking for accommodations. We cannot give it to them . . . as the trustees do not admit that the school is necessary. If we were free to do so, we would at once hire a house and teach the children while waiting to give them better accommodations." As a stop-gap measure, twenty schools were put on double session.[11]

CHAPTER 13

The Best Men
to the Rescue

WHILE THE DEBATE over the state of the schools became more agitated, the movement for political reform which began as Dr. Parkhurst's moral crusade gathered momentum. Members of the reform City Club, until then a fairly inactive association of well-bred gentlemen, decided to organize clubs in every assembly district to promote the cause of reform; four such clubs, called Good Government Clubs, were formed in 1893. The following year, before the mayoral campaign, another twenty clubs were established, which offered weekly lectures on such subjects as civil service reform, school reform, ballot reform, tenement-house reform, boss rule, police corruption, and so on. The Good Government men were dedicated to the principle of nonpartisanship in municipal government. Having decided beforehand that the clubs were not to participate in any spoils of office, they had to find another way to stimulate interest in the work of politics. This was done by assigning each club a particular department of the city government as its special study assignment. Good Government Club E was responsible for the public schools.

The motivations and assumptions of the reformers were inextricably related to their social origins. For the most part, the reform movement—later the Progressive Movement—was Yankee Protestant. They were men—and women too—of good families and good education, who, with few exceptions, were not of recent immigrant stock. They were upper-class, or at least upper-middle-class. They were appalled by Tammany Hall, by the slums, by the torrent of immigration, by corruption in public life. They felt that certain democratic excesses had brought the city to the point of ruin. Though they always declared themselves on the side of democracy, in this fight as in many others they

134

often acted on elitist assumptions. They believed that the best men were not elected to office, and that the best men were not appointed to office. Men reached high public place for all the wrong reasons, through the spoils system. City government was in the hands of Tammany, and Tammany was in the hands of first- and second-generation people who entered politics for *reward*. It was a tenet with the reformers, reflecting their own comfortable status, that men should regard public office as a public service, not a job; and that any public official who actually needed the income of his position was very likely not qualified to serve. Underneath this set of attitudes was a marked disdain for politics and politicians. There was a presumption that anyone who actively sought political preferment or called himself a politician was not to be trusted. Time and again, reform organizations congratulated themselves that they were "above politics," they were nonpartisan, they sought no reward other than the triumph of civic virtue. Their sense of class superiority was so deeply ingrained that it was entirely unconscious. They wanted to turn out the "others" who had used power to despoil the city, and put in their place—*themselves,* the best men. They sought power, but never admitted even to themselves that it was power they were after. In the genteel world of the Yankee reformers, power-seeking was uncouth. But "good government" was a worthy ideal, and the effort to take "politics" out of city government became for many of the city's elite a worthy, exciting and fashionable undertaking.

The president of Good Government Club A, the first organized, was an exemplary figure of the reform movement. He was only twenty-five, a recent graduate of Yale University. He belonged to excellent social clubs and was a member of the Society of Colonial Wars and the Society of the War of 1812. Club A's clubhouse, which was as much a social club as a political meeting-place, stood at 58th Street and Lexington Avenue; it was handsomely furnished with parlors, reading rooms, a billiard room, and a library. The club entered the political wars in its first year. Its leaders attempted to persuade the local Republican organization to run "a man of high character and standing in the community and pledged to non-partisan and business-like methods in the administration of city affairs," as a candidate for the assembly. These were code words which meant that the Republicans were asked to run a candidate acceptable to the reformers. Not surprisingly, the Republicans were not prepared to give up their partisanship to please the reformers; they entered a Republican candidate. Club A nominated a reformer, causing a three-way race. The result was that in a usually Republican district, the Tammany candidate won. Club A considered itself unique: "Never before has such a political organization, composed entirely of voluntary workers, actuated by patriotic principles, existed in this city." [1]

The reform movement was heartened in early 1894 by the creation of a state committee to investigate the city's police department, as a result of Reverend Parkhurst's revelations. The legislature was then Republican-controlled, and Republican legislatures always were more than willing to embarrass Democratic-run New York City by investigating corruption. The investigative committee, chaired by State Senator Clarence Lexow, consisted of five Republicans, one independent Democrat, and one Tammany Democrat. The legislature passed a $25,000 appropriation to finance the committee's work, but the funds were vetoed by Democratic Governor Roswell P. Flower, who feared that the probe was calculated to embarrass the city's Tammany administration. Despite the Governor's veto, the Lexow committee was determined to have its investigation. The New York Chamber of Commerce advanced funds to the committee to proceed, and the investigation began in March 1894.

Starting with little in the way of firm evidence, the Lexow committee was remarkably successful in uncovering massive, systematic corruption in the police department. Over the next eight months, until election day 1894, damaging testimony about the relationship between the police and politicians captured newspaper headlines. Witnesses testified that many members of the police force belonged to Tammany clubs, and that affiliation with Tammany was a prerequisite to advancement on the force. Some officers admitted that they paid for their appointment to the force and subsequently for promotions. There were admissions that the police were paid to protect prostitution, policy-making, and other illegal activities. Some witnesses claimed that Tammany Hall dictated the selection of justices of the police courts. The police commissioners themselves were accused of participating in bribes and the protection of vice. The counsel to the Lexow committee, John Goff, later wrote that "the primary business and aim of the department" was public plunder. The police force had become "an established caste. They were not of the people. . . . They were not the guardians of the peace or of society. They were the violators of the peace and the oppressors of society." [2]

The sensational nature of the hearings aroused public indignation. Looking ahead to the November 1894 elections, the anti-Tammany movement cast about for other issues, and inevitably saw the potential of the unsettled situation of the educational system. Reformers applauded the bill prepared by the 1893 commission, particularly the section which stripped the ward trustees of their all-important power to appoint and promote teachers. They were certain that teachers' jobs were patronage, and ward trustees were Tammany politicians. The solution seemed simple to the reformers: eliminate the patronage belonging to ward trustees, and only high-minded citizens would want the job in the future.

The appropriation of school reform as an anti-Tammany issue became possible only with the sudden death, in January 1894, of Adolph Sanger, president of the Board of Education, Tammany stalwart, and the Board's chief spokesman for reform. He was a seemingly vigorous man of fifty; his death robbed reform of the impetus it had achieved within the Board and gave the school-reform issue to the anti-Tammany forces.

When the Board of Education considered the sweeping proposals of the 1893 commission, it decided to restore to the trustees their power of appointment. This compromise was designed to improve the bill's prospects in Albany, where heavy opposition was expected from the trustees. The most the Board hoped for was the power to redistrict the school lines and thereby to cut down the power of some local boards. But despite the Board's concession, trustee opposition to the bill was so strong that it never reached the floor during the 1894 session.

In the same session, however, the local trustees managed to win passage of the Ahearn bill, which gave trustees the power to appoint principals and vice-principals. Up until that time, the Board of Education appointed principals from the trustees' nominations; the Ahearn bill reduced the central board to a simple veto, and permitted the trustees to resubmit their choice and deadlock the issue. Passage of the Ahearn bill was a humiliating setback for the central board, which had intended to limit the powers of the trustees, armed with the authority of a state commission and the support of the major newspapers.

Though emasculated by the Board of Education and buried by the state legislature, the 1893 commission report was by no means dead. The original bill, stripping the trustees of their powers, became a part of the reform campaign for the mayoralty in November. Stephen Olin, the reform member of the 1893 commission, blasted the public school system in the June 1894 issue of Nicholas Murray Butler's *Educational Review;* his most withering remarks were directed at the concept of local control of the schools:

> The advocates of the present system of governing the schools do not, as I understand, claim that greater efficiency or economy is thereby secured, or that better teachers are obtained, or the uniformity of discipline and management increased. For all these purposes the methodical administra-

tion of a central body is to be preferred. But they say that the present system increases the local interest in the schools. I confess that this argument does not seem to me a very strong one. If there are men who will not accept the office of trustee unless they can appoint teachers and employ tradesmen, there are other citizens no less worthy of respect, who have no taste for the distribution of spoils but are interested in education. . . . The community cannot afford to purchase local interest in the schools at the price now paid for it. . . . Mulberry Bend may not control its own police, nor Murray Hill assess its own taxes, nor Hell's Kitchen select its own health inspectors.[3]

To the reformers, the issues were clearly drawn. On one side stood Tammany, local control, the ward trustee system, corruption, favoritism, nepotism, inefficiency, patronage, and backward schools. On the other were the reformers, centralized control, professional supervision, businesslike administration, scientific pedagogy, honesty, efficiency, and modern schools. The reformers saw themselves as the guardians of progress, while their critics saw them as intolerant, self-righteous, and arrogant.

The reform campaign to win public support was aided by the publication of a survey of the school buildings, prepared by a member of the Board of Education, Charles Wehrum, in the spring of 1894. He visited every school in the city and compiled a matter-of-fact record of dismal conditions. He found classrooms with neither desks nor any other furniture. He visited numerous buildings which were badly ventilated, where rooms were cold, draughty, and dark, where teachers could not be heard because of the noise from trains or factories. He reported frequent overcrowding. Many of the buildings were permeated by foul odors emanating from badly maintained water closets. Many schools were still burning gas, instead of using electricity, which filled the rooms with noxious fumes. Some schools had two or three principals, each in charge of a separate department, each drawing full salary. In many classrooms, three children shared a desk meant for two.

Typical reports were the following:

Grammar School No. 36, No. 710 East Ninth Street—Eleventh Ward. 1,653 pupils, 35 classes. Exits poor; two rooms in the Primary dark, one room very obnoxious, urinals on both sides; air in this room bad; sickness prevailing. Two rooms no desks.

Grammar School No. 39, No. 235 East 125 Street—Twelfth Ward. 2129 registered, 40 classes. Babel let loose! School crowded, four classes being taught in assembly-room without division; ventilation in front room bad; street noisy; old building in use since 1849, annex since 1881; closets old; sinks smell badly.

In yet another school, at Fordham Road and Webster Avenue, Wehrum found; "Building on low ground; playground in cellar damp and chilled; closets unclean and dark, no sewer connection; stream of foul water running through yard; children sick with diphtheria." [4]

With the opening of school in September 1894, the pressure for seats was greater than ever. A strike during the summer delayed the opening of several new schools. All of the upper ward schools were filled to capacity, as were the schools of the Lower East Side. There was an increase of 15,000 students over the previous year. At some of the predominantly Jewish Lower East Side schools, the crowd of parents with children demanding admission was so great that the police had to be called to maintain order.

FUELED BY THE LEXOW DISCLOSURES, the reformers began their campaign for the mayoralty. On September 1, 1894, at the initiative of members of the Chamber of Commerce, 3,500 "select" citizens met at Madison Square Garden to form a citizens' movement "entirely outside of party politics and solely in the interest of efficiency, economy and the public health, comfort and safety." Out of this organizing meeting grew the Committee of Seventy, a new group dominated by wealthy merchants and professional men, which intentionally adopted the same name as had the group that ousted Boss Tweed twenty-three years earlier. The Committee of Seventy charged that the Gilroy administration was guilty of corruption, inefficiency, ignorance, and extravagance. The Tammany mayor had ignored sound business principles and had relied instead on the execrable motives of personal and political advantage. The chief principle of the Committee of Seventy was nonpartisanship. "Municipal government should be entirely divorced from party politics and selfish personal ambition or gain. The economical, honest and businesslike management of municipal affairs has nothing to do with questions of National or State politics." The committee promised to manage the affairs of the city "as a well-ordered household, solely in the best interests of its people."

Its platform called for strictly nonpartisan appointments to public service at top levels, and appointment according to civil service principles for subordinate positions; the creation of small parks in crowded areas; better enforcement of health laws, more public baths and lavatories; a better system of street cleaning and garbage disposal; increased rapid-transit facilities; and greater home rule for the city. Its statement on the public schools was a "demand that the quality of the Public Schools be improved, their capacity enlarged, and proper playgrounds provided, so that every child within the ages required by law shall have admission to the schools; the health of the children be protected, and that all such modern improvements be introduced as will make our Public Schools the equal of those in any other city in the world." [5]

The Committee of Seventy chose one of its members, William L. Strong, as its candidate for mayor. Strong was an independent Republican, a wealthy, sixty-seven-year-old merchant; he had come to New York from Ohio in 1853 to seek his fortune and had succeeded. To make the ticket appealing to anti-Tammany Democrats, the Seventy filled the remaining places with Democrats.

To get Republican support for its nonpartisan ticket, the Seventy approached Thomas Platt, a shrewd upstate legislator who dominated the state Republican party. Known as "Boss Platt," he financed the party with contributions from corporations who wanted immunity from unfavorable legislation. His organization was one of the most powerful Republican machines in the country. In the fall of 1894, Platt was preoccupied with securing the Republican gubernatorial nomination for his candidate, Levi P. Morton, a former vice-president of the United States and a socially prominent businessman. After succeeding, he turned his attention to the situation in New York City, where a rebellion against the Republican organization had been brewing for most of the year within the New York County party. Leading business and professional men had set up a committee of thirty to oppose the frequent deals between Boss Platt and Tammany boss Richard Croker. Most of these Republican rebels were members of the reform Committee of Seventy.

Platt had not been consulted in the selection of the Committee of Seventy, nor in their designation of Strong. He resented the emergence of this new force in city politics, which owed him nothing. Platt was especially suspicious of Strong, who was part of the anti-Platt "committee of thirty." But with the gubernatorial election pending, Platt was anxious not to split the party. At the very least, he hoped to pick up some anti-Tammany Democratic support for Morton by endorsing the candidate of the reformers. An emissary from Platt interviewed Strong. The interview went satisfactorily, and the Republican organization consented to endorse the candidate of the Committee of Seventy.

The state Democratic party was in disarray. A severe economic depression in 1893 contributed to a large Republican victory in the off-year election. The 1894 gubernatorial and mayoral elections threatened disaster. The Democratic governor, Roswell P. Flower, declined to run again, which was just as well since the party leaders did not want him anyway. The Democratic state convention nominated an upstate Democratic boss, David B. Hill, for governor, but failed to quell the rebellion within the party's ranks. Party dissidents nominated an outspoken reformer as their own candidate.

With the state party split, the city party was not much better off. Boss Croker cannily induced a highly respected and popular business-man, Nathan Straus, to accept the Tammany nomination for mayor. Straus pledged a nonpartisan, businesslike administration and hoped to win back the support of some anti-Tammany Democrats, but they rejected him because of his Tammany support. The day after the Republicans endorsed Strong, Straus withdrew from the race. Tammany substituted the faithful former mayor, Hugh J. Grant.

On the surface, the campaign against Tammany was a fight for good government, yet the atmosphere was charged with class and religious antagonisms. The city's social leadership barely contained its distaste for the common Irishmen who were powers in the city; even more basic to the good Anglo-Saxon stock of New York was its ingrained aversion to Catholicism. With Irish Catholics controlling Tammany Hall, and Tammany Hall controlling the city, reformers saw the election as a crusade to save the city.

At the November election, with the unified support of the Committee of Seventy and the Republican organization, Strong trounced Grant. Republicans swept to victory in state and national elections, as voters blamed the Democrats for the devastating 1893 depression.

The reformers could scarcely believe what they had wrought. Their hopes for the future were extravagant. The election of November 6, 1894, was deemed "The Great Political Revolution" to commemorate the day when reform forever destroyed Tammany's hold on New York City. The history of this great movement was published only a year later. E. L. Godkin, editor of the *Nation* and the *New York Evening Post*, wrote in the campaign history that the true significance of the election was not just the triumph over corruption or the election of an honest man or the prospect that municipal offices would be filled by honest men instead of thieves: "It is the final adoption of a new agency in the work of municipal administration." What was "new" and therefore "final" was the non-partisan character of the winning combination. This, Godkin wrongly prophesied, heralded a new era in municipal politics.[6]

IMMEDIATELY AFTER THE ELECTION VICTORY, the school reformers were ready to seek new legislation. The Committee of Seventy had appointed subcommittees to prepare reports on specific problems; the subcommittee on the public school system was chaired by Stephen H. Olin, late of the 1893 commission. His committee members were Henry L. Sprague (former member of the Board of Education), John B. Pine (trustee of Columbia College), William Ware Locke (a minister), and Professor Nicholas Murray Butler. The subcommittee was firmly committed to centralization and professionalization and against local control.

The subcommittee report followed the original recommendations of the 1893 commission by urging the abolition of inspectors and the elimination of most of the trustees' powers, "both being relics of a system long since outgrown." The only reason that the trustees had held on to their powers so long was that "the dispensers of patronage have been loath to give it up and have successfully resisted all attempts to take it away from them." The report promised that once a new system of administration was established, it would be possible to appeal for the extension of kindergartens, for manual training, for more and better supervision, for more and better accommodations, for playgrounds, for vacation schools, and for improvement of sanitary conditions within school buildings. The reform bill which they were preparing, the subcommittee modestly submitted, "may fairly be deemed an expression of the deliberate will of the entire community." This remarkable conclusion was reached in a roundabout way. They reasoned that their bill was a recapitulation of the 1893 commission bill, which had been drawn by a commission which was created by the legislature and appointed by the mayor; and that the 1893 bill with one minor exception had won the unanimous approval of the Board of Education. "The bill, so far as we are aware, met with no opposition except from those who have a personal interest in the continuance of existing abuses." This claim to represent the popular will stretched the truth. The one "exception" which the Board of Education failed to approve was the heart of the bill: the removal of the appointment power from the ward trustees. The assertion that local control of the schools had no friends but spoilsmen was equally exaggerated. The prose style of the report was unmistakably that of Nicholas Murray Butler, who as editor of the *Educational Review* had set up the ward-trustee system as a dragon that he was sworn to slay.[7]

A legion of school reformers set to work to build public support for the Olin subcommittee's bill. At the center of the lobby were Good Government Club E and its ladies' auxiliary, which was first called the

Women's Association for Improving the Public Schools (WAIPS); this latter group later became independent and in October 1895 was renamed the Public Education Association (PEA). Its chief advisor was Nicholas Murray Butler. The school reformers filled the press with stories and letters that dramatized the plight of the schools.

Two bills to reorganize the schools were introduced in Albany in the spring of 1895. One, sponsored by the reformers, was the Pavey bill; the other, the Bell bill, proposed to create a paid commission to run the schools. The Board of Education was divided in its support for the two. To the chagrin of the reformers, most of Mayor Strong's new appointments to the Board were not unswerving proponents of centralization. One Board member warned that to turn the management of the schools over to employees who were accountable to no one "would be to establish a bureaucracy, which would be repugnant to American institutions and traditions." The Board of Education voted to support the Pavey bill, but only if it were amended to restore the trustees' power of appointing teachers. To unify the reform forces, the two bills were merged, retaining most of the Pavey features; the Board of Education, under intense pressure from reformers, endorsed the compromise measure.[8]

The trustees, inspectors and teachers rallied in hostile opposition to the proposed legislation. The teachers called a mass meeting to demonstrate their resistance to outside interference in the schools. Teachers and supervisors sent delegations to Albany to testify against the bill.

The compromise measure managed to pass the assembly, but was badly beaten in the state senate, where Boss Platt was in firm control. The reformers had been divided and ineffectual; the defenders of the status quo had been certain and angry. And Boss Platt was determined to punish Mayor Strong for appointing anti-Platt Republicans to high office in the city. The Platt organization got even less patronage from Mayor Strong than it had received from Tammany Hall. Consequently, most of Mayor Strong's legislative package was stalled in Albany.

Because of the Platt–Strong quarrel, school reform was dead for 1895. But the reformers refused to be discouraged. So certain were they that they were in the vanguard of history and progress, that defeat barely slowed them. If anything, it led them to redouble their efforts to sweep the forces of reaction out of the school system.

CHAPTER 14

Professor Nicholas "Miraculous" Butler

THE FIELD MARSHAL of the school reform movement was Nicholas Murray Butler. He was involved to some degree in virtually every organization that joined the fight for school reform. Every important reform document about the public schools was either written or edited by Butler. Many, if not most, of the reformers never attended public schools nor did their children, so they tended to rely heavily on Butler's authority as an educational expert.

Like most of his reform friends, Butler was a white Anglo-Saxon Protestant of good family. After graduating from Columbia College in 1882 at the age of twenty, he was on his way to a brilliant academic career; he earned his M.A. and his Ph.D., he studied abroad, and—by the time he was twenty-four—he returned to Columbia to receive an appointment as an instructor in philosophy. Only four years later he became a full professor and dean of the faculty of philosophy. But that was not all. In 1887 he organized and was the first president of the teacher-training institution that later became Teachers College of Columbia University. In 1891 he founded the *Educational Review*, one of the leading journals of the field, and became known by educators around the country as an incisive and outspoken advocate of the professionalization of education.

Butler made the issue of reforming New York's schools his own. His unbending determination to alter the system was as strong as that of Bishop Hughes a half-century earlier. Hughes' crusade was specifically religious, but animated by the hostility of the poor towards their rich patrons. Butler's crusade to replace laymen with experts was religious in its intensity and was animated by Butler's perception that power in the school system was in the keeping of those who were socially and professionally inferior. Both Hughes and Butler were consummate politicians; each used every available political avenue in his struggle to oust the insiders, but their tactics were necessarily different. Hughes, knowing he represented an isolated minority, turned his people into a disciplined political force. Butler knew that he did not have the kind of support that could win a referendum or the sort of followers who would attend mass meetings. However, Butler did have access to the highest seats of power, as well as a carefully conceived propaganda program.

Butler's critics considered him an arrogant and elitist meddler. Not all the opposition to Butler's scheme to abolish local lay control came from self-seeking politicians, as Butler charged. There was a strong case to be made for the existence of local control of the schools, but Butler completely disparaged it.

One recent student of this period, David Hammach, emphasized the sharp class differences between the centralizers and the advocates of local control. The centralizers, like Butler, tended to be part of an economic and social elite; when Butler formed a school reform lobby in 1896, 92 of its 104 members were listed in the New York Social Register. Hammach described the reformers as aggressive modernizers who had taken part in the centralization of great corporations and the raising of standards in the professions, civic leaders who were as zealous about efficiency and nonpartisanship as they were about moral reform and traditional Protestant values. Like Riis, the reformers saw the school as a moral antidote to the ills of communities which differed from their own.[1]

Opposition to centralization existed, but was overwhelmed by the sophisticated tactics of the reform forces. Support for local control and home rule was traditionally expressed through Tammany Hall, which was controlled by the Irish and commanded the loyalty of the poor ethnic communities. However, Tammany's defense of the ward-trustee system was of no value in 1895 and 1896, since the legislature was dominated by upstate Republicans. The reformers' ill-concealed disdain for Catholics, immigrants, poor people, and neighborhood leaders (whom Butler called "unknown" and "insignificant") must have aroused antagonism to their plan to take control of the public schools. Yet so many of the schools were unsatisfactory that support for the status quo was shallow. As in 1842, the greatest resistance to change came from those who were directly involved in the operation of the schools.

The objections of the staff were articulated in the weekly newspaper *School*, edited by H. S. Fuller. Where Butler's *Educational Review* was a journal directed at pedagogical professionals, covering the latest scientific and philosophical comment both here and abroad, *School* was written for the New York school teachers. *School* was chatty and informal; it reported on news of the schools, personal notes about teachers and principals, meetings of the Board of Education, and any other news that might affect the interests of its readers. Fuller was frequently critical of both the central board and the local boards, but cautioned against any radical changes in the schools.

Other journalists were awed by Butler's fierce intellect and high professional standing. Some of the leading dailies sent writers to Butler for "background briefings," as if they were consulting an objective observer rather than the leader of one side. But Fuller was not at all impressed by Butler. Early in 1895, he dubbed him "Professor Nicholas Miraculous Butler" and contemptuously referred to him as a "continental expert." Fuller wrote that Butler had "publicly intimated his entire confidence in his own ability to reorganize, single-handedly, the public school system of New York City. It is not often that even Columbia graduates a genius who attains such dazzling heights." He added mockingly: "There is little doubt but that if the Higher Powers would entrust him with the task of constructing a new universe, Professor Nicholas Miraculous Butler would enter upon that undertaking with equal confidence, unabashed and unaided." [2]

Fuller believed that local control was both desirable and necessary in a democratic form of government. While the reformers were disparaging the trustees as tools of Tammany, Fuller insisted that they were an honest body of officials and argued that the faults of the system could be remedied by redistricting and equalizing the duties of the trustees.

Attacking the central provision of the reform legislation, which gave the powers of the trustees to a body of twenty professional superintendents, Fuller maintained that there was quite enough supervision already, by some 300 principals; what was needed, he believed, was that each principal be made the uncontested supervisor of his own school. He attributed the Pavey bill to "a coterie of educational faddists." [3]

School had little influence on public opinion and none at all on the thinking of reformers. But it did echo the sentiments of the thousands of people who worked in and around the school system. Just as schoolmen had belittled Joseph Mayer Rice as an outsider, so did *School* view the reformers as a handful of self-important men and women with no previous experience in the schools who were determined to destroy the system in order to try out their theories.

At the height of the 1895 drive for centralization, Fuller saw the fight as a sub rosa social battle. Teachers wrote him that the struggle was "between five men who stand for the Bill, and 4,000 men and women in the service of the schools." (The "five men" referred to the five members of the Olin subcommittee, who had prepared the reform bill.) Fuller went even further: "It is an issue between fully 1,400,000 residents of the city, and a sentimental fad of a remnant of the Four Hundred." He charged that the reformers, Butler chief among them, deliberately misled the city's press into thinking that their own views represented educational authority and the wishes of a majority of the city.[4]

During the legislative battle, the Women's Association for Improving the Public Schools (which changed its name to the Public Education Association [PEA] in October 1895) invited the teachers of the city to a meeting, to describe the advantages of the Pavey bill. About 700 teachers attended. At the meeting's close, a large majority voted to oppose the bill. Not only did the teachers fail to see the value of adding another level of superintendence over their heads, but they resented the patronizing role of the society ladies who wanted to educate them as to their best interests.

Butler wrote in the *Educational Review* that the teachers were stirred up by "ringsters" who were "fighting desperately to retain their power and patronage." He warned that defeat of the reform bill would "demonstrate clearly that politics has a foothold in the New York schools that is firmer by far than the reformers have ever asserted it to be."[5]

In his early years as a reformer, Butler's criticisms of the schools were reasoned assaults dealing with administration, curriculum, and methodology. But in the heat of the battle, he threw moderation to the winds. The tone of his writing became intemperate and accusatory. The schools were not only inefficient, they were "riddled with politics." The trustees were "as a rule, utterly unknown and insignificant persons, often active local politicians, and not always, it is believed, above preying upon the schools." He defined the opposition as incompetent teachers and principals "who fear the effects of competent supervision," and power-hungry trustees.[6]

The daily press never displayed any sustained interest in school affairs, particularly when the issues were as dull and uninteresting as internal organization and curriculum. But charges of corruption and political hanky-panky were good copy. The schools, as an institution, were peculiarly vulnerable to attack because of their utter defenselessness. The central board was not about to defend the trustees, nor were the trustees of a mood to support the central board. If a board of trustees spoke out in self-defense, it looked foolish; one board called in news-men to meet the principals of its district, who then vowed to the news-men that they had not received their jobs for political reasons. Everyone involved looked silly. It was not uncommon to read in newspaper editorials that "nobody wishes to see the present system preserved ex-cept the Tammany rascals and incompetents who have either adminis-tered or profited by it." [7]

After the defeat of school reform in 1895, the reformers decided to wait and see whether the Board of Education would develop its own reform legislation. Quite sensibly, they believed that if they could work together with the Board, their joint bill might pass in 1896.

The Board assessed public sentiment in the fall of 1895. Public hear-ings were held during October, and the trustees were polled. The poll revealed that the trustees were not nearly so intransigent as the reform-ers had painted them. They had a clear idea of what their powers should be, and it was not at all unreasonable. Of twenty-four trustee boards, twenty thought that the city should be redistricted. Most boards thought that they should continue to have the power to make repairs; most thought that they should retain the power to promote and transfer teachers. At the same time, the majority were willing to accept an eligi-ble list for teacher appointments; most also agreed that the central board should have the power to purchase sites and erect schools without the involvement of local boards. A small majority opposed an eligible list for principals; almost every board opposed turning the schools over to a full-time, paid "Board of Experts."

The Committee on School System, headed by Commissioner Strauss, was given the task of writing the Board of Education's own school-re-form bill. This committee, even though its chairman was associated with Tammany, had been the source of numerous innovative policies in the three years since the late Board President Sanger organized it. The com-mittee recommended a bill that incorporated proposals from previous reform bills, almost doubling the number of school districts, but leaving the trustees' powers virtually intact. The local boards would still have power to appoint teachers, but only those who had been examined by the city superintendent and admitted to the eligible list.

As soon as Butler read about the Board's bill in his morning *Tri-bune*, he dashed off letters to the president of the Board, Robert Maclay, and to the leading reformer on the Board, Charles Bulkley Hubbell. He advised them that the Committee of Seventy would fight the Board's bill if it did not abolish the trustees. He wrote to another Board member that the reformers "believe ourselves able, whether we can pass a bill of our own or not, to defeat any other bill that does not wipe out the trustees entirely." Butler wrote to a friend: "What a mess the Board of Education's committee proposes to make of school reform." [8]

Butler was not afraid to fight the Board of Education, because the school reformers had never been stronger. He knew he had considerable resources to draw on if a showdown in the legislature became inevitable. Good Government Club E was filled with zeal for its work, as were the good ladies of the PEA. The anti-Tammany newspapers, particularly the *Times* and the *Tribune,* wrote stories and editorials along lines he suggested. The *Evening Post* published frequent columns calling for radical reform; Butler was the writer's informant.

Another good break came Butler's way just at the time he learned that the Board intended to retain the trustees. He received a letter from one of the key members of PEA, Mrs. Schuyler van Renssalaer, who was also a writer and a friend of Joseph Pulitzer. She wrote Butler that "Mr. Pulitzer has promised me that the *World* shall be run just as we want it to be in regard to the school question, and that he will publish any editorials and as many of them, as I choose to send on." Mrs. van Renssalaer asked if he would mind editing whatever she wrote: "The subject is as yet so new to me that I am afraid of sending in anything of an editorial character until it has been read by a more competent eye than my own." Butler, who relished the role of behind-the-scenes manipulator, had complete control over the *World*'s statements on the public schools.[9]

Mariana Griswold van Rensselaer worked as hard for school reform as any member of the PEA. In her class origins, she typified the well-bred woman of the era, but in her personal accomplishments she was unique. Born in New York City in 1851, she traced her American ancestry to 1639. She was educated by private tutors and by travel abroad, married at twenty-two to wealthy and prominent Schuyler van Renasselaer, and widowed at 33; her only son died tragically in 1892. After her son's death, she turned to political activities by writing a pamphlet which strongly opposed female suffrage and by throwing herself into the school-reform fight. From 1899 until 1906 she served as president of the PEA, then returned to scholarship and produced a monumental, two-volume history of seventeenth-century New York City, as a result of which Columbia University (whose president was then Nicholas Murray Butler) awarded her an honorary Litt.D. in 1910.

DESPITE THE EXERTIONS of Butler and his friends, the Board of Education would not satisfy their demands. The Board approved the bill prepared by the Strauss committee and dispatched it to Albany for the 1896 session of the legislature.

Butler was outraged that the Board ignored the single issue which he considered of greatest importance: the elimination of the local trustees. He could hardly believe that some of Strong's appointees signed the report endorsing "Strauss's preposterous school bill." He made much of the fact that Strauss was a member of Tammany Hall.[10]

Anyone less intrepid than Butler would have bowed to the wishes of the Board of Education, for the entire Board had either been directly appointed or continued in office by reform Mayor William Strong. Board President Maclay was Strong's most trusted advisor on school affairs. Butler's remarkable feat was to appropriate the imprimatur of the "reform movement" on behalf of his particular ideas about the public schools, without the help of the movement's mayor. Butler did this through his domination of the subcommittee on school system, which wrote the Committee of Seventy's school legislation. It was a manifestation of Butler's political genius that the banner of reform stayed with his five-man committee instead of passing to Mayor Strong's Board of Education.

In January 1896 Butler called together a dozen friends who had worked for school reform the year before, "to consider ways and means of defeating or amending the Strauss bill." He and his friends agreed to form a new organization, composed of 100 leading citizens, to press for the abolition of the ward trustees. Within days a letter went out to a list of prominent men, inviting them to join a new Citizens Committee for Public School Reform. The organizing group had five members: Butler and Olin; former Mayor Abram Hewitt, Elihu Root, and J. Kenedy Tod. All were top-drawer members of New York society.[11]

While the Citizens Committee was organizing, Butler wrote members of the legislature, urging them to remain uncommitted until the opposition to the Board of Education's proposal "had time to crystallize." He informed the Republican legislators, most of whom were not from New York City, that "the Strauss bill" was written by a member of the Law Committee of Tammany Hall and would be opposed by all the city's newspapers.[12]

The reform bill which bore Butler's stamp of approval was sent to Albany under the sponsorship of the reform City Club. It contained only one provision: abolish the ward-trustee system. Butler shrewdly realized that it was important that school reform appear to have a broad base, so he attempted to play down his own role in the preparation of the new reform bill. He pointedly referred to the bill as a City Club bill and reminded his friends in the legislature and the press that he was not a member of the City Club. Nonetheless, it was Butler, not the City Club, who organized the campaign for the passage of the bill. The secretary of the City Club wrote Butler to apologize for its inability to send a delegation to the hearings on the school bill; he confessed that the City Club's main interest in the 1896 session was ballot reform.[13]

The City Club bill went to Albany with the endorsement of the Good Government Clubs, the Citizens Committee for Public School Reform, and the PEA. An officer of PEA wrote to Butler that their Executive Council had approved the bill unanimously: "There was absolutely no discussion of the measure. . . . You see what entire confidence we have in our Advisory Board!" Butler of course was their chief advisor. She added a postscript: "I have tried to write in a very calm and judicial way but I am really tremendously excited over being allowed to publicly take a hand in this fight." [14]

In dealing with legislators, Butler had the considerable advantage of his proclaimed "expertise," in an era when experts were revered. The legislators knew little of the New York schools and less of pedagogy. Butler wrote them with all the authority of the modern, scientific, pedagogical expert. Several legislators admitted openly that they relied on his judgment. Hamilton Fish, the speaker of the assembly, wrote him: "I am inclined to believe you are right in your views as to the school bill from what I can learn on the subject, although I must confess my knowledge is quite limited." Another assemblyman, in reply to Butler, wrote: "I am not a school expert, and I do not consider myself competent to judge as to which is the better bill. . . . I am certainly open to conviction." When Butler bombarded the chairman of the Cities Committee in the assembly with editorials and articles about the schools, the assemblyman sent the following message. "In view of the fact that you have promised to acquaint me with the particular features of the proposed bill, I have not investigated the matter carefully." [15]

Butler was a superb political maneuverer. He knew that his greatest strength lay in proclaiming that his goals were right for purely educational reasons, while professing to have nothing whatever to do with politics. At the same time, he stressed to the practical politicians that what he advocated was also the desire of "the vast majority of intelligent citizens," as well as a unanimous press. To a Democratic assemblyman who was not anti-Tammany, Butler wrote: "I am particularly anxious that all political questions and all questions as to the fitness or unfitness of particular school trustees, should be kept entirely out of the discussion. They are really not germane at all. The question at issue is simply one of efficient educational administration, nothing more." [16]

Actually, Butler's disagreement with the Board of Education's plan

was political, not educational. The Strauss bill provided numerous educational reforms. Butler knew that the only difference between the two
proposals was the fate of the trustees and local control. Strauss and the
majority of the Board of Education thought that the trustees kept the
people in touch with the schools; Butler thought that laymen should
have nothing to do with educational administration. His position was
unreservedly snobbish, not only on professional grounds, but socially as
well. Those who agreed with him were "enlightened" and "intelligent"
citizens, while those who disagreed were "barbarous" and self-seeking.
He often wrote that the "right" kind of people were not appointed to
serve as inspectors and trustees; what he meant was that *his* kind of
people were not appointed. It was not really the patronage system that
he disliked, for Butler himself gathered names of people whom *he* considered the right kind of people and submitted them to the mayor for
appointments. When he wrote that school trustees were insignificant
people who were utterly unknown, he meant that they were unknown to
him and his circle; it was of no moment to him whether they were
known and respected in their own neighborhood.

Butler's contemptuous attacks on the ward trustees in the daily
press and in the pages of the *Educational Review* revived sour memo-
ries of the days when elected local school boards contained many tav-
ernkeepers and minor ward politicians. But in the post-Tweed reorganization, the rules of the Board of Education specifically excluded liquor
dealers from appointment to the local school boards. And though Butler
neglected to account for the inconsistency, the chairman of the Board
committee which selected local trustees was the leader of the reform faction, Charles Bulkley Hubbell. The fact was that the school trustees of
1896 were solid, if unrenowned, middle-class citizens. Contrary to the
charges about Tammany control, nearly 40 percent of the city's trustees
were Republicans. More than half were merchants; others were lawyers,
doctors, judges, and bank officials. It had become customary practice by
the Board to accord representation on the local school board to every
sizable minority group within a district, to the extent that it was possible on a small board. The Board recognized the right of representation
of all sects and nationalities. This informal system produced boards
which reflected their districts, but which were not composed of the kind
of educated, altruistic people whom reformers admired. The Strauss bill
aimed to retain these boards, but to submit their power of appointment
to civil service principles, thereby eliminating the chief cause of complaint against them.

Boss Platt's Republican organization at first distrusted the reform legislation, because so many school reformers were insurgent Republicans. Butler wrote an ally of Platt: "You can assure everybody of the fact that no bill that this Legislature will pass will do more for the Republican organization than our school bill. This will not happen because it is framed in the interest of that organization, but simply because it deals a terrific blow at Tammany Hall. Anything that weakens that body strengthens its opponents." [17]

While Butler maintained a public stance of the nonpartisan, objective educational expert, he launched a brilliant statewide lobbying effort on behalf of the reform bill. He identified the key votes in both houses of the legislature and tracked down people who could pressure them to favor reform. He wrote to the legislators' best friends, even their priests and ministers. He corresponded with school officials in remote cities and towns, asking them as fellow professionals to apply pressure to their local assemblyman and state senator. He explained that the battle was entirely "nonpolitical," that it was a simple, though epochal, confrontation between good and evil: "In this school fight it is civilization against barbarism and your influence will be most potent." He contacted college presidents, school superintendents, textbook publishers, the state superintendent of education (who replied that he was too busy to get involved and would follow Butler's lead), even the United States commissioner of education. He advised all his correspondents that the Board of Education's proposal (which Butler always called "the Strauss bill") was a Tammany bill, while the City Club legislation was the vehicle of good government and modern pedagogy.[18]

Simultaneously, Butler orchestrated the school question in the New York press. He advised the *Evening Post* in March, a month after the centralization plan was introduced, that its support "should be pretty continuous from now on until the matter is settled. I am also asked not to embarrass some of the Senators who are supporting our bill vigorously by attacking Platt in connection with it. They would like to have the public press confine itself to criticism of Tammany and of the inefficient system." Butler wrote a letter to the editor of the *Tribune* and advised him to print it as a news article "because what I want to effect is public opinion at Albany and letters have no weight with them. Besides, I should far rather have it appear without my name. It will do far more good." [19]

As time for a vote drew nearer, Butler got rougher in his tactics. He confided to his Republican allies who controlled the legislature that the protests against centralization and professionalization were organized by Tammany Hall. He wrongly claimed that the Board of Education had split along partisan lines, all Democrats being for the Strauss plan, all Republicans against. He tried unsuccessfully to prevent hearings by the state senate; he did, however, convince his supporters in the assembly that no more hearings were necessary. He did not want his opponents to have any outlet for their views.

The opposition to centralization was trapped by Butler's rhetoric. With the press solidly arrayed against them, their defense of local control sounded like apologies for Tammany domination of the schools. The much-maligned Commissioner Strauss, in a newspaper interview, said that the criticisms of the Board of Education's proposal were "wholly unjust and indefensible." He decried the labeling of it as the "Strauss-Tammany bill," since he was the only member of the Board of Education who belonged to Tammany Hall. He pointed out that only two members of the seven-man committee which drafted the plan were Democrats, and that the bill had the support of ten members of the Board who were either Republicans or independents.[20]

The weekly newspaper *School* came to the defense of the ward trustee as "the representative of the public in school government." It predicted that the reformers aimed at "the centralization of power and its removal from the people." The cry for centralization "comes from those who do not attend the schools; who have no children in them; who have never had any experience, interest or sympathy with them. . . . The larger number of persons who can be kept in touch with and interested in the public school system, the better it will be for the schools." *School* mocked Butler's blue-ribbon Citizens Committee for Public School Reform, asserting that the school trustees were at least as honest and certainly more representative of New York than the 100 socially prominent men who were listed as members of the Citizens Committee.[21]

At the hearings before the Senate Committee of Cities, the President of the Board of Education, Robert Maclay, gave the Board's case for retaining the ward trustees:

> New York is a peculiar city. It is a cosmopolitan city. If you do away with the trustee system you do away with the people's schools. The trustees are in touch with the schools, and none others are or can be but those who live in the locality of the schools. We have a peculiar population, made up of all nationalities. They are people whose children we want to get in our public schools. There is a fear on the part of these people that we are going to interfere with their religion. If we have ward trustees representing all classes, confidence will be maintained.[22]

The reformers turned out in force for the hearings, led by the indefatigable Nicholas Murray Butler. A large delegation of upper-class women, which included the governor's wife, appeared to testify as members of the Public Education Association. The women were new to the ways of the legislature, but thrilled to be entrusted with a role in the forefront of the battle. Being women accustomed to giving orders and being entirely certain of their cause, they took hold of the new situation. Mrs. Edward Hewitt, daughter-in-law of the ex-mayor, chided Board President Maclay for his lack of knowledge of the public schools. Mrs. Schuyler van Renssalaer, recently appointed a school inspector by Mayor Strong, testified to the senate hearing that the schools were so inefficient that "no one who can afford to send his children to a private school would think of sending them to a public school." The activities of the society women in Albany led an amused press to dub their proposal the "pink tea school reform bill." [23]

The Senate Committee on Cities, at the conclusion of its hearings, reported out a compromise which grafted some reforms of the Board of Education plan onto the single demand of the reformers, abolition of the ward-trustee system. The reformers were delighted. Governor Levi P. Morton urged the legislature to pass the compromise measure promptly.

In New York, the teachers and trustees realized that the question was no longer which of several bills would pass, but how to stop centralization. The teachers called a mass protest meeting. The circular for the meeting was titled: "PUBLIC SCHOOLS IN DANGER!! ARISTOCRACY VS. THE PEOPLE." The *Herald* and the *Sun,* both supporters of reform, reported that 3,000 to 4,000 people attended. Butler quickly wrote reassurances to the Republican floor leader, saying that he himself had viewed the meeting, that no more than 1,800 persons participated, that many of those were teachers' escorts, that the meeting had been ignored by the press (except for the "imaginative reporter" from the *Herald,* who had inflated the head count), and that teachers had given out tickets to school parents.[24]

The reformers decided to confine their lobbying efforts to small meetings. The PEA women, doing the thing they knew best, organized a series of teas for public school teachers. Their mission to educate the teachers about the true needs of the schools was blatantly patronizing. One teacher wrote to *School* that she did not quite understand the sudden interest of certain "distinguished ladies" in the public schools. "Have we teachers done anything remarkable? Why should we be invited to their 'teas' and waited on by these estimable ladies? Mrs. Ben Ali Haggin's children do not come to our schools. We do not expect that Mrs. Levi P. Morton will drive up to our schools in the governor's open barouche to take us on a ride in the Park of an afternoon. We have no ambitions to be admitted to the select social circles of the city." [25]

The reformers' command of the daily press obscured the extent of opposition to their legislation. Not only was centralization opposed by the trustees, the teachers, and the principals, but also by a majority of the Board of Education and a majority of the city's representatives in the state legislature. There were even reformers who questioned the propriety of dissipating all local control for the sake of "efficiency." Joseph Mayer Rice, author of the article in 1893 which launched the drive for school reform, took a position against the reform legislation because of its complete abrogation of local control.

With the vote in Albany only a few days off, the last gasps of the opposition were recorded. At a mass meeting at 150th Street, speakers warned that if the trustees were abolished, the uptown wards would insist on running their own school system. One speaker said, "We must be represented by persons who live in the district. I don't know if any member of the Board of Education would know where to get off the trolley car if he came to visit the schools in this district." [26]

School warned that if the school system were centralized, the schools would become "a great perfunctory machine in which the individual parent or teacher is lost," and that sound character and good citizens cannot be produced by a highly centralized, aristocratic system. It charged that Governor Morton had succumbed to pressures of "social caste and sectarian influence." [27]

Despite the flurry of protests and petitions, the reform measure passed both houses of the legislature with ease, by 31–13 in the senate, and by 88–43 in the assembly. Republican Boss Platt apparently took a hands-off position, allowing the overwhelmingly Republican legislature a free vote. Most of the Republican legislators were not from New York City. They saw the issue as an opportunity to rebuff Tammany Hall, while satisfying the insistent demands of a unanimous press in the city, educational experts from all over the state, and the most prominent and prestigious people in the city, including the Republican governor and his wife.

After passage in the legislature, the bill then went to Mayor Strong for his approval. Strong at once became the focal point of diverse pressures. He decided to hold public hearings, alloting one day to each side. Every association of teachers and supervisors spoke against centalization. Fifteen members of the twenty-one-man Board of Education signed a statement opposing it. The Board held that the legislation violated home rule, since the people of the city were denied a referendum; they also felt that the bill violated sound principles of civil service, since the new Board of Superintendents would have the power to examine applicants as well as to appoint and promote them. These functions, the Board felt, should be separated. And lastly, the Board feared that the immense concentration of power in the Board of Education increased the danger of political interference in the schools.

A former school commissioner wrote the mayor to remind him that the fact that the police department was corrupt should not lead to the assumption that the schools too were corrupt. "No wrongdoing, no scandal, no abuse has been shown which cries for a remedy. . . . Who cares for this radical change? Not the teachers. . . . Not the Board of Education. . . . Not the parents. . . . Who wants the change then? A small number of excellent persons whose restless activity and earnest anxiety for the public good lead them to forget that the great law of the natural . . . world is that growth must be slow to be sure and gradual to be effective." [28]

A number of trustees and inspectors sent Strong a petition against the measure which revealed their perception of the reformers' social bias: "Who are these reformers? A few men and women with more zeal than knowledge, led by the wife of the Governor himself. . . . In this agitation, the reformers have held meetings behind locked doors, have never allowed public debate, have controlled the Press. . . . The Trustees are branded as ignorant Tammany heelers, as Catholics mainly and saloon and bar-keepers. Why? Because the reformers . . . say we are such people." [29]

A petition with 100,000 signatures opposing the abolition of local boards was delivered to Strong. He listened to both sides without revealing his intentions. Even Butler was worried. Meanwhile, the counter pressures on Strong to sign the bill came from his own social circle and political allies. The school reformers were, after all, an appendage of the Committee of Seventy which elected Strong. Strong was independent enough to keep his friends guessing about his predilection, but they were, in the end, his friends. He was of their milieu. And he had no fear of making an unpopular decision because he had no further political ambitions.

Whatever Strong's predispositions, he could not help but be impressed by the arguments in favor of centralization. The reform case was baldly restated in a confidential analysis of the school situation presented to Strong by a member of his staff. The writer stated as plain fact that the management of the schools had been "deplorable and inexcusably bad," because of "the impossibility of locating responsibility." He predicted that, under the new law, "We shall no longer be obliged to hunt around for the persons responsible for a failure to furnish adequate schooling facilities for the children of the masses; there can be no shifting of the blame upon the Trustees, or by the Trustees upon the Board of Education, or by both upon the Board of Estimate and Apportionment." He held that localities had no more right to manage the schools than they had to manage their own police force, particularly since local management had been so demonstrably incompetent. It was not a good thing in a city

> largely impregnated with foreign influences, languages and ideas that the school should be controlled locally; for in many locations the influences that would control would be unquestionably un-American. In some districts there are vast throngs of foreigners where one scarcely hears a word of English spoken; where the mode of living is repugnant to every American idea. The best interests of the city demand that the children of such a population be brought under American influences and instruction.[30]

On April 22, 1896, Mayor Strong signed the reform bill and centralized the school system of New York City.

BETWEEN
THE WARS

1896–1913

CHAPTER 15

New Education for the New Immigration

WITH THE PASSAGE into law of the reform bill, local control in the New York public schools was entirely eliminated. The reformers won a total victory. Power over the educational work of the schools was lodged in the hands of professional educators, and there it would remain for almost seventy-five years.

Control of the school system was divided between the lay Board of Education and a new, powerful Board of Superintendents. The Board of Education, still twenty-one members appointed by the mayor, assumed authority over the business aspects of the system; they were empowered to establish, discontinue, or consolidate schools, to purchase sites, erect buildings, and order repairs. The Board of Superintendents took charge of the educational side of the system, absorbing most of the power that formerly belonged to local boards. It was their responsibility, subject to the approval of the Board of Education, to hire and promote teachers and principals; they prepared eligible lists of personnel based on competitive examination, in accordance with civil service principles. Though the job of inspector was retained, its powers were insignificant and vague.

The great "school war," as Butler dubbed it, was very nearly finished. All that remained were a few skirmishes to make the triumph complete. The first of these was an attempt, led by Butler, to oust John Jasper from his post as city superintendent. It seemed an affront, in Butler's view, to maintain this visionless representative of the old order in a post created for an educator of stature. Butler corresponded with school superintendents in Cleveland, St. Paul, and Denver, encouraging them to apply for Jasper's job. Butler's idea of the proper powers of the city superintendent was revealed in his description of one of the candidates: "The Cleveland law makes Mr. Jones an autocrat. He appoints, removes, transfers and promotes all principals and teachers at his own will and with no appeal to any other authority." Jones had a life appointment, removable only for cause. Butler remarked that Jones had never abused this great power. Butler's faith in professional authority was as unbounded as his distrust of lay authority.[1]

But none of Butler's candidates garnered a majority, and John Jasper was re-elected to a six-year term. Butler accepted this setback grudgingly and turned his attention to the Charter Commission, which was going to determine the system of education for all of Greater New York. The old city of New York was scheduled to become the borough of Manhattan on January 1, 1898. On that date, New York, Brooklyn, and numerous towns and villages in the Bronx, Richmond, and Queens counties were slated to become one consolidated city.

Republican Governor Levi P. Morton had appointed a Charter Commission for the new city. Butler appealed to every influential acquaintance to urge the governor to appoint a reformer to the education committee of the Charter Commission. He was actually lobbying for his ally and superior, Seth Low, president of Columbia University. Low did receive an appointment and became the only educator on the three-man committee. All three members were from Brooklyn, which was significant in view of the antagonism between the proponents of the centralized system of New York and the defenders of the still highly decentralized system in Brooklyn.

The New York reformers, strong and confident after their recent victory, devoutly believed in the rightness of centralization and professionalization; they believed their ideas represented scientific advance in the field of education. Naturally, they were anxious to spread their superior system throughout the rest of Greater New York. But Brooklyn had a strong tradition of local control and was equally anxious to maintain it despite the pressures of the New York reformers.

The Brooklyn system contained all the practices which New Yorkers had routed from their own schools. In Brooklyn, the central board was large (forty-five) and weak. Each school was managed by a three- or four-man Local Committee; each member of the central board sat on at least one Local Committee. Friends of the system claimed that it kept the schools close to the people. But its enemies, who included Brooklyn City Superintendent William Henry Maxwell, charged that the local committees were thoroughly political and undermined the efficiency of the system. Superintendent Maxwell was a cofounder with Nicholas

Murray Butler of the *Educational Review* and shared Butler's views.

The education committee of the Charter Commission worked out a compromise between centralization and local control. The new charter established a borough school system, comprised of four strong borough boards and a weak central board. Brooklyn was permitted to retain its peculiar local-committee system. But the school systems of the other boroughs were centralized within each borough. In Queens, thirty-five independent school districts were merged into one; in the borough of Richmond, twenty-nine districts became one. The New York system was continued as the Borough Board of Education for Manhattan and the Bronx. Each borough board had its own borough superintendent and complete control of the instruction within its own schools. The central board was supposed to secure a uniform financial system, as well as set minimum standards for hiring teachers. The education chapter of the new charter also created a semiautonomous Board of Examiners to examine applicants for school jobs; this was a proposal of Butler's which was designed to end complaints that examination and appointment powers should not be concentrated in the same authorities.

The New York reformers were dismayed that Brooklyn held on to its provincial prerogatives, but were buoyed by their undaunted faith in the steady march of progress. Their optimism seemed already to have been repaid in Manhattan, where improvements in the schools were apparent by the end of 1897, only eighteen months after the schools had been centralized. The construction-and-repair bottleneck was broken once authority was concentrated, and the number of new sittings rose dramatically. The reformers justly claimed credit for streamlining and simplifying the process. Perhaps even more important, the schools at last had full political backing, thanks to the Strong administration. The Strong administration supported the needs of the schools in the Board of Estimate and Apportionment and promoted the passage of new schoolhouse bonds in the legislature.

In 1897, New York's last year as an independent city, the reform faction finally won control of the Board of Education, signified by the election of reformer Charles Bulkley Hubbell as its president. The reform Board's first act was to establish three high schools. But their next major decision sowed bitterness which would linger on to plague the new borough system. The reform Board passed a resolution requiring applicants for the three new high school principalships to have had experience as high school principals. This proviso automatically disqualified everyone working in the New York system, since the system had no high schools. The three positions were placed outside civil service so that the Board could appoint its choices without asking them to submit to a New York examination. New York teachers and principals were outraged; they saw this action as a confirmation of the reformers' disdain for people employed by the system, an attitude encountered frequently during the bitter legislative battles of 1895 and 1896.

The teachers' anger fed into a large tide of public disenchantment with reform. Strong's administration had been honest and efficient, but the voters apparently were bored with reform. In the vital election of November 1897, with the patronage of the entire metropolitan city at stake for the first time, Tammany candidate Robert Van Wyck, a little-known judge, easily defeated reform candidate Seth Low. Low was defeated by the refusal of the Republican organization to support his candidacy. The Republicans, knowing it would cause a Tammany victory, entered a separate ticket and split the anti-Tammany vote. Boss Platt had had his fill of stiff-necked reformers. He preferred to let Tammany win; he was used to working with them.

In 1898, Robert Van Wyck started his four-year term as the first mayor of Greater New York, and the new borough school system of the metropolitan city was inaugurated. Since the Manhattan-Bronx area contributed the most taxes, it controlled a majority of votes on the central board. Each borough board sent representatives to the central board, and the Manhattan-Bronx majority elected Charles Bulkley Hubbell as president of the central board. The reform majority then elected Brooklyn City Superintendent William Henry Maxwell as the first city superintendent.

Maxwell, a Protestant of Scotch-Irish descent, immigrated to the United States in 1874 at the age of twenty-two. Trained as a teacher, he lacked the political patronage to get a job in the New York or Brooklyn schools. He worked on a newspaper for several years, and finally, because of a good series that he wrote about the needs of the Brooklyn schools, he obtained an appointment to teach in the Brooklyn evening schools. By 1882, he was an associate superintendent in the Brooklyn system; five years later, he was elected Brooklyn superintendent. After his election as the first city superintendent of Greater New York, Maxwell served for twenty years. Throughout his tenure in office, Maxwell worked to remove all partisan influence from the schools. He was close to Nicholas Murray Butler and was a coeditor of the *Educational Review* from 1891, when it was founded, until 1896.

After Van Wyck's election, the reformers feared for a time that he would use his power-of-removal authority to replace them, just as reform Mayor Strong had replaced some Tammany men when he took office.

But Van Wyck did not remove any of the reformers before the expiration of their term of office. Within eighteen months after his taking office, Van Wyck's new appointments eliminated the reform majority. Van Wyck did punish the Board of Education in other ways. He castigated it for initiating costly experiments and regularly cut the education budget in the Board of Estimate and Apportionment.

Meanwhile the new borough system became more deeply entangled in red tape with every passing day. All the boroughs complained that the others were receiving an undue amount of funds. Borough boards attacked the central board for interfering in local affairs. *School* complained that the borough authorities were not able to direct their schools without coming to central headquarters "for every single pencil or broom that is required." [2]

The relationship between the borough boards and the teachers was openly hostile. When the Manhattan-Bronx board adopted a salary schedule which required teachers to pass examinations before obtaining a pay increase, the teachers got the legislature to establish a salary schedule based on length of service, not examination scores.

Relationships among the borough boards themselves were unpleasant and at times chaotic. The Brooklyn board sued the other boards to prove its right to set higher teachers' salaries, even though it meant reducing the money available to the other boroughs. The Court of Appeals upheld Brooklyn's right to set its own salaries, but directed Brooklyn to levy a special tax to repay the other boroughs. When the Brooklyn board decided to drop the issue, the Brooklyn Teachers Association brought 2,500 individual suits against the city for the promised salary increases. The corporation counsel threw the entire issue back to the unhappy central board.

By the end of 1899, the central board was in a deep muddle. The Queens and Richmond boards declared that they lacked the money to pay teachers for the last quarter of the year. The teachers demanded the minimum salaries they had won in Albany. The Manhattan-Bronx teachers sued the city for their salary increases, and a court ordered the city to issue bonds to pay them. Mayor Van Wyck, knowing that the education authorities were required by law to expend more for teachers' salaries, assailed them for their extravagance, slashed the schools' budget for the coming year, then wrote every board member in every borough that board members would be personally liable if they could not stay within the appropriation allotted them by the Board of Estimate and Apportionment.

As if all these problems were not enough, the city superintendent was publicly at odds with the central board and some of the borough boards. When Maxwell called a city-wide examination of teachers, the Manhattan-Bronx board, no longer dominated by reformers, told its teachers not to appear. And when Maxwell presented his first annual report, castigating the Brooklyn local-committee system, the central board refused to print it, on the grounds that it was offensive to school officials.

When the 1900 legislative session opened, numerous bills to reorganize the New York City schools were submitted. One bill would have abolished the central board and established autonomous borough school systems. Others sought further centralization. *School* reported that Mayor Van Wyck and Republican Governor Theodore Roosevelt favored a single central board. *School* urged a bill which would increase the powers of local school inspectors, warning that the schools were "without local supervision or local interest. They are being forgotten by the voter and the taxpayer except as part of a perfunctory bureau." If further centralization should occur, *School* warned, "the public schools of the entire city will pass into the direction and control of a vast bureaucracy." [3]

Early in the legislative session, a bill passed which rescued the city's schools from their immediate financial crisis. Though the Board of Education objected strenuously to Albany's interference with home rule, the legislature enacted a four-mill tax and a system-wide salary schedule.

It was universally acknowledged that the school system needed more than money. The borough system, in its two years' trial, proved to be unstable and weak. Governor Roosevelt appointed a new Charter Revision Committee in April 1900 to correct the defects of the city's charter. The educational section was assigned to men identified with New York reform. Several months later, they reported back with a plan which abolished the borough boards and centralized the entire professional staff under a single city superintendent. Lay control was retained in a single, central Board of Education of twenty-three members, but at Brooklyn's insistence, the legislature increased it to forty-six. To quiet widespread criticism of total centralization, local school boards were established in each of forty-six districts. The local school boards were supposed to inspect the schools in their districts and give voice to local opinion, but had no power to interfere with the administration of the schools. Each local board was composed of five lay members appointed by the borough president. In theory, the local boards were supposed to have close communication with the Board of Education, since each central board member was assigned to sit on a local board. But since the local boards had no meaningful power, they had no impact on school policy.

The legislature passed the Revised Charter in 1901. Friends of centralization were delighted to see the professional staff concentrated in one headquarters; friends of local control were pleased that the local boards were revived.

Shortly after the new organization took effect, it became evident that the locus of power was in the professional bureaucracy. The most distinctive feature of the new system was that every significant educational decision was made by the Board of Superintendents; the lay Board of Education held final authority, but only to approve or veto the prior decisions of the Board of Superintendents. The Board of Education with its forty-six members was too large and unwieldy a body to assert leadership.

The highly centralized, highly professionalized educational system mandated by the Revised Charter of 1901 proved to be highly durable. After being the center of political storms for almost a decade, the schools settled into an institutional form which lasted for almost seventy years more, with relatively minor changes.

THE MOST IMPORTANT IMPACT of centralization was a remarkable burst of energy and innovation, unparalleled before or since in the New York schools. The reformers had a program, a social-work approach to children's lives, and the new concentration of authority made it possible to introduce massive changes in the schools. The first victim of change was the traditional notion that school was a place to learn reading, writing, and arithmetic, and that students who failed had only themselves to blame. In its place was substituted what was then called "the new education" and later known as "progressive education"—amorphous terms which covered a range of antitraditional approaches. The development of this movement in education has been described by Lawrence Cremin in *The Transformation of the School* as "a many-sided effort to use the schools to improve the lives of individuals." The school was charged with responsibilities which previously belonged to the family, the settlement house, and the community. Its functions were expected to expand to include "direct concern for health, vocation, and the quality of family and community life." Advocates of the new education were influenced by recent research in psychology and the social sciences; their desire to merge science and democratic idealism led them to the conclusion that education must be adjusted to the needs of the many different kinds of children who were the clients of public education.[4]

The new education was both a political and a humanitarian response to the attainment of nearly universal education in an economically and ethnically diverse society. The restriction of child labor and the enforcement of the compulsory attendance law corralled into the school a new kind of pupil, a child who in previous years was a vagrant or a worker. This realization of the democratic ideal of universal education meant that the schools had to create programs for those children who had neither the interest nor the ability to take part in a formal academic curriculum.

These developments coincided with, and were clearly stimulated by, the massive migration from southern and eastern Europe. The public schools were quickly overwhelmed by the children of these immigrants. In the first fifteen years of the century, the average school register nearly doubled, rising to almost 800,000 pupils. Since Maxwell had declared, as a tenet of progressivism, that there must be a seat for every child, many schools were forced to schedule double sessions.

The public schools were faced with a serious plight: within their confines were many children who could not speak English, children of widely varying abilities, children who would leave school as soon as they were old enough (fourteen) to get working papers, and children with special physical disabilities. In 1904, Maxwell tabulated the first age-grade statistics ever compiled by an American city and found that 38 percent of all pupils in the eight-year course were overage for their grade. The *New York Times* reported "that the boys and girls receiving instruction at the city's expense are dull, and that from 25 to 50 percent of the pupils are from two to six years too old for the classes in which they are being taught." Maxwell, admitting that he found the figures more startling than anything he had imagined, urged principals to set up special classes for the overage. The fact that so many children seemed unsuited for the traditional academic curriculum brought progressive pressure to "meet the needs" of all children by diversifying the schools' offerings. During this period the system established courses which emphasized manual, vocational, and industrial training; special classes were set up for the mentally defective, crippled children, deaf children, blind children, tubercular children, and anemic children.[5]

The social work impulse of the progressive movement brought into the schools programs which had originated in settlement houses. Evening trade high schools and industrial schools of elementary grade were opened, primarily for young immigrants and Negroes who worked at manual labor during the day. The public schools inaugurated a medical inspection of all school children. Libraries were established in every school building. Special classes were created for those who were overage or slow learners; some of these classes were for non-English-speaking pupils, others for children who intended to leave school at age fourteen to go to work. Summer classes were opened, as well as evening study classes in recreation centers. And, in a complete break with the past, the school buildings were made available to their surrounding communities for after-school use for academic, recreational, and social programs. Over a million people patronized the Board of Education's evening recreation centers in a single year, and a similar number attended evening lectures on specialized subjects; so popular was the lecture program that it evolved into an organized system of adult education. Vacation playgrounds, where there were sometimes musical performances, averaged a daily attendance of over 125,000.

At the time, progressive educators saw this diversification of the schools' role as the democratization of education. But not everyone within the system applauded Maxwell's changes. One principal complained to the *New York Times* in 1905 that "fads" like sewing, drawing, nature study, and music took time away from the basic studies and caused academic failure. "Do good to the greatest number. The weaklings must fall. This is the keynote of the situation forced upon the teachers by the involved course of study they are expected to follow," the principal complained. "It might be possible to find good in the course of study mapped out for our use if there were classes of fifteen or twenty children. With classes numbering from thirty-five to seventy, the teacher is obliged to rush on without a single halt. She cannot turn back for a second to help a straggler. . . . The system makes the backward children." [6]

Half a century later, Richard Hofstadter decried the movement in the education profession "to exalt numbers over quality and the alleged demands of utility over intellectual development"; this trend was typified by the replacement of the ideal of "mastery" of subject matter with the ideal of "meeting the needs" of all children:

> Far from conceiving the mediocre, reluctant, or incapable student as an obstacle or a special problem in a school system devoted to educating the interested, the capable, and the gifted, American educators entered upon a crusade to exalt the academically uninterested or ungifted child into a kind of culture-hero. They were not content to say that the realities of American social life had made it necessary to compromise with the ideal of education as the development of formal learning and intellectual capac-

ity. Instead, they militantly proclaimed that such education was archaic and futile and the noblest end of a truly democratic system of education was to meet the child's immediate interests by offering him a series of immediate utilities.[7]

Maxwell was the man in the middle; he was torn between two worlds, committed to the traditional aims of education ("the ordinary school studies which all civilized peoples grant are essential,—reading, grammar, arithmetic, history, geography, drawing, penmanship, spelling"[8]), but conscious of the high rate of failure in schooling the masses by traditional studies. Maxwell was the leading agent for the radical transformation of the New York public schools in the early years of the twentieth century. In many ways, he embodied the contradictions of progressivism, particularly its curious combination of elitism, paternalism, and democratic idealism.

Like many of his progressive peers, Maxwell was an autocrat committed to egalitarian goals. He wanted to see the public schools become a ladder "from the gutter to the university," an expression which aptly typified both his democratic idealism and his patronizing manner. He held that "that state or that city is on its downward career which does not constantly recruit its force of able men and women from the ranks of the poor as well as from the ranks of the well-to-do. . . . No school should ever say to its pupils: 'Thus far shalt thou go, and no farther.'" It was a source of great pride to him that any pupil who entered a New York public kindergarten could continue his education at no cost as far as his ability would carry him. Ironically, it required autocratic school administration to open up educational opportunities on a democratic basis.[9]

Underlying Maxwell's democratic faith was his conservative nature. He believed that it was the duty of the school to provide equal opportunity, but the responsibility of every individual, according to his ability, to achieve or fail. "The best corrective of the evils generated by the accumulation of wealth," he wrote, "is not anti-trust laws or other repressive legislation, but a system of schools which provides a training for all that is equal to the best which money can buy. . . . The trained man will demand and will receive, in due time, his due share." Though he had lived through economic depressions, Maxwell would not concede that training and education might not be sufficient to protect men from massive economic dislocations. "Education is a chief cause of wealth and the certain corrective of its abuse. In a community in which every man has been trained to his highest efficiency, monopoly and poverty would be alike impossible." This overweening belief in the leveling power of education typified the views of Horace Mann and William Seward, as well as most other major educational reformers. Education, it suggests, can overcome all obstacles and inequities, whether economic or social. One implication of this analysis is an acquiescence in the social and economic status quo, abjuring governmental intervention and relying on the power of education to eliminate the abuses and injustice among men, some day. Modern historians such as Rush Welter, in *Popular Education and Democratic Thought in America,* and Henry J. Perkinson, in *The Imperfect Panacea: American Faith in Education, 1865–1965,* have shown that Americans have tended to rely on schooling as the antidote to problems that were social, economic, and political in nature.[10]

Public school officials before Maxwell had a narrow vision of the school's role; they worried about refining the curriculum and enlarging the capacity of the schools, but thought little about the schools' relationship to society. Schooling was a given; society was a given. There was no impulse to change one or the other.

But William Henry Maxwell, unlike his predecessors, saw the public school as a potential agent for the uplifting of society. In his 1904 annual report to the Board of Education, Maxwell approvingly quoted Robert Hunter's *Poverty,* a muckraking exposé of social conditions; Hunter criticized the schools' ignorance of the city child's life and held that "the charge of neglect must rest upon the community and the school, and not, in most cases, the parent." Totally reversing the reproachful attitude of his forebears in the Public School Society, Maxwell agreed:

> There is only too much truth in Mr. Hunter's rebuke. We have not studied our industrial and social life sufficiently. We have been too much absorbed in purely educational questions. We are not adapting our work sufficiently to our environment. Our duty to the home, our duty to society, is to increase rather than to diminish the number of hours per day during which the school will take charge of the child.[11]

The social-work concept of schooling—that is, the school as an institution that "takes charge" of children in lieu of an incompetent family and a disintegrated community—was very well suited to a man as autocratic and paternalistic as Maxwell. Though his interest was more in reshaping children than in reshaping society, Maxwell nonetheless placed the public schools on the side of social concern and social betterment.

In his first report to the Brooklyn Board of Education, when tens of thousands of children in Brooklyn and New York were not in school because of insufficient accommodations, Maxwell wrote that the first obligation of the school was to provide a seat and a desk for every child. "Our next is to employ the best, and only the best teachers that can be found, and, after this has been done, to extend our system upwards as far as the people desire it to go. No constitution forbids the extension of educational facilities." An outspoken foe of political involvement in the hiring of teachers, Maxwell championed the merit system of examining, hiring, and promoting teachers and principals; to this end, he encouraged the development of teacher-training institutions in the city.[12]

Throughout his tenure, Maxwell inveighed against rote learning and old-fashioned "show" exercises, in which a student wrote a sentence or two over and over until it was perfect to show to visiting superintendents. He urged teachers to de-emphasize memory-work and to turn their efforts towards teaching students how to think.

Maxwell achieved his greatest impact after the Revised Charter took effect in 1902, concentrating enormous power in the Board of Superintendents, which he dominated. He raised the standards for teachers' licenses, which quickly caused teacher-training institutions to improve their offerings. He worked for higher teachers' salaries, as well as for the equalization of salaries paid to men and women professionals. In 1898, the city had only a few high schools, of erratic quality: within a few years, high schools were an integral part of the school system, available to every student. Maxwell believed that the specialized high school, emphasizing either academic or commerical or manual-training work, was more economical and more efficient than comprehensive high schools; this decision set the pattern of development for the city's high schools.

The task facing Maxwell was prodigious. An English educator visited the New York schools in 1900 and reported on the wide gap which existed between the self-conscious intellectual ferment of leading schoolmen and the still stultifying practices in the classroom. He described the dim outlines of a new and distinctly American education, based on faith in "right environment as the corrective of an evil heredity." The sociologists among schoolmen looked to education "as a moral antiseptic." Whereas the previous aim of the school had been to teach mental discipline and useful knowledge, the new American school aimed "to train the individual will to recognize and to respond sympathetically to the larger will of society." American education, observed the writer, saw itself intimately bound to the social problems of the nation. Idealistic school officials believed that the schools could be truly democratic, that they could be self-governing and self-disciplined. The schoolmen believed they could purposely encourage "the creation of a right taste, oc-

cupation of the hands and minds of children in useful ways which
stimulate to industry or in directions which appeal to their love of
beauty or of use; the development of the sense of wonder at and sympathy with nature . . . reverence for the beautiful, the good, the true. . . ."
"We hope to reach the day in our public schools," said one member of
the New York Board of Superintendents, "when they will be practically,
if not nominally, run by the pupils, i.e. the teacher will float on the interest which the pupils manifest, because of the superior order of instruction and the superior personality and character of the teacher."
How remote the schools were from this ideal was recorded by the same
British writer. He visited schools and was astonished by the military discipline which predominated. Never before, he wrote, had he "witnessed
such stillness and rigidity of posture as he saw during the opening exercises in a large school in this city." He visited a class where the teacher
regulated the class routine like clockwork, "time being kept between
question and answer by clapping the hands." [13]
   A few years later, a journalist from one of the leading progressive
magazines observed several New York public schools. She found some
schools where modern methods were followed, but others where teaching
was harsh and repressive. In one school, "I did not once hear any child
express a thought in his own words. Attention was perfect. No pupil
could escape from any grade without knowing the questions and answers of that grade." [14]
   Between aspiration and reality, between theory and practice, the
disparity was enormous.

WITH THE OPENING OF THE TWENTIETH CENTURY, European immigration
to the United States rose sharply. Whereas immigration had averaged
less than 500,000 people annually in the last fifteen years of the century,
the number of immigrants exceeded 800,000 in 1903 and hovered around
1,000,000 in each of the next ten years. However, it was not numbers
alone that aroused hostility to continued immigration, but the fact that
by the turn of the century the "new" immigration—from southern and
eastern Europe—began to be greater than the old immigration—the
Teutons and Anglo-Saxons from Northern Europe.° In the 1870s, the

   ° The new immigration came from Austria, Hungary, Bulgaria, Greece,
Czechoslovakia, Italy, Yugoslavia, Poland, Russia, Finland, Spain, Portugal,
Rumania, and Turkey; in New York City, most of the new immigrants were
Jews and Italians. The old immigration came from the United Kingdom, Ger-
many, France, Belgium, Denmark, the Netherlands, Switzerland, and Sweden.

new immigration represented less than 10 percent of all immigration to the United States, in the 1880s it was about 20 percent, while in the 1890s it was slightly more than 50 percent; but in the first twenty years of the new century, it accounted for about 75 percent of all transatlantic immigration, a vast human tide. Of some 13 million immigrants to the United States in the years from 1897 to 1914, fully 10 million of them were immigrants from southern and eastern Europe.

Prejudice against the new immigration was so intense and widespread that Congress finally adopted restrictive legislation in 1924 which virtually closed the nation's doors. Future immigration was tied to a quota system based on the national census of 1890, a sure way of keeping out those national groups which had arrived in great numbers between 1890 and 1915.

The old immigration was held to be desirable because it was

> of the same stock as that which originally settled the United States, wrote our Constitution and established our democratic institutions. The English, Dutch, Swedes, Germans, and even the Scotch-Irish, who constituted practically the entire immigration prior to 1890, were less than two thousand years ago one Germanic race in the forests surrounding the North Sea. Thus, being similar in blood and in political ideals, social training and economic background, this "old immigration" has merged with the native stock fairly easily and rapidly.[15]

But the new immigration was supposed to be so different from native Americans as to cast doubt on its assimilability. It was widely believed that their inferior genetic traits would dilute the strong, industrious, pioneering spirit of the older racial strains and that this infusion posed a threat to the future of the nation. Many popular and scholarly articles advanced the idea of a relationship between nationality and innate intelligence and argued for exclusionary immigration policies. Not only were the new immigrants charged with being ignorant, poor, dirty, immoral, and overly fecund, but their eagerness to work at any wage caused American workingmen to accuse them of stealing jobs, undercutting unionization, and depressing the American standard of living. Worse, the Jewish and Italian immigrants were suspected of harboring anarchist, socialist, and syndicalist philosophies, and it was doubted that these particular groups were capable of appreciating the genius of American institutions.

Many progressive reformers joined the movement for immigration restriction. They saw their efforts at amelioration eroded as the slums grew larger with each incoming boatload of immigrants. One scholar,

John R. Commons of Wisconsin University, wrote in 1908 that a line drawn across Europe which separated Scandinavia, the British Isles, Germany, and France on one side from Russia, Austria-Hungary, Italy, and Turkey on the other would divide not only distinct races, but distinct civilizations:

> It separates Protestant Europe from Catholic Europe; it separates countries of representative institutions and popular government from absolute monarchies; it separates lands where education is universal from lands where illiteracy predominates; it separates manufacturing countries, progressive agriculture, and skilled labor from primitive hand industries, backward agriculture and unskilled labor; it separates an educated, thrifty peasantry, from a peasantry scarcely a single generation removed from serfdom; it separates Teutonic races from Latin, Slav, Semitic and Mongolian races. When the sources of American immigration are shifted from the Western countries so nearly allied to our own, to Eastern countries so remote in the main attributes of Western civilization, the change is one that should challenge the attention of every citizen. Such a change has occurred.[16]

The poverty and problems of the new immigrants were not substantially different from those of earlier immigrants; their numbers, however, both annually and cumulatively, were so much greater than any previous mass immigration that restriction became a crusade among many different groups. Oscar Handlin has shown how racist, pseudoscientific studies were used to buttress popular prejudices and to justify closing America's doors on a selective basis. The political coalition which wanted to halt the influx from southern and eastern Europe was composed of Yankee nativists, Southerners, former Populists and Progressives, anti-Catholics, anti-Semites and the organized labor movement; as Handlin notes, each had its own reasons, but together they achieved their common goal.[17]

Before immigration was decisively restricted, the population of New York City was dramatically altered. In 1890, the Irish and the Germans were more than half the city's 1.5 million inhabitants. By 1920, the consolidated city of Greater New York contained 5.6 million people, and its largest groups were the Jews (about 1.5 million) and the Italians (about 800,000).

Like previous immigrant groups, the Jews and Italians settled mostly in distinct neighborhoods, where their language was spoken, and where they had their own grocery stores, newspapers, places of worship, and social clubs. The Italians moved into what had been Irish wards in the downtown section; the area bordered by Canal Street, Houston Street, the Bowery, and Broadway came to be known as "Little Italy." Another Italian quarter grew up around East 110th Street. The Jews settled in a previously German neighborhood on the Lower East Side; the section bordered by Division Street, Rivington Street, Norfolk Street, and the Bowery was called "New Israel."

Many who worried about the assimilability of the new immigrants turned to the public school as the appropriate institution to cope with the newcomers' social, economic, and physical problems. In the nineteenth century, the public school was only one of many acculturation devices; it was expected to provide a seat for the child, lecture him on good habits, and drill into him enough schooling to get him into the world of work on a competitive, if minimal, basis. But in the early twentieth century the public school was transformed into a vast, underfinanced, bureaucratic social-work agency, expected to take on singlehandedly the responsibilities which had formerly been discharged by family, community, and employer. Of course, the immigrants continued to be acculturated through work, politics, religion, the press, and their own communities, but the idea took hold that the public school was uniquely responsible for the Americanization and assimilation of the largest foreign immigration in the nation's history.

The public schools reflected the residential concentration of ethnic groups. A 1905 study found that a school district below Houston Street with 64,000 pupils was 94.5 percent Jewish; ten of the district's thirty-eight schools were 99 percent Jewish. The homogeneity of the schools did not disturb the parents of the children. In 1904, the Board of Education proposed shifting 1,500 children from overcrowded Lower East Side schools to less crowded schools on the West Side in Hell's Kitchen. A protest meeting was called at a settlement house, and 2,000 irate Jewish parents turned out. The chairman of the evening said, "Of all the serious problems that have confronted the people of the east side in the last twenty years, there is not one that has met with such general opposition as this measure. We are patient almost to a fault. We have stood for a good many municipal evils, for dirty streets, for grafting politicians, and overbearing policemen. But we will not stand for this." Speakers complained about the dangers of a long cross-town trip and expressed fears that their children would miss morning prayers and afternoon religious school and would encounter hostility in a non-Jewish neighborhood.[18]

Ethnic characterizations of the school children developed quickly into stereotypes. The poor academic performance of large proportions of the immigrant children was seen not as an indictment of the school, but as confirmation of popular attitudes about the new immigration. In an 1896 series, the *New York Tribune* said that the public school teachers

> have to deal largely with pupils who are offensively dirty and densely ignorant, both by inheritance and by force of circumstance. In fact, it has been suggested, in joke, that the Board of Education establish a washing school as an annex to each of the primary schools, and that the first exercise in each school be to give the pupils a bath. From a sanitary and moral point of view this is not a bad idea, and the suggestion that was started in a jocular way has so far attained serious consideration in the Board of Education that some of the new schoolhouses projected for the East Side have been designed with a number of shower baths.[19]

The pupils in a predominantly German public school were commended for their neatness and cleanliness: "Their parents take a pride in sending the children to school in their best clothes and in a thoroughly washed-up condition. In some instances, the girls have been known to attend school in white kid gloves, but this is the exception, not the rule."

The *Tribune* descriptions of two predominantly Jewish public schools typified conflicting stereotypes of the Jewish pupil. To the despair of their teachers, the pupils at Primary School No. 20 on Broome Street were especially clannish and resistant to Americanization:

> Not that all immigrants are ignorant, or that they all find it difficult to grasp the American idea of self-government, but . . . a certain class of foreigners have settled in the neighborhood of this school and, by flocking together, have become stubborn in their ignorance. Many immigrants take a pride in acquiring American ways, and endeavor to mingle among the native citizens, but those living in this part of Broome Street have formed a little colony by themselves. . . . The population here is made up of Hebrews of the poorest class. The children are sent to the public school with an honest desire to learn English and something about American ways, but the process of teaching them is an unusually tedious one. . . . Most of the children attending Primary School No. 20 . . . come from families in which English is seldom spoken at all, and where good manners and cleanliness are decidedly at a discount.

But then there were the ambitious pupils of Grammar School No. 75, which was called "the Hebrew public school of New York." At No. 75, the principal and teachers spoke "in the highest terms of the ability and good behavior of the pupils." Many of the children worked after school as newsboys; the worst punishment was to keep a pupil after school, which would make him late for work or religious school. The

Tribune reporter observed that "it seems that a college education is more prized by these pupils than a commercial education," since the college course drew more pupils than the commercial course. Despite their poverty, many Jewish families were willing to make sacrifices, to forego immediate rewards, in order to keep a promising student on the path to a profession.

It was said of the Italian children that they showed little aptitude for traditional academic work, but excelled at drawing and manual activities. Italian children left school to work and help support their families in greater proportions than did Jewish children.

Comparing ethnic groups by intelligence tests or by school-retardation rates became fashionable in the first quarter of the twentieth century. A 1908 study of fifteen New York elementary schools ranked ethnic groups on the basis of the proportion of students retarded in at least one grade; German pupils had the least retardation (16 percent); then came American-born children (19 percent); Russians (23 percent),* English (24 percent), Irish (29 percent), and Italians (36 percent). Another study in 1911 ranked ethnic groups according to the proportion of pupils who graduated from high school; the rank order was Italians (0 percent), Irish (0.1 percent), native whites (10 percent), British (10.8 percent), Germans (15 percent), Russians (16 percent). The groups who had the least school success, then, were the Italians, who were both recent arrivals and non-English-speaking, and the Irish, who spoke English and were part of an earlier immigration.[20]

School officials in New York blamed educational retardation on a variety of factors: language problems, physical and mental handicaps, late enrollment in school, part-time schooling, large classes, non-English-speaking homes, the strain of after-school employment, the effect of poverty on family and community life. All these conditions were influential, to be sure, but the consistent differentials in the comparative studies, crude though they were, indicated that cultural factors were key to successful adaptation to American schools.

Whether a group or individuals came from a city or from a rural area was one such important cultural determinant of successful assimilation. Those immigrants who came from cities and towns appeared to have significantly greater school success than those who had been peasants in their former homelands. Another key element was the immigrant group's attitude toward schooling, which influenced motivation in school. The overwhelming majority of Italians who came to New York

---

* Russian-born immigrants were largely Jews.

around the turn of the century were Southern Italians, mostly peasants, mostly unskilled laborers, and mostly illiterate. In Italy, wrote Glazer and Moynihan, the Southern Italians were "considered inferior, hardly civilized." The schools they had known in the old country were terrible and compulsory; few peasant children managed to complete primary school, and fewer still attained a secondary or higher education. "Education was for a cultural style of life and professions the peasant could never aspire to. . . . 'Do not make your child better than you are,' runs a South Italian proverb." Schools, then, were not thought of as a route to opportunity or economic success. There was instead suspicion by closely knit families that the schools would pull their children away from the traditional way of life and displace traditional values; the schools' assimilationism was seen as a potential threat to the continuity of family and community. Interestingly, however, the Italians preferred public schools to parochial schools; a government report in 1908 showed that some 60,000 Italian children were in public schools, as against 8,000 in parochial schools (slightly more than half the city's Irish children were in public schools at the same time).[21]

Like the Italians, the East European Jews came to New York in search of economic security. But the Jews also came to flee oppressive political regimes which condoned anti-Semitic pogroms, denied Jews basic rights, and severely limited their educational opportunities. Italians knew that they could return to Italy, and sometimes did. Jews knew that theirs was a one-way migration, that there was no returning to the increasingly brutal existence in the ghettoes and *shtetls* of Eastern Europe.

The Jews had their tradition of education, a tradition which survived in synagogues when Jews were excluded from secular schools in the old country. In the new country, and especially in New York City, the educational system was not only free, it was completely open to Jewish children. The opportunity to get an education overshadowed any other factors, like the crowded, deteriorated condition of many slum schools. Reformers complained about the deplorable state of the Lower East Side schools, the dark and dank classrooms, the noxious fumes from open gas jets, the din from adjacent factories and passing elevated trains. One reporter inspected an Allen Street school which was 90 percent Jewish and recorded that "the impression one gets on leaving this building after having visited all the classrooms is that he has been in a tomb." Yet this "wretched ramshackle structure" was so much in demand that children were turned away for lack of accommodations. Far from protesting school conditions, Jewish parents rose up in anger only when their children were not admitted because of overcrowding; in 1897, the parents of 500 Jewish children nearly rioted when their children were turned away from Grammar School No. 75, which already had 2,000 pupils in a building meant for 1,500.[22]

The educational aspirations of the Jews were noted in a government report in 1901:

> The poorest among them will make all possible sacrifices to keep his children in school; and one of the most striking social phenomena in New York City today is the way in which the Jews have taken possession of the public schools, in the highest as well as the lowest grades. . . . The city college is practically filled with Jewish pupils, a considerable proportion of them children of Russian or Polish immigrants on the East Side. . . . In the lower schools, Jewish children are the delight of their teachers for their cleverness . . . obedience and general good conduct, and the vacation schools, night schools, social settlements, libraries, bathing places, parks and playgrounds of the East Side are fairly besieged with Jewish children eager to take advantage of the opportunities they offer.[23]

The Jews' dedication to schooling produced economic results in a relatively brief period. While the Jews of the new immigration made their living primarily in the needle trades or small shops, their children and grandchildren were represented in disproportionately large numbers in professional occupations. More than half the city's doctors, lawyers, dentists, and public school teachers were Jewish by the 1930s. For the impoverished Jews who crowded into the city's slums in the early twentieth century, the public schools were indeed a ladder "from the gutter to the university."

CHAPTER 16

Problems of Centralization

ONCE THE NEW YORK CITY SYSTEM was centralized and professionalized, the reformers thought that their fight for the schools was won. Firm in their conviction that progress was inevitable, they congratulated themselves and turned their attention to recapturing the mayoralty from Tammany Hall. In 1901, reformers allied with the Republican party behind Seth Low and succeeded in beating the Tammany candidate. Unfortunately, the Revised Charter cut the term of the mayor elected in 1901 to only two years, which gave Low little time to accomplish change.

Seth Low typified the "best man" tradition of New York reformers. His credentials were impeccable: fine family, good education, successful business, unimpeachable integrity. Low was born in Brooklyn in 1850; his father was a merchant and clipper-ship owner who was descended from an old Massachusetts family. Seth Low was educated in private school and by travels abroad; he finished his secondary education at the Brooklyn Polytechnic Institute and was a member of Columbia College's class of 1870. He joined his father's prosperous firm, married the daughter of a U.S. Supreme Court Justice, and became active in Young Republican politics in Brooklyn. He was devoted to municipal reform, especially to the separation of local from national politics. In 1881 and again in 1883, he was elected mayor of Brooklyn. He introduced the merit system into Brooklyn's municipal service. In 1889, not yet forty years old, Low became president of Columbia College, where he energetically centralized the graduate organization, bringing Teachers College, Barnard College, and the College of Physicians and Surgeons into association with the university.

Low was an honest and efficient mayor, but he was defeated in the 1903 election because of his cold, upper-class demeanor. As mayor of New York, Low supported fully the needs of the public schools. But his

181

term in office was short, and the financial demands of the schools were enormous.

With an unsympathetic Tammany administration in office, Maxwell had to fight a battle on two fronts. On the one side were the intractable problems of the schools; on the other were the politicians, blaming him for his inability to solve them. He found himself in a familiar dilemma: enrollments increased, but the Board of Estimate cut the education budget; meanwhile the mayor attacked him and the Board of Education for being extravagant.

Under Low's successor, Mayor George B. McClellan, Jr., the Board of Estimate and Apportionment created a committee to investigate the extravagance of the public school system. The committee recommended the elimination of such "inessential" subjects as sewing, music, drawing, and physical and manual training, as well as the school library system. Maxwell countered by accusing the Board of Estimate of crippling the efficiency of the schools. After years of on-and-off warfare, the Board of Estimate ordered a professional inquiry in 1910. City officials complained that the schools expected to receive huge appropriations without submitting proper budget documents. They could not understand why, after so many millions had been spent, there were still so many thousands of children either overage in their grade or receiving only part-time instruction. The resulting study, called the Hanus Report, was prepared by a panel of university experts over a three-year period, and it was on the very ground where Maxwell sought his greatest accomplishments that the Hanus Report most severely criticized the schools.

Maxwell considered that one of his greatest achievements was the standardizing of the curriculum throughout the city, so that all children attended school for the same number of years and each grade studied the same subjects. He believed this to be necessary in a mobile population where children changed schools frequently. The end result, he thought, was to bring weak schools up to the standard of the strong. Practical education, like manual training, commercial and industrial education, had been introduced into the schools under Maxwell's leadership, and the schools had gone far towards opening their facilities for community use. He recognized that his goals had not been fully achieved, but he knew how far the schools had come in a relatively short period.

The Hanus Report criticized the uniformity of the New York curriculum, calling it too inflexible to serve the needs of a diverse population: "We view this uniformity of prescription as vicious in principle and injurious in practice. It is undemocratic, unsocial, unpedagogical." The committee chided the schools for not giving sufficient emphasis to commercial, industrial, and vocational subjects. It held the curriculum partly responsible for the fact that fewer than 42 percent of all pupils finished the eight-year course.

The report recommended that the curriculum of each school be planned every five years by its principal and teachers, based on the "needs and desires of the community in which their schools are located, and . . . the dominant interests . . . of the pupils that enter their schools." The course of study should be built around human problems, emphasizing important aspects of commercial and industrial life, and incorporating the customs, activities, and pursuits familiar to their pupils. Each school should be a true neighborhood center, aware of local needs and identified with local interests.

The principal, the report recommended, *should* be the real head of his school, but the existing supervisory system destroyed his independence. The principal was found to have no authority over the content of studies or methods of teaching in his school. Nor did the teachers have any part in designing curriculum or choosing texts; they were directed what to do and how to do it. "The teachers, as a rule, are conscientious and energetic. . . . In respect to their profession, they are static and depressed. . . . No one in the system is discussing aims and principles with them and showing how these should affect their teaching."

The report called for the abolition of the powerful Board of Superintendents and the reduction of the ineffectual forty-six-member Board of Education to eight members. It recommended that the central board devolve some of its authority on local boards, "to work out courses of teaching and activities suited to the nationality, industrial conditions and character" of the neighborhood.

> The form of these local boards should not be provided by state laws. It should rather be left to the Board of Education. For such a scheme can only be worked out by experiment. Different methods might be tested out in different boroughs or in different sections. The powers of teachers, principals, district superintendents, as well as local authorities should be determined by by-laws of the Board of Education.

Lest their last recommendation be construed as a call for a return to community control of the schools, the authors emphasized that "nothing could be worse for the school system than to have the authority for school administration dispersed throughout the forty-six districts." Local board members should be "watchful lay guardians of the educational in-

terests of the people" and should not interfere with administration, supervision, or classroom work.[1]

When the Hanus Report was made public in 1913, city officials
were interested in it chiefly to find justification for placing budgetary
controls over the Board of Education. Its findings were largely ignored,
except for its proposal to decrease the size of the Board of Education.
Laws aimed at reducing the size of the board were defeated in the legislature in 1915 and 1916; finally in 1917, a law passed which changed the
board from forty-six to seven members. Not until 1944, when still other
studies had criticized the administration of the schools, were most of the
executive functions of the Board of Superintendents transferred to the
direct control of the city superintendent.

While the Hanus Report had little immediate impact on the problems, it did publicize the restlessness and discontent of the city's teachers under their stern-willed chief executive. Maxwell's admiring biographer, Samuel Abelow, wrote that "Dr. Maxwell, because of his
dominating personality, had developed a form of benevolent despotism
which was hampering the freedom of the individual teacher." Some of
the lesser supervisors aped Maxwell's style, without his saving benevolence. In 1913, the Board of Education elected a president who sided
with the teachers. In 1916, Maxwell took a leave of absence because of
ill health; not recovering, he retired in 1918. Two years later he died.[2]

William Henry Maxwell and Nicholas Murray Butler, more than
any other men of their time, were responsible for the metamorphosis of
the New York schools. Butler led the revolution, Maxwell consolidated
it. To be sure, the cult of efficiency, the worship of the expert, the admiration of centralized organization, were the shibboleths of the day. Science supposedly would provide the answer to everything; social institutions could be run as efficiently as businesses. Get the politicians out,
put the experts in. It was, though unconsciously, an elitist philosophy.
The old decentralized system, the reformers said, was not accountable,
it was too complex to fix responsibility at any point. In its place they
substituted a closed, bureaucratic system of experts, not because they
thought there would be more accountability, but because experts should
be able to get the job done. Experts never have to be accountable to the
public, as politicians do; Butler and Maxwell thought that this made
them stronger, more impartial, more efficient, better able to do the right
thing regardless of popular opinion.

When he retired in 1918, Maxwell wrote to Butler that throughout
his twenty-year tenure as city superintendent there had been many obstacles, "but through it all, I was steadily supported in everything I did
or advocated by the feeling that I was trying to do what you would ap-

prove. For this purpose I made a careful study of everything I could find that you had published. . . . So, if I have done fairly well, you are chiefly the cause." [3]

Butler assumed the presidency of Columbia University after Seth Low was elected mayor in 1901; he remained in that position until his retirement in 1945. He never ceased to be an energetic and controversial figure. He interested himself in the cause of international arbitration of disputes and served as president of the Carnegie Endowment for International Peace from 1925 until 1945; his efforts were recognized in 1931, when he shared a Nobel Peace Prize with Jane Addams. His political ambitions were never satisfied; in 1912, he became the Republican vice-presidential candidate on William Howard Taft's losing ticket, replacing the regular candidate, who died before the election. In 1920, he campaigned for the Republican presidential nomination unsuccessfully under the slogan, "Pick Nick for President and Pic-Nic in November." As an educator, he was repelled by the tendencies of the progressive education which he had helped to launch; he deplored "the present day notion that an infant must be permitted and encouraged to explore the universe for himself as if everything were at its beginning and there had been no human experience whatever." He became a thorough conservative in his politics, opposed to socialism, bolshevism, the income tax, high inheritance taxes, child-labor laws—anything that he construed as governmental interference with individuals. He denounced New Deal tax programs as a "steal-the-wealth" plan. Upton Sinclair called Butler "the intellectual leader of the plutocracy." As the busy decades rolled by, it was forgotten that Butler had ever participated in a reform movement or that he had crossed swords with the old guard as a young man and won. [4]

The great work created by Butler and Maxwell endured through wars, political vicissitudes, the Great Depression, and mayors of all stripes. The New York City school system, as framed in the Revised Charter of 1901, was highly centralized, highly professionalized, and highly bureaucratized. As Butler had hoped, there was minimal provision for public involvement in school policy. In its first flush, centralization brought with it the benefits of efficiency; gone were the days when local boards could veto the central board, when teachers had to curry favor with local trustees, when narrow local interests could stymie city-wide reforms. The experts made the decisions. The teachers were examined by an impartial board and appointed on the basis of test scores; even principals and other administrative employees of the Board of Education were part of the civil service system.

Each era recoils from the excesses of its predecessor. The reforms of the centralization movement would, a half-century later, become the evils which would stir a new reform movement. The very concepts for which Butler and Maxwell labored—professionalization, centralization, the merit system—would later become the targets of public school critics. In the waning days of the tightly centralized system, it became clear that its major weakness was its remoteness from popular control; popular control, in the form of elected school officials, does not guarantee efficiency, but it does guarantee political legitimacy. This was the system's single greatest flaw, but it would not become evident until midcentury.

THIRD
SCHOOL
WAR

The Crusade
for Efficiency

CHAPTER 17

Disillusioned Progressives

PUBLIC SCHOOL REFORM, so astutely engineered into law by Nicholas Murray Butler, antedated the general upsurge in progressive sentiment in the city and the nation. As the progressive passion for institutional reform swung into high gear in the early years of the century, the New York City schools were accorded virtual immunity from hostile scrutiny; under Maxwell's leadership, they were thought to be in the vanguard of social progress.

By the time the Hanus Report was issued, more than fifteen years had passed since Butler's great legislative and political coup. The new generation of progressives and reformers did not feel that the school system was their creation, and they began to wonder whether the public schools had been sufficiently reformed. Despite all the schools' loudly proclaimed pioneer social programs, the slums were as bad as ever. If anything, the problems of the city and its schools were mounting. Ironically, progressive rhetoric contributed to a revolution of expectations, making the schools more vulnerable to charges of failure than in the past; their severest critics, as the Hanus inquiry demonstrated, were other progressive educators. So long as the schools had functioned solely as that agency in the community which disseminated literacy to children, as in the nineteenth century, critics were limited in their complaints to such things as the sanitary conditions of the school buildings and the system's inability to accommodate everyone of school age. But once the schools accepted responsibility in the fight against the slums, then they became the primary culprit when the slums refused to disappear.

The slums, in fact, continued to grow in the second decade of the century, as immigration continued unabated. The schools, beleaguered by the pressure of numbers, burdened by their stock of antiquated buildings, hampered by periodic municipal economy drives, struggled just to keep a normal program operating for as many children as possible. Maxwell often spoke proudly of the special programs for children with unusual problems or physical defects, but his critics complained that these programs were too limited and too piecemeal in their application.

The accommodations problem, scourge of the previous school system, remained unsolved under Maxwell. Despite his best efforts, numbers were against him. Between 1898 and 1915, the pupil register nearly doubled. Though more than $100 million was spent during this period for new school sites and buildings, it was not possible to build schools fast enough to keep pace with enrollment. Rather than turn children away from school, as had been done before his tenure, Maxwell chose the alternative of part-time schooling in crowded areas. Maxwell's long-held credo, "a seat for every child," seemed even further from realization than it was when he first became city superintendent. By 1914, more than 100,000 children of a total enrollment of almost 800,000 were either on part-time or in double session classes.

Another major concern of the school system was the large number of children who were overage for their grade, almost 40 percent of the entire register in 1904. Almost a decade later, the Hanus survey complained that there were pupils in the second grade who should be in the eighth grade; almost 16,000 fourteen-year-olds were in the sixth grade or lower. Many of these pupils were non-English-speaking or had started school late. Maxwell was convinced that much of the overage or retardation problem was due to remediable physical defects and worked to establish regular physical examinations in the schools. But the limited number of special programs and the constant influx of new immigrant children into the schools kept solution of the problems out of reach.

Added to the overage problem was the schools' difficulty in holding onto their pupils. Among children below age fourteen, there was a high rate of truancy; and when school children reached fourteen, the working-paper age, nearly 50 percent left school. Even in high school, 30 percent or more of the students dropped out each year. Critics saw these statistics as an indictment of the schools' curriculum; they believed that if the schools offered courses of interest and relevance, pupils would stay in school longer. These critics demanded vocational and industrial training in the schools, both to keep pupils in school and to prepare them for employment.

Maxwell disagreed. He believed that for economic reasons pupils would continue to drop out as soon as they were old enough to work. And while he favored manual training, such as drawing and cooking, he opposed the introduction of vocational training into the elementary grades. He felt that the academic studies of the grades should not be diminished and feared that businessmen were trying to convert the schools into cheap labor-training programs.

As problems mounted, the reform coalition which always supported Maxwell began to dissolve. The schools came under fire from two different reform factions, each with its own set of priorities. One group was the social reformers. They were, for the most part, social workers, philanthropists, and progressive intellectuals, clustered in the settlement-house movement and in organizations like the Public Education Association (PEA). The other faction was political reformers.

The social reformers, appalled by the baleful influence of slum life, pressed the schools to lengthen the school day, week, and year. The longer the child was removed from the street and, although it was rarely admitted outright, the slum home, the more chance the school had to work right influences on him. These progressives looked to the school to do "preventive social work." In the school, the child was supposed to learn the attitudes, the values, and the skills that his home environment was too poor to provide. Teachers inspected children's heads for lice and chastised them about the importance of cleanliness and sound nutrition. With the major burden of cultural assimilation thrust upon the school, and, at Maxwell's insistence, with the school accepting that burden willingly, it became a matter of urgency to hold onto the school's captive audience for as long a period of their lives as possible.

While the social reformers urged costly new programs on Maxwell, the political reformers loudly criticized the heavy expenditures of the school system. The political reformers were the forces of good government: merchants, lawyers, bankers, and wealthy industrialists who had fought Tammany Hall for almost fifty years on behalf of honesty, economy, and efficiency in municipal government; they held that better services could be rendered by intelligent, honest, nonpartisan officials at less cost to the taxpayers. These reformers were appalled that the schools, with more than 20,000 employees, consumed more than 20 percent of the municipal budget without apparent success in their role. They were sure that so large and expensive an operation, while not necessarily corrupt, harbored waste and inefficiency.

Despite the progressives' antipathy toward giant trusts, they joined in the general public admiration for the technological success of big business. They saw in business organization and techniques the potential solution to the vexatious social problems of the city. As in the 1890s, reformers demanded that municipal affairs be conducted in a "businesslike" fashion by efficient, nonpolitical technocrats. It was in this spirit that the political reformers came to embrace the wisdom of the efficiency expert and to anticipate a new era based on the scientific management of human problems.

Maxwell had managed to hold his disparate coalition of supporters together through several different city administrations. Under reform Mayor Seth Low (1901–1903), the schools enjoyed unstinting financial and political support. During the two terms of Mayor George B. McClellan, Jr. (1903–1909), a sometime Tammany man of scholarly bent, Maxwell maintained a working relationship with the city administration. But after Mayor William Gaynor took office in 1909, Maxwell was locked in battle with the Board of Estimate. Ironically, the source of Maxwell's opposition was not Gaynor, elected as a Tammany candidate, but the reformers who controlled the Board of Estimate.

In the election of 1909, Gaynor was the only city-wide Tammany candidate to win. The other elected city officials ran on a reform Fusion ticket. The leading Fusion official was the president of the Board of Aldermen (a post analogous to president of the city council), John Purroy Mitchel. Mitchel was a bright and socially prominent young man, only thirty in 1909. Though he was both Irish and Catholic, he had little in common with the large numbers of poor Irish Catholics in the city. Mitchel's grandfather, John Mitchel, was an Irish national hero, a writer whose provocative attacks on England caused the English to banish him from Ireland. His father was fire marshal of the city of New York, who married Mary Purroy, a public school teacher whose father had been president of the Board of Aldermen. John Purroy Mitchel was born in a comfortable home in Fordham, attended a private preparatory school, and graduated from Columbia College and the New York Law School.

Mitchel made his mark as a crusading lawyer in the corporation counsel's office during the McClellan administration. His investigations as McClellan's commissioner of accounts led to the removal from office of the borough presidents of Manhattan and the Bronx. He was handsome and charming, earnest and forthright. The forces of Fusion, impressed with his record as a corruption-fighter, selected him as their candidate for president of the Board of Aldermen in the campaign of 1909. Otto Bannard, Fusion's candidate for mayor, was defeated while the rest of the Fusion ticket was elected, as a result of publisher William Randolph Hearst's entering the mayoral race and splitting the anti-Tammany vote at the top of the ticket.

As president of the Board of Aldermen, John Purroy Mitchel was a voting member of the Board of Estimate, the body whose approval was necessary for the expenditure of city funds. Mitchel, a great admirer of efficiency experts, headed the Board of Estimate committee which sponsored the Hanus survey. True to the good-government tradition, Mitchel traced many of the city's ills to mismanagement. Ex-Mayor McClellan wrote years later that when Mitchel investigated a city department with the aid of the efficiency experts and found no graft, he was sorely disappointed: "That was Mitchel's weakness, as it is the weakness of most so-called reformers. He was always looking for corruption, always suspecting officials of being grafters." McClellan thought that Mitchel's point of view "was entirely that of a prosecutor, for he lacked fairness and could not see both sides of a question." [1]

Like other reformers, Mitchel was committed to the principle of home rule. As applied to the schools, home rule meant that the schools should be entirely subject to the budgetary controls of the Board of Estimate. The struggle over the fiscal independence of the schools had been a recurring theme since the Tweed era, when politicians, under the banner of home rule, converted the Board of Education into a subordinate department in order to control its vast funds and jobs. Mitchel had no such devious intentions. He thought it administratively improper, for instance, that the legislature should regulate teachers' salaries. He believed that so long as the teachers could lobby for higher salaries in Albany, the city would be unable to control its own expenditures.

Because of his experience in the politics-ridden Brooklyn system, Maxwell instinctively resisted political interference in educational policy-making. In his annual report in 1912, he called "the emancipation of the Department of Education from municipal control" the most vital need of the schools. (The phrase "Department of Education" was an anachronism which persisted after the brief period from 1871 to 1873 when the system had been a subordinate Department of Public Instruction.) He charged that the subserviency of the school system to the Board of Estimate and the Board of Aldermen had

wrought nothing but evil during the past ten years. It has caused unreasonable delay in the purchase of necessary sites and in the erection of necessary school buildings; it has led to the purchase of unnecessary sites; it has crippled the schools by preventing the appointment of needed teachers and will cripple them still more next year. . . . What is needed is a sufficient income raised by taxation that shall be at the absolute disposal of the educational authorities, subject only to the most vigorous accounting.[2]

While reformers were divided in their ideas about the schools, they united unequivocally in their determination to defeat Tammany in the 1913 mayoralty race. A Fusion committee picked John Purroy Mitchel as its candidate. Mitchel was in the good graces of President Woodrow Wilson and had the important backing of William Randolph Hearst. Hearst's support was significant not only because of the power of his mass-circulation newspapers, but because he ran his own political party, the Independence League.

Tammany nominated Edward E. McCall, chairman of the Public Service Commission, whose name was little known to the public. Mayor Gaynor launched an independent campaign for re-election, but died in mid-campaign.

Gaynor's death united all the anti-Tammany forces behind John Purroy Mitchel and assured his election as mayor.

CHAPTER 18

The Solution to
New York's Problems

THE FUSION PLATFORM on which Mitchel was elected pledged an administration dedicated to economy, efficiency, and home rule. One of Mitchel's goals was to relieve the city of the burdensome problems of the public schools. He was concerned equally about the obsolescence of their curriculum, which led to massive educational failure, and about the obsolescence and overcrowding of the buildings, which led to a seemingly inexhaustible demand for new schools. Maxwell and Thomas Churchill, president of the Board of Education, girded themselves to protect the system's independence and its budget.

Mitchel pursued three courses of action regarding the schools. First, he sought legislative authority to control teachers' salaries; second, he supported legislation to reduce the size of the Board of Education; and third, he tried to reorganize the curriculum to conform to the latest progressive thought.

Mitchel's efforts in Albany to gain control of teachers' salaries were unsuccessful. He and the city controller, William Prendergast, made it clear that they would like to reduce teachers' salaries, which mobilized teacher opposition. When organized teacher groups lobbied against the city's bill, Mitchel criticized the teachers for engaging in political activity; he said caustically that the teachers "seem to think at times that they are the Board of Education." Throughout his tenure, Mitchel and the teachers were at odds.[1]

The city's forty-six member Board of Education was an anachronism, a political compromise inconsistent with the current, efficiency-oriented trend in educational administration. Other major cities, such as Cleveland (1892), St. Louis (1897), Milwaukee (1897), Baltimore (1898),

Philadelphia (1905), Boston (1905), Pittsburgh (1911), Detroit (1916), and Chicago (1917) had reduced the size of their education boards. Since small boards were an integral part of progressive ideology, Mitchel received the enthusiastic support of the Public Education Association, the Citizens Union, the City Club, and other reform groups. His legislation was defeated in 1915 and 1916, but enacted in 1917. The winner of the 1917 mayoralty election would have the power to appoint a new seven-member Board of Education.

The existing Board of Education fought the plan to decrease its number. Teacher organizations, knowing of Mitchel's desire to cut their salaries, did not want him to have the power to appoint an entirely new board. Board President Thomas Churchill detected a class issue in the question of the size of the new board. He felt that the large board was more democratic than a small one, that it permitted the expression of a wide range of opinion and the participation of people from many different communities. He feared that the work of a small board would be full-time, and no one could work full-time at an unpaid job but men of great wealth, "and they are the very worst type to put in control of the schools." [2]

Churchill was one of the chief targets of those who wanted a small, reform Board of Education. He was not one of them. He had graduated from public schools and City College; he taught in the public schools for fourteen years and then practiced law. He had been elected Board president in 1913, before Mitchel was mayor. As Board president, he made a practice of speaking to meetings of parents and teachers around the city. He was a Tammany man, and his name was often mentioned as a possible nominee on a city-wide ticket. He sided with the teachers in their desire for more status and more income and was popular with them. One teacher wrote the *Times* on the day before Churchill was re-elected board president in 1915: "He brought a human quality into a machine, a democracy into a bureaucracy dominated by an autocracy. . . . Mr. Churchill has a singular insight into the needs of the great mass of the common people. . . . He is not a bureaucrat, an aristocrat or an autocrat—he is a democrat in the best sense of the word." [3] His reform opponents, however, considered Churchill a politically motivated obstructionist and a blind defender of the ineffective existing system. Since only one-fifth of the Board came up for reappointment each year, Mitchel was not able to control a majority of the board for the first two years of his term and had to work with Churchill.

It was not only for reasons of economy that Mitchel wanted to control the Board; he was equally interested in the programmatic aims of his friends in the PEA. On his first day in office, he appointed one PEA insider to the Board and reappointed three others. The PEA did not share the mayor's enthusiasm for paring the school budget, but welcomed the opportunity to have direct access to the city's chief executive. Since the PEA's days as an adjunct to Good Government Club E in the 1890s, it had grown in power and ambition. It was no longer a women's organization. Its president, Charles P. Howland, was a Mayflower descendant, a prominent lawyer and philanthropist. After 1911, according to Sol Cohen's recent history of the PEA, "only the rich and the powerful, bankers and corporation lawyers, possessors of private fortunes, or those who had access to fortunes, were sought out." PEA launched many experimental programs, all with the same long-range goal: "The transformation of the public schools into child welfare agencies." [4]

Convinced that it was possible to make the schools both better and more economical at the same time, Mitchel invited Board President Churchill and several other city officials to join him in a trip to the Midwest to view innovative school programs. The Mitchel group visited Gary, Indiana, a city which reform-minded educators were saying had a remarkable school system. Mitchel may have heard about the Gary schools from his friends in the PEA; two months previously, two observers from the PEA visited Gary to study the schools and returned to New York filled with enthusiasm.

Mitchel's visit to Gary was a fateful one for the New York schools and for Mitchel. The Gary plan seemed to offer solutions to many, perhaps all, of New York's worst school problems: overcrowding, part-time classes, an outmoded curriculum, insufficient practical education, truancy, poorly-equipped schools. Mitchel's efforts to install the Gary system in the New York schools ultimately turned into one of the major controversies of his term and plunged the schools into a bitter political fracas.

Dr. William Wirt, the superintendent of the Gary schools and the author of the Gary plan, had been a student of John Dewey at the University of Chicago. He developed his unusual plan of instruction in his first superintendency, at Bluffton, Indiana. From there he had been hired to create a school system for the new city of Gary, a factory town of the United States Steel Corporation. Wirt, a man with strong convictions, a limited budget, and a free hand, began in 1908 to organize a distinctive school plan. By 1911, the Gary system attracted a steady stream of visitors—teachers, administrators, and social reformers—who spread its fame in educational circles. Wirt's plan first received national attention in 1912 at the Second National Conference on Vocational Guidance, held that year in New York City. As the Wirt plan gained national prominence, it was variously referred to as the Gary plan, the Gary system, the duplicate school plan, the platoon school, and the work-study-play school.

The Gary plan was many things to many people, but above all it was an effort to put progressive ideology into practice. The Gary school was a children's community, and it was a community school. It offered an enriched curriculum, superb facilities, and a fully utilized school plant—and all at a saving to the taxpayers. The plan was based on a simple premise: full use of the school facilities. Wirt had noticed that in the traditional school, the auditorium and playground were used for brief periods during the day and were left empty while classrooms were usually overcrowded. Under his scheme, students spent half the day in class, and the other half in the playground, workshops, auditorium, laboratories, or community facilities. To put the plan into effect, a school had to be equipped with extensive shops, laboratories, and recreational facilities. Once a school was properly outfitted, it could accommodate twice as many students as the traditional school, because the Gary school had two platoons. Group X spent the morning in the classrooms and the afternoon in the work/play program, and Group Y had a reverse schedule. In effect, there were two separate schools in one building, using all the schoolrooms and programs all day. Wirt discarded the old progressive tenet of "a seat for every child" as an inefficient use of classroom space. He relied on departmental teaching; instead of having one teacher instructing the same children all day, each class moved to different classrooms for various subjects.

The elaborate workshops in the school were key to Wirt's conception of the school as a children's community, in which each child made an active contribution to the school society. Each shop was conducted by a skilled workman, rather than a teacher; instead of following a dry course of study, the shop work was based on the needs of the school. The domestic science classes helped to run the school kichen; the carpentry shop built and repaired school furniture. Students in the print shop prepared the blanks, forms, reports, and other materials which the school required. Similar practical contributions to the school community were made by students in the plumbing shop, the electrical shop, the machine shop, the laundry, the paint shop, etc. Students not only learned by doing, like apprentices, but saw the practical value of their work. In recognition of the importance he attached to shop work, Wirt placed the shops alongside regular classrooms and laboratories and installed large windows to the corridor, so that students could see the interrelatedness of work and study.

Each department found ways to contribute to the school community. For example, botany students helped to care for the school grounds; commercial students managed a school banking system, with student savings accounts, and took charge of the school's regular accounting and secretarial work, under professional supervision.

Wirt believed in the unit school, a single building housing all grades from kindergarten through high school. Children worked alongside others of different ages. Wirt felt that the younger ones drew inspiration from the older, and the older ones gained confidence by helping the younger. Seventh and eighth grade students received regular assignments as classroom helpers in the elementary grades.

In the spring of 1915, Randolph Bourne published a series of articles in the *New Republic* which projected Wirt and the Gary plan into the forefront of the progressive education movement. Bourne saw the Gary schools as the ideal realization of John Dewey's philosophy of democratic education. He expanded his observations into a book entitled *The Gary Schools* in 1916, which won an even larger audience for Wirt's plan. Bourne hailed the Gary plan as a modern, humane answer to the problem of educating children in an industrial society. John Dewey had realized that since the city child could not receive his practical education in the home or community as his forefathers had, the school was called on to fill the vacuum. The hours spent by the city child on the streets were detrimental to his character, wrote Bourne, therefore the school must keep children fruitfully occupied for most of the day doing things that were meaningful to them as children. Bourne wrote that the Gary school was not only a preparation for life, it was "a life itself, as the old household was a life itself."

> The ideal school plant . . . carries out a belief in educating the whole child, physically, artistically, manually, scientifically, as well as intellectually. Mr. Wirt believes that by putting in the child's way all the opportunities for varied development, the child will be able to select those activities for which he is best suited, and thus develop his capacities to their highest power. This can be done only in a school which provides, besides the ordinary classrooms, also playgrounds and gardens, gymnasiums and swimming pools, special drawing and music studies, science laboratories, machine shops, and intimate and constant contact with supplementary community activities outside the school.[5]

Each of the schools in Gary was situated on ten to twenty acres of land, which afforded good space for playgrounds and parkland. Half of the schools' grounds consisted of a public park, open for public use all the time; the schools' recreational facilities, when not in use by the school, were available to the community. After school hours, the schools became a community resource, like any other public property.

The schools opened in the evenings for an adult education program which Randolph Bourne called "a people's university"; the bulletin of courses read like a university catalogue. The playing fields, gymnasia, and swimming pools were kept open in the evenings and on weekends as well. The school auditorium was available for political meetings, social gatherings, lecture groups, or any other use that the community might wish to make of it. The school plant, wrote Bourne, "thus becomes in the fullest sense a social or community center." Unlike other cities, where school boards act "as if they were trustees of private property," the Gary schools belonged to the people of Gary:

> They are public in the same broad sense that streets and parks are public. . . . In such a community and such a school education would never be finished. . . . The child would not "graduate," "complete" his or her education, but would tend to drift back constantly to the school to get the help he or she needed in profession or occupation, or to keep on enjoying the facilities which even the wealthy private home would not be able or willing to afford.[6]

Bourne reported that Wirt looked forward to the day when the public school would include a year or two of college in the same building. In this one plant, which offered schooling from the nursery through college, education would be a continuous process, thereby breaking down "those artificial barriers by which we measure off 'education,' and make it easy for people to 'finish' it. The Wirt school seems definitely to forecast the day when the public school will have swallowed the college, and the 'higher education' will have become as local and available as the three Rs." When Bourne discovered that one of the Gary Schools had an art gallery, it suggested to him that the school might one day also absorb museums and galleries.

Teachers tried to enliven academic courses by relating them to contemporary life. Bourne gave this example: While the city of Gary was considering whether to create a new waterfront park, some history classes were studying a syllabus on the city from ancient times to the present; the course brought together history, sociology, and geography and turned into a study of city planning through the ages. Other history classes took their cue from the World War to delve into the history of the Balkan nations and modern Europe. Newspapers and magazines were as important as textbooks; anything that might make the subject more lively was considered an appropriate resource. Another history class compared Athenian with Gary education: "This is another illustration of that constant effort to make the pupils realize the meaning of what they are doing and what is around them. The effort of the Gary education is to make the child acquainted with the purpose of his school. He is not taught as an inferior who must take without question wisdom from immensely superior teachers, but as an equal and democratic citizen of his school community, learning whatever and whenever he can."

Other classes found ingenious ways to relate their studies to the school's life and the life of the larger society. One Gary school had its own zoo, which became a zoology laboratory. "One class in zoology last year made an illustrated booklet descriptive of the school zoo," Bourne wrote. "The text was written by the pupils, the photographs prepared by them, and then the booklet was tastefully printed in the school printing-shop by the pupils themselves." The physics classes used the school's lighting, heating, and ventilating systems as their textbook. Chemistry students tested the city's water and the foods delivered to the school, learning of their composition and studying the physiological effects of different foods. "The children are practically deputy food-inspectors, and make their reports on the official blanks. It is said that the result of this sort of inspection is that in a prosecution for violation of pure-food laws in Gary a case has never been lost." The chemistry students also used their skills to test the materials sold to the schools, such as coal and cement, to see whether they came up to official specifications. The children were not only using their environment as a textbook, "but they are able to turn their knowledge immediately into work which is immediately beneficial, not only to themselves, but to the whole community."

In a typical eight-hour day, a student would spend two hours in the classrooms, two hours in shop, laboratory or studio, one hour in the auditorium, two hours in recreation, and one hour at lunch. Dr. Wirt insisted that work, study, and play were activities of equal importance and that no invidious distinction was to be made among them.

Discipline was not the problem in Gary that it was in schools which stressed silence and quasi-military order. Wirt believed in a relaxed, informal school atmosphere, which was encouraged by his emphasis on laboratories, shops, and "application" work in the classrooms. Activity and conversation were as natural to the Gary school as compulsion was to the traditional school. Wirt explained, "When I was a youngster, I was punished for whispering—talking to another boy, because I had something I wanted to say to him. What barbarism! Why, if children want to talk, let them talk. . . . Put them on their honor. Make them see with their own eyes and understand with their own brains what is best for them." [7]

The atmosphere of the Gary school struck Bourne as the highest form of character-training: "Self-activity, self or cooperative instruction, freedom of movement, camaraderie with teachers, interesting and varied work, study and play, a sense of what the school is doing, social introspection—all combine to give an admirable moral training and to produce those desirable intellectual and moral qualities that the world most needs today. Not obedience but self-reliance does such a school cultivate."

Despite its amazing resources and its long hours, the Gary school cost no more than a traditional school. Wirt insisted that the more people use a public facility, the cheaper it becomes for everyone. The plan's "unique contribution" to the economics of education, wrote Bourne, was "to treat the public school as a public service, and apply to it all those principles of scientific direction which have been perfected for the public use of railroads, telephones, parks and other 'public utilities.'" He quoted Wirt as saying, "You can afford any kind of school desired if ordinary economic public-service principles are applied to public-school management. The first principle in turning waste into profit in school management is to use every facility all the time for all the people." Wirt referred disparagingly to the insistence of New York school officials that every child is entitled to his own seat and desk, and that each teacher is entitled to the exclusive possession of a classroom. This ideal was as outrageous and uneconomic as proposing that every person was entitled to have his own private streetcar seat or his own individual seat in a public park. The New York ideal presumed that every person in school wants to do the same thing at the same time. But, said Wirt, "We are willing to have some one else use our public library, look at our pictures in our public museum, walk in our public park, sleep in our Pullman berth or in our hotel bedroom, or travel in our steamboat when we are otherwise engaged." Thus, the great economies of the Gary plan came from its doubling the capacity of a single school building, a feat which was accomplished only by providing shops, laboratories, playgrounds, gymnasia, and other rich resources. The lavish outfitting of the school was not an extravagance, but a necessary part of a money-saving reorganization.

The Gary plan seemed to incorporate everything that reformers and progressives had demanded for years. Even John Dewey himself extolled the virtues of the Gary system in his book *Schools of Tomorrow*

RARE OLD PRINT MADE IN 1840. THE LANCASTER MONITORIAL SCHOOL—An educational idea imported from England, in which a thousand pupils sat in one great room under the control of one teacher, assisted by monitors who passed up and down the aisles to maintain order. There was no Public School system, supported by taxation, in America's greatest metropolis, New York, until 1844. As early as 1805, a system of non-sectarian schools was established and amalgamated with the Public School system in 1853

PRINT OF AN INFANT SCHOOL IN NEW YORK IN 1825—The infants sat on a flight of wide stairs. on the plan of the ancient amphitheaters. The teacher stood on the floor below and taught by word of mouth. The pupils were marched to and from the school room in lockstep with much formality and rigid discipline. These schools were conducted by a charitable organization which was known as the Infant School Society

First public schools in New York, courtesy of the
New York Public Library Picture Collection

BOOK MANUAL.

THE pupil should stand erect,—his heels near together,—toes turned out,—and his eyes directed to the face of the person speaking to him.

Fig. 1. Represents the Book-Monitor with a pile of books across his left arm, with the backs from him, and with the top of the page to the right hand.

Fig. 2. The Book Monitor, with the right hand hands the book to the Pupil; who receives it in his right hand, with the back of the book to the left; and then passes it into the left hand, where it is held with the back upwards, and with the thumb extended at an angle of forty-five degrees with the edge of the book, (as in fig 2), until a further order is given.

Fig. 3. When the page is given out, the book is turned by the thumb on the side; and, while held with both hands, is turned with the back downwards, with the thumbs meeting across the leaves, at a point judged to be nearest the place to be found. On opening the book, the left hand slides down to the bottom, and thence to the middle, where the thumb and little finger are made to press on the two opposite pages. If the pupil should have thus lit upon the page sought for, he lets fall the right hand by the side, and his position is that of fig. 3.

Fig. 4. But, if he has opened short of the page required, the thumb of the right hand is to be placed near the upper corner of the page, as seen in fig. 4; while the forefinger lifts the leaves to bring into view the number of the page. If he finds that he has not raised enough, the forefinger and thumb hold those already raised, while the second finger lifts the leaves, and brings them within the grasp of the thumb and finger. When the page required is found, all the fingers are to be passed under the leaves, and the whole turned at once. Should the Pupil, on the contrary, have opened too far, and be obliged to turn back, he places the right thumb, in like manner, on the left hand page, and the leaves are lifted as before described.

Fig. 5. Should the book be old, or so large as to be wearisome to hold, the right hand may sustain the left, as seen in fig. 5.

Fig. 6, 7. While reading, as the eye rises to the top of the right hand page, the right hand is brought to the position seen in fig. 4; and, with the forefinger under the leaf, the hand is slid down to the lower corner, and retained there during the reading of this page, as seen in fig. 6. This also is the position in which the book is to be held when about to be closed; in doing which, the left hand, being carried up to the side, supports the book firmly and unmoved, while the right hand turns the part it supports over on the left thumb, as seen in fig. 7. The thumb will then be drawn out from between the leaves, and placed on the cover; when the right hand will fall by the side, as seen in fig. 2.

Fig. 8. But, if the reading has ended, the right hand retains the book, and the left hand falls by the side, as seen in fig. 8. The book will now be in a position to be handed to the Book-Monitor; who receives it in his right hand, and places it on his left arm, with the back towards his body, the books are now in the most suitable situation for being passed to the shelves or drawers, where, without being crowded, they should be placed with uniformity and care.

In conclusion, it may be proper to remark, that however trivial these minute directions may appear to some minds, it will be found on experience, that books thus treated, may be made to last double the time that they will do, under the usual management in schools. Nor is this attainment of a correct and graceful mode of handling a book, the only benefit received by the pupil. The use of this manual is calculated to beget a love of *order* and *propriety*; and disposes him more readily to adopt the habit generally, of doing things in a methodical and systematic manner.

Book manual, courtesy of The New York Historical Society, New York City

Bishop John Hughes,
courtesy of the
New York Public Library
Picture Collection

William Henry Seward,
courtesy of the
New York Public Library
Picture Collection

William M. Tweed

Nast cartoon depicting
Tweed Ring's ban
on Harper Brothers' textbooks,
courtesy of The New York
Historical Society,
New York City

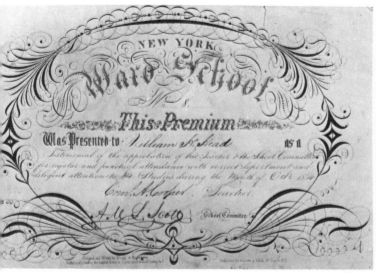

Late nineteenth-century
ward schools issued
merit tickets to students,
courtesy of the Museum
of the City of New York

Protestants continue to
fear Catholic subversion
of the public schools
throughout the
nineteenth century,
courtesy of the New York
Public Library
Picture Collection

Fire drill in a typical public school in the 1870s, courtesy of the New York Public Library Picture Collection

Jacob Riis photograph of a class in the condemned Essex Market School in the 1890s, courtesy of the Museum of the City of New York

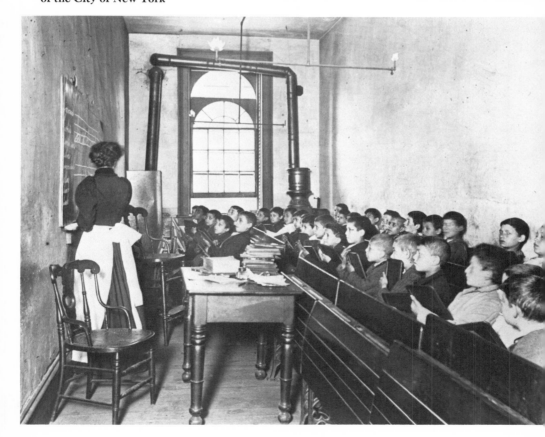

The young Nicholas Murray Butler,
courtesy of Teachers College

Dr. William H. Maxwell,
courtesy of Brown Brothers

ABOVE: Mayor John Purroy Mitchel,
courtesy of the New York
Public Library Picture Collection

UPPER RIGHT: Mayor John Hylan
and Mrs. Hylan
reached Palm Beach just
thirty-one hours late,
but they're happy as young,
carefree honeymooners,
courtesy of Wide World Photos

LOWER RIGHT: Dr. William A. Wirt,
originator of the Gary Plan,
courtesy of Wide World Photos

White parents protest bussing plans, 1964, courtesy of Wide World Photos

Milton Galamison leads a demonstration favoring integration, 1964,
courtesy of Wide World Photos

Albert Shanker, UFT president,
courtesy of United Press International

Rhody McCoy, unit administrator
of the Ocean Hill-Brownsville district,
courtesy of Leroy Henderson

(1915), which he wrote in collaboration with his daughter, Evelyn. He titled the chapter on Gary, "The Relation of the School to the Community." Dewey said that the Gary schools were cultivating a social spirit in children, that they were learning to see their part in the welfare of the community. It could teach democracy because the school itself was a democratic community.[8]

Even before Bourne and Dewey placed the progressive stamp of approval on Gary, New York reformers were captivated. The Gary plan was the perfect articulation of the school as the mechanism of social progress. It created a "right" environment. It accepted responsibility for children's moral and practical education. It was an education for the *whole* child, consciously taking the place of home and community. The Public Education Association made its importation to New York a crusade.

Mayor Mitchel needed little convincing to be attracted to the potential of the Gary plan. Like Wirt, Mitchel believed that the principles of scientific management should be applied to schools. The Gary scheme of organization was especially appealing, because it made possible the accommodation of nearly twice as many students in one school building as there were classroom seats. While the initiation of the plan would be expensive, alteration of old buildings was considerably cheaper than building new schools. And the buildings which were equipped for the Gary plan would have a swimming pool, an auditorium, shops, laboratories—elaborate facilities which no school building in the city, old or new, could presently afford.

Mitchel was intrigued by the prospect of ending the part-time problem, enlivening the curriculum, and at the same time saving huge sums of money in the school construction program. The social reformers in the PEA and related organizations were excited by the Gary plan because it represented the embodiment of progressive educational ideals.

Soon after Mayor Mitchel's visit to Gary in the spring of 1914, he offered Dr. Wirt a contract to establish his system in New York on an experimental basis. Wirt agreed to work as a consultant to the New York schools, at $10,000 a year for one week's work each month. New York school officials kept quiet, but resented the arrangement. To begin with, Mitchel completely by-passed Dr. Maxwell in his dealings with Wirt. And not only was Wirt an outsider from the provinces, but the mayor had agreed to pay him the same salary for part-time work that Superintendent Maxwell received for full-time service. Wirt's salary seemed particularly outrageous to New York schoolmen since Wirt received only $8,000 for a year's work in Gary.

Dr. Wirt arrived in New York in the fall of 1914. His first assignment was to reorganize PS 45, a badly overcrowded school in the Bronx. Ninety percent of its 3,000 children were on part-time. Though PS 45 lacked the extensive shops and other facilities that Wirt considered vital, he set up a duplicate school plan, lengthened the school day, and put all the children on a full-time schedule. The school day was divided into ten forty-minute periods. While half the students were in regular classrooms, the other half were either in the auditorium, on the playground, on excursions, in manual-training class, or working on gardening, pottery, carpentry, sewing, cooking, or printing. Gardening classes worked on beautification projects in Bronx parks.

There was another alternative available for students not engaged in regular class work: released time for religious instruction. This was part of Wirt's plan in Gary, and it worked well in that community. But New York, with its many religious groups, was another story. The pastor in the neighborhood of PS 45 promptly established religious classes, as all but a few hundred of the students were Catholic. Protestant and Jewish leaders in other parts of the city viewed the practice with alarm; the religious antagonisms that simmered below the city's surface threatened to boil up. Catholic officials felt compelled to deny publicly that the Gary plan was a Catholic plot.

By March 1915 the Board of Education felt sufficiently satisfied to appropriate $50,000 to Wirt to reorganize another crowded school, PS 89 in Brooklyn. Dr. Maxwell, who had been unceremoniously ignored by advocates of the new system, asked what the money was being appropriated for. When told that it was to be used for the same plan that was in PS 45 in the Bronx, Maxwell said, "Well, I visited that school the other day, and the only thing I saw which could be called the Wirt system was a lot of children digging in a lot." [9]

Maxwell's skepticism reflected the barely concealed hostility of much of the supervisory staff. Wirt encountered passive resistance instead of cooperation, in the form of "bureaucratic delays, deliberate foot-dragging, deliberate modifications" in his plans. The schoolmen's anger was directed as much at Mitchel as it was at Wirt and his plan. They deeply resented the hiring of an unknown "expert" from a small town, particularly since he had been imported by a politician who cut the schools' budget at every opportunity.[10]

School officials were further embittered when the Board of Estimate rejected a request from the Board of Education for $67,000 to pay for summer-school teachers, and another request for $276,000 for evening-school teachers. Controller Prendergast said that teachers ought to be willing to volunteer to staff the summer and evening schools and denounced teachers for their lack of dedication. He complained that they worked a short year, a short week, and a short day, without making any effort to produce more labor for the same pay. He said, "Their efforts so far as the public records show have been uniformly centred upon getting increases of salary for themselves without increase of service." [11]

Three months after the Wirt plan was funded at PS 89, the chairman of the Board of Education's Committee on Vocational Schools wrote a letter to the *Times* to say that "even the prospect of giving better education at cheaper rates would not justify us in rushing forward without getting the opinions of our experts." But Mitchel and Prendergast had no interest in waiting for "the opinions of our experts"—that is, the regular supervisory staff. A week after this letter appeared, the Board of Estimate voted $620,000 to install the Gary plan in twelve additional schools in the Bronx.

Wirt explained that the money would be used in the twelve schools to build pools, auditoriums, workshops, and new classrooms. He predicted that this expenditure would actually save money by eliminating the need for new school buildings in a crowded district. A news story noted that "in the Department of Education there are critics of the system, but they are not at the present time receiving much attention. They may be said to be in a state of 'watchful waiting.'" [12]

At the beginning of the fall 1915 term, Controller Prendergast issued a report which ignited an angry controversy. He urged the immediate extension of the Gary system into schools throughout the city. On the basis of the Gary plan, he recommended the following changes:

1. Lengthening the school day from five to six hours
2. Lengthening the school year from forty to forty-four weeks
3. Shortening the common school course from eight years to seven, because of the longer school year
4. Allowing no salary increases for teachers whose work-day and year were extended, because their lengthy summer vacations were already "a public scandal"
5. Not allowing teachers on annual salary to earn extra money in summer or evening school, a practice he considered "vicious" and tantamount to "graft"
6. Not permitting new school construction, except in new districts or to replace obsolete buildings (but in no case to relieve crowded schools)
7. Making a 10 percent reduction in the teaching staff, based on Wirt's proposal to use only sixty-four teachers for every seventy-two classes, since other classes were pooled in the auditorium and playground

Prendergast's proposals, as well as his gratuitous attacks on teachers' character and motivation, won the Mitchel administration no friends among the city's 20,000 teachers. If anything, the open animosity of Mitchel and Prendergast towards the teachers spurred the activities of the Teachers League, a nascent organization trying to draw teachers into the trade union movement.

The day after Prendergast issued his report, the executive director of the PEA, Howard Nudd, wrote the *Times* to commend Prendergast. Prendergast, wrote Nudd, recognized that the Gary plan means a richer education as well as economy. Nudd predicted that Prendergast's proposals "should open the way for the school and city officials to get together immediately to formulate a policy for carrying out this recommendation promptly. There should be no delay. The situation is acute." [13]

A week after the Prendergast report, the Board of Superintendents conferred with Wirt. Associate Superintendent William Ettinger, who would succeed Maxwell a few years later, warned Wirt that "the child and not the dollar is what we think of in New York. . . . The plan in a certain degree is a great step backward. It lowers the status and the salary of the teachers."

The superintendents asked Wirt if he believed the plan should be extended to the rest of the city before waiting for the results of the demonstration in the Bronx schools. Wirt told them: "No, it is my idea to extend the plan gradually." Wirt insisted that the plan was not simply a double-session program aimed at saving money: "My idea is to have the children wholesomely busy in class, shops, auditoriums, and playgrounds, and so keep them off the streets, but this cannot be done by the schools alone. The public must help: The churches, settlements, libraries, museums and public playgrounds. My point is not how cheaply, but how well, the schools can be run." [14]

Three days later, Mayor Mitchel wrote to Board President Churchill that, because of the demonstrated success of the Gary plan, he wanted it installed in *all* the city's schools. Churchill balked. He urged the mayor to give the plan a full trial before attempting to graft it onto the schools of the city's "many and divergent communities."

Prendergast, who seems to have developed a genius for antagonizing the administration's opponents, amplified the case for the Gary plan in a magazine article. While the city wanted the best education for its children, it also wanted "full value for every dollar." Since the public school is "the factory for character building, it should have the child within its control a longer period than it has at present." The part-time problem had to be solved. The superintendents who opposed the Gary plan complained about "importing talent from outside," but offered no answers themselves. "The trouble is," wrote Prendergast, "that too much deference is paid to the 'mossbacks' in the educational system. . . . They should be laughed at; then they would be brought to their proper level."

Prendergast predicted that if the Gary plan were adopted, about 20 percent of the present buildings could be abandoned, and the city would profit by the sale of the land. He also predicted savings to the public based on a cut in the number of teachers and supervisors. The greatest opposition to the plan, he held, came from the top staff, because the issue was a clear cut battle between an "entrenched bureaucracy" and "useful modern ideas." [15]

Dr. Wirt set to work reorganizing twelve Bronx schools which were said to be the most congested in the city. The twelve schools contained 35,000 pupils, but only 25,000 sittings; 2,500 of the existing sittings were "unsatisfactory" and had to be replaced. The city had two new schools under construction in the area, at a cost of about $1 million each, but another four buildings would be necessary to end the part-time problem in the area. Wirt proposed to solve the part-time problem by introducing the platoon plan into the twelve schools. His proposal would create space in the twelve schools for as many as 46,000 pupils, an allowance which anticipated future population growth in the area.

Under the "New Organization," unsatisfactory annexes would be vacated and unsatisfactory classrooms would be used for auditoriums, playrooms, laboratories, and workshops (some of the unsatisfactory classrooms were auditoriums which had been partitioned into classrooms). Wirt proposed building modern annexes at four of the schools and rearranging and re-equipping the eight others. The total cost for his program, including the cost of land for the new annexes, was under $750,000; for this sum the part-time classes would be eliminated, the school day lengthened, and the twelve schools equipped with laboratories, shops, and play space.

The children would have a daily program which consisted of 220 minutes in the classrooms for academic work; 40 minutes in the gymnasium or play-yard; 40 minutes for general exercises in the auditorium; 60 minutes for lunch; 80 minutes for drawing-rooms, science laboratories, manual training, or workshops. In the lower grades, the last 80 minutes would be given to play, excursions, library, church instruction (released time), or home. Wirt wrote that "as a rule the children will have 380 minutes in school in addition to the luncheon hour, in place of the 300 minutes provided in the regular full-time school. Such a study-work-and-play school removes the children very largely from the demoralizing life of the street, and gives ample time for academic, physical, and pre-vocational training." [16]

With the politicians pressuring for quick results and the supervisory

staff waiting for him to fall on his face, Wirt struggled to achieve immediate implementation of the platoon system. Despite his labors, the
"New Organization" was only the bare bones of the Gary plan, lacking
the soul and spirit that excited visitors to Gary. Wirt's reorganization
put all the students in those schools on a full-time basis, but nothing
was said of a children's community or a school integrally related to its
community. Teachers and principals unused to any change found themselves reshuffled and their schools turned upside-down in what must
have seemed like an unwelcome experiment.

A certain amount of confusion was inevitable as the reorganization
took effect, but parents had little patience with the upheaval. Parents
worried that their children were getting short-changed, that there was
too much play and time-wasting. A mother whose child had been in PS
45 in the Bronx, the first to be reorganized, complained in a letter to
a newspaper that her child changed classes every forty minutes and
had to wear her coat all day to avoid losing it; she claimed that the
child spent the day listening to a phonograph, taking photographs, sit-
ting in assembly, dancing, or in similar nonacademic activities. At the
term's end, said the mother, she transferred her child to another school
and found that she was a full term behind in her studies.[17]

A reporter, Thomas S. Baker, visited the schools in Gary and came
back with impressions that reflected popular anxieties about the tendencies of progressive education. While acknowledging that the townspeo-
ple of Gary were very proud of their schools, the writer was critical of
what appeared to be the "cult of the easy." The aim of the system "is to
allow the child to study those things which he likes to learn in the manner in which he likes to learn them." He reported that business letters
from manufacturers were held up as examples of good writing; this was
evidence, he thought, that the Gary schools were overly utilitarian. He
wondered whether this "pursuit of happiness" could "make hard-working
and accurate scholars and produce thoughtful men."[18]

A school principal commented that he believed the Gary plan a
step backward educationally because it focused on the group rather
than the individual. The plan had nothing to offer educationally, he
warned. It was "simply a departmental schedule so drawn as to accommodate two groups of classes in one building." It creates such a huge
educational machine that "the individual is lost." The principal becomes
a business manager, and "the school becomes an educational factory."
He predicted that if the plan were adopted throughout the city, problems of retardation and overage would multiply as a result of the minimal contact between teachers and pupils.[19]

Bourne, who had seen the Gary plan in operation in Gary, responded to the skeptical principal that "no public school has ever made so determined an attempt to get away from mass teaching and uniform methods." But Bourne had not seen New York's adaptation of the Gary plan; there was every indication that the system was being implemented hastily in the cheapest way possible, without adequate teacher preparation or necessary physical facilities.[20]

Why the haste to install the Gary plan? The Mitchel administration had decided that it was the answer to the problem of overcrowded schools and had stopped the school-construction program. During the terms of McClellan and Gaynor, the city spent $5 million to $7 million annually for new sites and schools; Mitchel reduced that figure during his term to little more than $1 million. Since the city school register continued to grow each month, it was imperative that the Gary plan begin to function as quickly as possible.

At the end of 1915, the Board of Estimate slashed the school budget again, making cuts in funds for summer schools, night schools, and trade education, as well as requiring kindergarten teachers to teach two classes each day. The city officials ignored complaints of fire hazards in decrepit school buildings; Prendergast explained that these dangerous old buildings would be closed as soon as the Gary plan was in general use.

CHAPTER 19

Tammany and the
Bureaucracy as Allies

As 1916 OPENED, the Gary controversy entered a new stage. Superintendent Maxwell released an attack on the quality of the academic work in the city's Gary schools, which had been prepared by Burdette R. Buckingham, the Board of Education's chief statistician. But Mitchel had his own ammunition: he gained control of the Board of Education and ousted Board President Churchill. These two events intensified the ill feeling and determination on both sides.

The Buckingham report compared two Gary schools to six prevocational schools and eight traditional schools. The prevocational schools, called Ettinger schools for the associate superintendent who organized them on an experimental basis, operated as traditional schools until the seventh grade, when students could choose to take prevocational courses in addition to their regular school work. Academic tests were given to seventh and eighth grade pupils in the three sets of schools, first in March, then in June. Their "rates of progress" in the intervening period were compared, and it was found that the greatest improvement was made by students in the traditional schools, followed by the prevocational schools, then the Gary schools. The statistical methods used were crude: test scores for each group were added and averaged, with no attempt to balance other factors or to match the groups.

The questions on the tests elicited precise information of the kind which the traditional schools emphasized. "In what state is Pike's Peak?" "Name a volcano in Italy." "What circles bound the Torrid Zone?" "Give an example in a sentence of a predicate nominative and underline it." Buckingham acknowledged that memory-work "is a much discredited function in modern pedagogical doctrine," but asserted that

210

pupils who do well on information questions also do well on thought questions.

Buckingham predicted that the standards of the future would quantify educational goals; teachers might, for example, require students to write 125 letters per minute or to add so many digits in so many minutes. With obvious pleasure in his own contribution to the advance of scientific pedagogy, he concluded, "Thus the educational scientist becomes an ally of the educational philosopher."

Commenting on the Buckingham study, Maxwell pointed out that it was only fair to recognize that "we are comparing the two types of schools as to matters which one of them emphasizes and the other does not." But he couldn't resist gibing at those who scorned his leadership and jumped on the Gary bandwagon: "The proponents of the new plan —not its thinkers and philosophers, but its noisy partisans, gifted with enthusiasm, but little else—have betrayed it" by claiming that it was equal or superior to traditional education in *every* way.[1]

Reformers were not willing to take the Buckingham report as a serious indictment. The PEA's Howard Nudd, whose letters advocating the plan appeared with frequency in the press, wrote a scathing critique of the report in the *Journal of Education*. He pointed out that neither of the Gary schools was properly equipped and that the tests were administered during a massive reorganization in these schools. Besides, both schools had previously been severely overcrowded, so that their students were educationally handicapped from the beginning. He insisted that Gary pupils spent as much time in academic work as pupils in regular schools.

The editor of the *Journal of Education,* who wanted the plan to have a fair trial in New York, called the report a scientific outrage which tried to "average the unaveragable": "An automobile goes forty miles an hour, a horse goes nine miles an hour, a wheelbarrow goes two miles an hour. The average speed of automobiles, horses and wheelbarrows is seventeen miles an hour. This is a mighty uplift on wheelbarrows, but hard on automobiles. That sort of thing may go in the sporting world, but in education it is a tragedy." [2]

Maxwell usually relished a good controversy, but this time he was *hors de combat* because of poor health. In April 1916 he took a leave of absence and left the city, in hope of recuperating.

Mitchel was not interested in criticism of the Gary plan; he was totally committed to push ahead with it. Since he replaced one-fifth of the members of the Board of Education each year, he finally commanded a majority in his third year in office. Mitchel handpicked his choice for president, infuriating members who were unaccustomed to explicit mayoral intervention in the internal affairs of the Board.

The new president of the Board of Education was William G. Willcox, an exemplary citizen with a deep interest in education. Willcox was a wealthy and successful lawyer, son of a Congregational minister. Willcox, at one time president of the trustees of Tuskegee Institute in Alabama, had a brother who was a professor at Cornell and a sister who was a professor at Wellesley. He had been part of the Mitchel entourage to Gary in 1914.

Willcox embarked on a campaign to win support for the Gary plan. To the New York Board of Trade, he explained that the entire part-time problem could be solved by an expenditure of $12 million and the adoption of a duplicate school plan; to house the same children in new schools would cost $50 million. He tried to allay fears that a fad was being foisted on the schools. Even if the schools were to abandon the duplicate school plan, the improvements would still be of value.[3]

Willcox had to combat persistent charges that the Mitchel Board of Education was dominated by the Rockefeller interests. Insinuations to that effect began early in Mitchel's term. On his first day in office, Mitchel appointed Abraham Flexner to the Board of Education. Flexner was a brilliant critic who had worked for the Carnegie Foundation for several years, then joined the staff of the Rockefeller-funded General Education Board. When Mitchel appointed him to the Board of Education, he was assistant secretary of the General Education Board and a leader of the PEA.

Before Mitchel gained control of the Board of Education, Flexner had acted on the mayor's behalf in controversial situations. Flexner headed a three-man Board committee which urged a reduction in the size of the Board; this recommendation was welcomed by the PEA and the Board of Estimate, but not by the majority of the Board of Education. Flexner managed Willcox's campaign to win the presidency of the Board.

Opponents of Mitchel and Willcox claimed that the mayor wanted to vest control of the Board in an arrogant, upper-class clique. Flexner's activist role on the Board, coupled with his Rockefeller connections, gave rise to rumors that the Rockefeller interests were trying to take over the New York schools.

Mitchel disregarded these attacks and added fuel to the fire by appointing Raymond B. Fosdick, an employee of the Rockefeller Foundation, to the Board of Education in 1915. Fosdick had served as Mitchel's assistant when Mitchel was exposing Tammany malfeasance as commissioner of accounts. During the Gaynor administration, Fosdick had been the commissioner of accounts, responsible for investigating the records of other city agencies. To Mitchel, Fosdick was a man of unimpeachable integrity; to Mitchel's detractors, Fosdick was another agent of the Rockefeller combine.

After Fosdick's appointment, charges of a Rockefeller scheme came with increasing frequency from labor leaders, from Tammany men, and from the Hearst press, which turned against Mitchel only a few months after he took office. Several labor unions formed their own committee on vocational education and asked Mitchel to appoint a labor representative to the Board of Education. Their spokesman said, "It was no fault of ours, of course, that our list of names was lost in the shuffle or the wastebasket. But let me tell you now that if John D. Rockefeller gets another appointment on the Board of Education, this town is going to hear about it." The president of the State Federation of Labor wrote Mitchel in early 1916 that "Commissioners Flexner and Fosdick are trying to secure control of public education for the Rockefeller crowd"; he demanded their removal. About the same time, a Tammany state senator asserted in a resolution in Albany that "the Rockefeller Foundation and its auxiliary boards and associations contemplate an attempt to gain control of the Board of Education of the City of New York." [4]

Despite the high prestige of the Rockefeller Foundation in the philanthropic and academic worlds, its public repute was tied to the controversial image of John D. Rockefeller. Scarcely a day went by during this period when Rockefeller was not assailed for his attitude toward the working man generally and trade unions specifically. In January 1915 a member of the United Mine Workers' executive committee met with Rockefeller to discuss working conditions in the Rockefeller mines in Colorado; after the meeting, he publicly denounced Rockefeller for treating his workers like "human chattels." He charged that the millions spread around by the Rockefeller Foundation represented "showy generosity" derived from "money withheld from the wages of the American working class." On other occasions, labor spokesmen expressed the fear that the big foundations were in reality "industrial trusts" which would ultimately be used to fight the trade union movement.[5]

Tammany was only too glad to feed the fear and suspicion that were created by the appointment of Flexner and Fosdick to the Board of Education. At every opportunity, Tammany orators described Mitchel as a servant of the power elite, indifferent to the needs and cares of the common people. The powerful Hearst papers, the *New York American* and the *New York Journal*, assailed Mitchel through this perspective at every opportunity.

Though Mitchel publicly assured labor leaders that the schools would never "come under any special or private interest" while he was mayor, he finally felt compelled to underscore his words with action. Both Flexner and Fosdick resigned from the Board in early 1917, but their identification with the Rockefeller "interests" returned to haunt Mitchel during his campaign for re-election later that year.

The notion of Rockefeller domination seemed foolish and contrived to the reformers; they failed to understand that the charges were symbolic of a popular fear that the changes underway in the schools had a class bias. Misconceptions about the Gary plan were rife because its supporters did such a poor job of popularizing it. The reformers spent their efforts convincing others of their own class—in journals, in letters-to-the-editor columns of newspapers like the *New York Times*, at conferences and debates. In their campaign for the plan, they were hampered by their own narrow social identification with upper-middle-class and upper-class life in Manhattan. They argued for the Gary system in a rational and genteel fashion, while their opponents went into the neighborhoods and the schools and talked to the people who were most intimately affected, parents and teachers. The repetition of the canard tying Gary and Rockefeller together had a certain subrational validity: the Rockefeller name represented the quintessence of the power elite, which moved in ways that were undemocratic, arrogant, and self-serving. And, indeed, in that symbolic sense the Tammany orators were correct; the city's "best men," men and women who were educated and wealthy, had decided to remake the schools. The reformers' intentions, at least in their own eyes, were honorable, but their impatience and self-righteousness converted the Gary plan from an educational issue into a full-fledged school war, fought out in the political arena, as impassioned as those which took place in the 1890s and the 1840s.

Willcox did not comprehend that when he sold the business community on the economical features of the plan, he simultaneously aroused the suspicion of immigrants who worried that the value of their children's education would be cheapened. But it was not to immigrant or working-class parents that Willcox went in search of support and understanding. His language frequently borrowed the formulations of the businessman; the schools, he said, are "expected to pay large dividends in intelligent and efficient citizenship. Does the city get its money's worth?" He merged the aspirations of the social and political reformers:

> Every item of the curriculum should be fearlessly challenged to show its value in practical benefit to the child. . . . Too much time has been wasted under the specious plea of mental discipline. . . . Mental disci-

pline can be acquired in useful as well as useless tasks. . . . We are not only spending vast sums of taxpayers' money, but we are spending precious years of childrens' lives, and we must spare no effort to spend them wisely and fruitfully.[6]

The Gary plan, Willcox believed, made it possible to save money and save children at the same time.

Throughout 1916 and 1917, the reform movement gave the Gary plan its full rhetorical support, as the Board of Education began implementing it in various parts of the city. Whenever a critical voice was raised, the next day's *Times* was almost certain to contain either an editorial or a letter from Howard Nudd of the PEA in defense of the new system.

In mid-1916, the Women's Municipal League, an organization which was part of the city's interlocking directorate of reform leaders, issued a broadside arguing the need for the Gary plan. It likened the schools of Gary, Indiana, to "the schools in Utopia." School affairs in New York were at "a crisis never before reached in educational history." The richly equipped schools of Gary were compared with the aging school buildings of New York, many of them lacking a playground, a gymnasium, an auditorium, and other such facilities. The Gary plan had only two "fixed principles": first, that children must be busy "all day long at work, study and play under right conditions"; and second, that "all available facilities of the entire community, which make for child welfare" must be coordinated and in use all the time for the child's benefit.[7]

This pamphlet, while bowing in the direction of economy, was a restatement of the aims of the social progressives. (It is interesting that the economy-minded reformers were all men, while the female progressives and some men were advocates of costly social reforms, possibly because wealthy women, who could respectably show their concern for children's welfare, were not expected to be interested in such complicated matters as taxes and accounting procedures.) The two "fixed principles" represented what was most attractive in the plan to the social reformers. On the one hand, the eight-hour school day promised to keep children away from their deleterious environment and expose them only to "right conditions." This hope was, to be sure, a naive and behavioristic reliance on the power of the school to reconstruct children's lives in accordance with a socially desirable ideal. On the other hand, the reformers were captivated by the Gary plan's explicit proposal to weave together the many strands of community life into an harmonious whole.

Many reformers believed that alienation was at the root of the slum dwellers' misery, that the sense of community had been destroyed by the advance of industrialism. In this view, the school, in common cause with the settlements, the churches, the governmental and private agencies of the neighborhood, might become a regenerating force in the community.

Supporters of the Gary system formed an organization called the Gary School League. Certain of the inherent virtue of the plan, the reformers discounted the validity of opposition to it, which they blamed on political machinations of Tammany or a traditional fear of change among school personnel. They felt sure that support for the plan would quickly follow enlightenment and understanding.

But other rumblings began to be heard, from groups which refused to be enlightened. Labor groups worried that the plan was intended to undercut their apprentice programs or to produce scab labor. Teacher organizations were angered by the city's intention to lengthen their work day without increasing their pay; they believed that the city was plunging headlong into the Gary system "on the cheap," with only the most superficial attention to curriculum reform. The supervisors who worked alongside William Henry Maxwell were offended by the way the revered superintendent had been shunted aside by the new generation of reformers; it was even publicly rumored that Wirt was slated to succeed Maxwell on his retirement, a possibility which cannot have pleased Maxwell's deputies.

A. Emerson Palmer, secretary of the Board of Education and authorized historian of the New York schools, went so far as to liken the Gary plan to the Lancasterian system. He suggested that the Gary system's use of student helpers was reminiscent of the Lancasterian student monitors; and he saw a basic likeness in both systems' emphasis on the group rather than the individual. The mention of Lancaster schools called up images of a "public" school for children of the poor, stirring up fear that the Gary system was designed to put public school children in a vocational track and to make them incapable of competing with private school children.[8]

An even more ominous augury of the growing dissatisfaction with the spread of the Gary plan was the Board of Aldermen's veto of Wirt's $10,000 salary at the end of 1916. Mitchel overrode their veto. He had only one year left in his term as mayor, one year in which to make the Gary system the basic administrative pattern of the schools.

As the issue became increasingly polarized, it was clear that the November 1917 election would determine the immediate future of the school system. The Gary system was so closely identified with Mitchel that its fate depended on his; whoever was elected mayor would have the power to appoint the entire new Board of Education on January 1, 1918.

In June 1917, the PEA urged the Board of Estimate to appropriate enough funds to put the Gary plan into forty-eight more schools. Nudd criticized what he called a concerted effort to discredit the plan and to convert it into a campaign issue; he charged that carefully manipulated neighborhood meetings were being held throughout the city, where the Mitchel administration was blamed for the part-time problem because of its refusal to build new buildings. These protest meetings were called by newly organized Anti-Gary Leagues, which sprang up in different parts of the city. No well-known names were connected with the Anti-Gary Leagues; their membership was drawn from neighborhood associations and parent groups.

Despite strenuous opposition, the Board of Estimate voted funds to extend the Gary plan into all the public schools of the city. At the public hearing where this decision was made, Willcox led the fight for the Gary plan; Churchill led the opposition. Churchill said, "This ought not to be called the Gary system, but the vagary system." Representatives of parent and community groups bitterly attacked the plan. Said one: "There is no parent who is not absolutely opposed to the Gary system. Most of the people of the city believe that it does not fit in with their ideals of education. It is enslaving the youth of the city, fitting them to be the tools and slaves of the aristocracy." The vote on the Board of Estimate did not break along partisan lines; two Tammany borough presidents supported the Gary plan.[9]

But Tammany nonetheless detected a popular issue and incorporated opposition to the Gary plan in its platform, which was released in early August. The regular party organization, calling itself the "Democratic Fusion Committee," pledged that:

> We will drive politics and political influences out of the Board of Education. We deplore the attempt of corporations and foundations to control our school system. We will banish the imported Gary system, which aims to make our public schools an annex to the mill and factory. We declare it to be unAmerican and undemocratic.

The Gary plan, charged the platform, made distinctions between the children of the rich and the poor. The platform pledged to provide "sufficient seating accommodations for all children of school age." [10]

Three weeks after the Tammany platform declared all-out war on the Gary system, the Board of Education announced its budget proposals for 1918. Another $5 million was requested for playgrounds, gymnasiums, shops, laboratories, and other facilities for extending the Gary system.

CHAPTER 20

"Mr. Mayor, Hands Off
Our Public Schools!"

THE CAMPAIGN FOR THE MAYORALTY began in earnest in September 1917. Tammany leaders chose for their candidate an undistinguished Brooklyn judge, John F. Hylan. Hylan, a simple man of limited experience, was born on an upstate farm, son of a poor Irish Catholic immigrant family. As a young man he worked on the railroad as a track layer, a fireman, and an engineer. He attended law school at night and finally got his law degree at twenty-nine. He was appointed a magistrate in Brooklyn by Mayor McClellan. At the time of his nomination for mayor, Hylan was a judge on the Kings County Court, to which he had been elected in 1915 for a seven-year term. Tall and red-headed, he was called "Red Mike." Years later, city planner Robert Moses recalled Hylan as "self-educated, dull, plodding and transparent . . . a decent political hack, picked because he was the most pliable, respectable material that could win without effort." [1]

Hylan was designated by Tammany Hall after William Randolph Hearst, the enigmatic publisher, withdrew from the race. Tammany men were divided on the subject of Hearst: some admired his radical politics; some feared the enormous power he wielded through his newspapers; others hated him for his arrogance and unpredictability. Hearst was a brilliantly successful publisher, but a deeply frustrated and ambitious politician. After winning election to Congress in 1902 with Tammany's support, Hearst tried time and again for higher office, always with his eye on the presidency. He was a serious candidate for the Democratic presidential nomination in 1904 (at the age of forty-one), ran for mayor in 1905 as an anti-Tammany candidate, ran for governor in 1906 as the Tammany candidate, ran for mayor in 1909 as an anti-Tammany candidate. His standing as a credible candidate declined sharply after his state-wide loss in 1906, but he continued to be a serious factor in politics, both because of the mass circulation of his newspapers and because of the ever-present possibility that he might enter a separate ticket, drawing votes away from one faction or another.

Rumors that Hearst was under consideration as the Tammany candidate for mayor in 1917 were squelched by Hearst's enemies in the organization. Boss Charles Murphy bowed to Hearst's opposition within Tammany, but still desired Hearst's support and got it by picking Hylan, who was Hearst's choice.

Hearst's statement endorsing Hylan delineated the issues of the campaign ahead:

> I have no personal hostility to Mayor Mitchel. He is an amiable young man, but without character or principles. He has a silly ambition for social recognition and a weak willingness to place himself entirely in the hands of selfish and sinister interests and allow himself and his great public office to be used for the private advantage of these selfish interests and against the public welfare. . . . He has made the public schools the football of politics and the sport of faddists. He has put our children on half rations of education. . . . The half-time system in the schools, which all denounced as the most damning vice of Tammany, Mitchel has sought to make a virtue by putting it into general operation and giving it the high-sounding title "The Gary System." [2]

Mitchel's supporters saw Hylan as a bumbling, insecure puppet of Hearst and Tammany Hall. The Mitchel administration had been free of scandal and received the endorsement of every newspaper except those owned by Hearst. Given Hylan's obscurity and his lack of experience, Mitchel had every reason to believe that he would be re-elected.

New York City held its first direct primary in 1917. (In the past, candidates had been chosen by party conventions.) Hylan had no primary opponent, but Mitchel had one. Mitchel was an independent Democrat who had been elected on a Fusion ticket in 1913 with Republican support. Once again, he hoped to win as a Fusion-Republican nominee. Republican William Bennett, a former state senator who had been a key figure in Mitchel's 1913 campaign, opposed him in the Republican primary. Bennett's challenge did not alarm Mitchel, whose friends in the elite Republican Union League Club assured him of their continuing devotion to the principle of nonpartisan conduct of municipal government.

But Bennett took his candidacy seriously; before the mid-September primary, he vigorously campaigned in every borough. Bennett opposed the Gary plan and promised to appoint to the Board of Education men who did not send their children to private schools. He attacked Mitchel for not keeping open evening schools where foreigners learned to speak English, while at the same time spending money on a sunken garden in Central Park. He charged Mitchel with overcentralizing city government, placing too much emphasis on Manhattan, and ignoring the needs of the other boroughs.

Another candidate, labor lawyer Morris Hillquit, entered the field against Mitchel in the general election on the Socialist party ticket. Hillquit was counsel to the huge International Ladies Garment Workers Union, which had a large immigrant membership. The Socialists wanted to register a strong protest against American entry into the World War, which became official in April 1917.

Mayor Mitchel was in more trouble politically than he suspected. He had stirred up opposition on many issues other than the Gary plan. One especially damaging controversy involved his economy-minded commissioner of public charities, John A. Kingsbury. Kingsbury, a Protestant who believed in minimal public assistance, investigated private charitable agencies receiving city funds; a number of them were Catholic. Since most of the great philanthropies and foundations were Protestant-dominated, Catholics felt a strong reliance on public funds to deal with their social problems. Kingsbury's criticism of their agencies made Catholic leaders think that the city adminstration was harassing them.

Mitchel irritated other blocs of voters by his fervent advocacy of military preparedness. In the summer of 1915, Mitchel trooped off for a month of voluntary training, in company with hundreds of other prominent men. Mitchel's strident patriotism alienated German voters, Irish voters (who did not wish to see England saved by the United States), and others, particularly Jews who were pacifists. He adopted measures to suppress the civil liberties of antiwar dissenters, forbade anti-Ally speeches, and denied police protection to Irish firebrands who denounced England in street-corner harangues. For the first time, public school teachers were required to sign a loyalty oath. When faced with antiwar hecklers in early 1917, Mitchel warned that "The time is approaching when the people of America will be divided into two parties —Americans and traitors. Those who are not for America are against her, whether they are in private life or on the floor of the legislature." [3]

No one of these issues reflected badly on Mitchel's integrity or ability. But each of them fed a steadily building image, delineated by his detractors, of a socially ambitious, insensitive politician who catered to special interests while ignoring the needs of the common people. Newspapers reported that he socialized with members of high society; he received large political contributions from Rockefellers, Vanderbilts, Morgans, and others of the city's elite. A Hearst newspaper ran a cartoon of Mitchel, dressed in evening clothes and dancing to the tune of "Everyone's a traitor but me." [4]

After ignoring Bennett's campaign against him in the Republican primary, Mitchel was in for a rude shock. Bennett narrowly won the Republican primary, and Fusion was dead, in all but name.

Now Mitchel realized he was in a fight for his political scalp. While he was still listed on the ballot as the Fusion candidate, he had to run against a Republican, a Tammany Democrat, and a Socialist. Attacked from all sides, Mitchel seemed to become more convinced than ever that his candidacy was a test of New York's patriotism. Immediately after his primary defeat, leading Republicans and Progressives such as Charles Evans Hughes, Theodore Roosevelt, Oscar Straus, and Henry Morgenthau gave Mitchel their public endorsement. Roosevelt called him "absolutely the best mayor New York ever had . . . the absolute incarnation of Americanism." [5]

The *New York Times* called him the "public spokesman and symbol of . . . Americanism" and called his opponents "pro-German, traitors and seditionaries." The Republican County Committee of Manhattan voted overwhelmingly not to endorse Bennett and called patriotism the central issue of the campaign. The Union League Club urged fellow Republicans to support the Fusion nominee instead of Bennett.[6]

While the Fusion party disintegrated, Tammany worked to establish its candidate. Each time Hylan delivered a speech, he stuck to the written text, word-for-word. Hylan's advisors ignored the pro-Mitchel press, which gibed at Hylan's refusal to speak spontaneously to reporters. Hylan stayed on the attack, refusing to debate with Mitchel, and ignoring the response to his attacks.

Hylan opened his campaign with a blast at Mitchel's educational policies. He accused Mitchel of turning "domination of the board over to private influences which seek to impose their own theories of education upon the people. . . . His appointees have been pledged to accept the theories of the Rockefeller Foundation." He warned that the new board of seven must not be "appointed at the dictation of and wholly subservient to the Rockefeller Foundation or other private influence." He assailed the Gary plan as a nefarious scheme of the Rockefeller Foundation and concluded that, "We must not have one class of education for the children of the poor and another kind of education for those children of the rich who can afford to buy it in private schools.[7]

Board President Willcox called Hylan's statements absurd and demanded that Hylan retract or explain; a month later, Hylan responded to Willcox that the management of Willcox's election by Abraham Flexner was proof of Rockefeller control of board policy. The *Times* tried to ridicule the issue by calling Hylan "the Murphy Foundation candidate for Mayor," but Hylan continued to repeat the charge linking the Gary plan and the Rockefeller Foundation.[8]

The Gary school issue fit perfectly into the overall theme of the Hylan campaign. His candidacy, he asserted, was "a battle of all the people of all parties against the most powerful combination of special privilege seekers the people of the city have ever had to contend with." As he accepted the Democratic nomination, his speech rang with phrases like "these agents of great wealth," "that invisible government of which we have read so much," "privilege seekers," "special interests," "predatory wealth." Hylan promised, "I shall give no quarter to exploiters and oppressors of the people." The beauty of Tammany's appeal was that it capitalized on the very issues and emotions which had been stirred in the public consciousness by the muckrakers of the progressive movement.[9]

Knowing how Mitchel was alienating large numbers of German, Jewish, and Irish voters with his strident super-Americanism, the Tammany candidate warned Mitchel that he "must not seek to hide behind the flag. I have not found that any one man has a monopoly of patriotism." [10]

Hylan delivered a stinging attack on the Gary plan at the Brooklyn Training School for Teachers, an institution dear to Superintendent Maxwell's heart. Hylan said that the opposition of Dr. Maxwell, "one of the greatest educators in the world," should have been enough to kill the Gary plan in New York. The candidate unleashed a rhetorical salvo that summarized all the popular fears and misconceptions about the new system:

> I say to you, Mr. Mayor, to the Rockefeller Foundation, and to any other private interests—hands off our public schools. I say, Mr. Mayor, our boys and girls must have the best possible education and under decent and honest conditions. . . . I say to you, Mr. Mayor, our boys and girls shall have an opportunity to become doctors, lawyers, clergymen, musicians, artists, orators, poets or men of letters, notwithstanding the views of the Rockefeller Board of Education.[11]

Hylan had the emotional side of the issue and exploited it fully. As if to underscore the Democrat's charges, the Bronx County grand jury opened an investigation of the Gary schools in the Bronx, to study charges that they were an obstacle to Americanization and a menace to children's health, because of excessive climbing of stairs between class periods.

Mitchel defended the Gary plan, insisting that its adoption would improve school facilities so markedly that public school children would have "the same advantages that had been enjoyed hitherto by the children of the well-to-do in private schools—what may well be termed a democratization of public education in the city of New York." But his defense of the new system was weak and vague. He claimed that certain superintendents diverted money intended for Gary schools to the establishment of prevocational schools for older children; he complained that there had been "deliberate attempts to wreck this system from inside of the school system." His supporters obtained a letter from Ida Tarbell, a muckraker famed for her articles on Standard Oil; she called the attempt to link the Gary plan to the Rockefeller interests "an absurd and disreputable trick." None of these defensive positions stopped Hylan, however, who knew he had a good campaign issue.[12]

While Hylan continued to denounce the Gary system and the Rockefeller Foundation, Mitchel counterattacked by impugning Hylan's loyalty. He called his campaign a battle against "Hearst, Hylan and the Hohenzollerns." Mitchel spoke as though he were single-handedly fighting the enemies of America; in mid-October, he charged that "every disloyal element in this city is arrayed solidly behind the Hearst candidate." [13] He tried to connect Boss Murphy's name to those of Irish sympathizers with the Central Powers. Tammany orators at once charged that Mitchel considered any persons with German and Irish names automatically suspect as traitors.

Less than three weeks before the election, rioting broke out at PS 171, on Madison Avenue near 103rd Street, a school where the Gary plan had recently been introduced. About 1,000 demonstrators, almost all of them boys under fifteen, smashed windows. It was reported that the rioters were trying to promote a general strike in protest against the Gary plan and that they had been "agitated by opponents of the Gary system, who have been speaking against it from street corners near schools in Harlem and the upper East Side."

The demonstrations and rioting spread throughout the poor neighborhoods of the city, occurring mainly in and around the thirty-two Gary-ized schools. At one school, 5,000 children marched with anti-Gary signs like "Can the Gary System and Mitchel"; some young marchers carried Hillquit placards, but the Socialists denied having anything to do with the outbursts. Demonstrations broke out in high schools, in part to protest a newly lengthened school day, which interfered with students' after-school jobs, but also to protest military training, which was made compulsory by a new state law. A city-wide high school strike was planned.

For ten days the anti-Gary demonstrations continued, breaking out in one locality, then another. Hundreds of mothers milled around the schools in Yorkville and East Harlem, complaining angrily about the Gary system. Riots broke out in the Williamsburg and Brownsville sections of Brooklyn; schools were stoned, and police tires were slashed by the young rioters. Schools on the Lower East Side and in the Bronx reported large-scale demonstrations. Of thirty-two Gary schools, nineteen reported riots and demonstrations; attendance was down at the others, and some schools closed because of the disorders.

The city-wide high school student strike was snuffed out when the Board of Education's Committee on High Schools refused to negotiate with protesting students and teachers: "The high school boys of this city are not going to run the schools. Rather than have the boys go on strike, make seditious speeches, and incite children to destroy property, make declarations against the war, and say that the Board of Education withholds information from the people, we will close." However, the Board did reverse itself and returned to a school day ending at 3:00 P.M., instead of 4:00 P.M.[14]

The anti-Gary demonstrations continued through the end of October, with little explanation of their origin. School officials blamed Socialist agitators; Mitchel blamed Murphy and Tammany Hall. Most of the riots took place in poor immigrant neighborhoods; most of the boys who were arrested were Jewish. There were widespread fears in the poor Jewish communities, played on by street orators, that the Gary plan would close the doors of opportunity, as Hylan so often threatened; immigrant parents worried that Gary schools were industrial schools. Whether the street orators were from Tammany Hall, the Socialist party, or the local parents' association, their message about the Gary plan struck a responsive chord in their audience. Just the fact that the plan came from Gary, Indiana, seemed menacing to those who were already fearful of their ability to advance in life. Gary, Indiana, was a town created by the United States Steel Corporation and named for Elbert Gary, chairman of the company, and in the eyes of a poor man and/or a Socialist, one of the nation's chief exploiters of the American workingman. Given the fact that the plan came from a factory town like Gary, how could it be anything but a program to subjugate the children of the masses to the needs of capitalism? The failure of the reformers to try to convince the object of their benefactions, the immigrants, contributed to the school disorders, since there was so little understanding and so much suspicion in the immigrant communities.

In the last emotion-filled days of the campaign, Mitchel accused Hylan of "aiding and abetting the pro-German propaganda in this country" and "of being the associate of the paid agents of Germany in this country." He produced the letterhead of an organization called the Friends of Peace Society, and Hylan's name was listed as an honorary vice-chairman. The Society, declared Mitchel, was pro-German. If reelected, Mitchel pledged to protect the treasury from Boss Murphy and the city from enemy sympathizers.

On the same day, a Tammany official revealed that a fund for popularizing the Gary plan had received contributions from a member of the Vanderbilt family and other well-known upper-class people. He charged that "the Gary system is a system by which the Rockefellers and their allies hope to educate the children of this and coming generations in the doctrine of contentment—another name for social serfdom." [15]

On election day, Hylan won with the largest majority accorded any candidate for mayor since the Greater City was formed in 1898. He received 297,000 votes and carried every borough. Mitchel, who had won 57 percent of the vote in 1913, received only 25 percent, or 147,000 votes. Hillquit, the Socialist candidate, ran behind Mitchel with 132,000. Bennett, the Republican, polled 53,000 votes. Mitchel carried the upper-income areas; Hillquit carried Jewish districts and a few German and Irish districts; Hylan swept everything else.

The day after the election, Hylan announced that the Gary plan would be eliminated at once from the schools where it had been established. He promised that the newly elected Board of Estimate would launch a major school building program.

The editor of the *Journal of Education*, assessing the New York election, commented that "it looks like a dark day for the name 'Gary' educationally." He saw two major reasons for the debacle: first, that Wirt's large salary was unpopular with the voters; and second, "too much stress [was] laid upon the economics of the situation, and much more was claimed for it by the administration than it could make the public believe." [16]

The day after his election defeat, John Purroy Mitchel applied for a commission in the military. He received a commission as a major in the air force. In July 1918, while on a training flight at Lake Charles, Louisiana, Mitchel crashed and was killed. He was thirty-nine years old.

MITCHEL'S ADMINISTRATION has been described by historians as "a legend of well-nigh perfect government." Theodore Roosevelt said of Mitchel during the campaign that he had "the invaluable quality of choosing and being able to command the services of the finest and best men of the city." A dissenting view of Mitchel's accomplishments was registered by Robert Moses, who began his own governmental career during this period. Moses said, "I know of nothing actually accomplished by the sterile Mitchel administration" other than a new zoning system. "Almost all the rest of the Mitchel proposals were pipe dreams. It was an honest outfit committed to saving rubber bands, using both ends of the pencil and similar efficiency devices, and to the impossible promise of making vast physical improvements without spending more money." [17]

The anti-Tammany coalition, shattered during the election, never recovered. America's entry into the World War created a new political and social climate; the era of genteel reform came abruptly to a close. Like the rest of the country, New York City was plagued by postwar inflation and mindless radical-hunting. The prewar progressive combination which had elected Mitchel in 1913 could no longer find a common ground politically.

While Hylan's rout of Mitchel spelled political death for the Gary plan in the New York schools, an evaluation of the plan which appeared in 1918 destroyed its reputation among progressives as "the schools of Utopia." Because of the attacks on the plan in New York, Wirt asked the General Education Board to evaluate the schools of Gary in 1916. The survey, conducted by Abraham Flexner and Frank P. Bachman, was sharply critical of the academic work of the Gary students. It found that many of the nonacademic programs were unsubstantial, that the quality of supervision was crude, and that students were becoming "habituated to inferior performance." [18]

Flexner observed in his memoirs that this survey "was disastrous to the exploitation of the Gary system. Education is far too complicated a process to be advertised on the basis of inadequate supervision and without definite accounting—mental, moral, and financial. With publication of the report the school world was relieved, and the Gary system ceased to be an object of excitement. It disappeared from discussion as suddenly as it had arisen." (Actually, the platoon system was adopted in more than two hundred cities in forty-one states during the 1920s, not because of its progressive tendencies, but because of the money that was saved by doubling pupil enrollment in a single building.) Interestingly, the Flexner-Bachman survey was finished well before the 1917 election, but not released until the fall of 1918. It was ironic that the study which deflated the Gary bubble was financed by the Rockefeller Foundation and carried out by a man who had served as Mitchel's representative on the Board of Education.[19]

Dr. William Wirt, having enjoyed for a few brief years the adulation of the progressive world, returned to Gary, where he remained as superintendent of schools for the rest of his career. As he grew older, he became a staunch conservative in politics, a fate he shared with a number of other progressives. In the 1930s, he became obsessed with the idea that Franklin D. Roosevelt was "the Kerensky of an American revolution" and that the New Deal was a covert attempt to overthrow democratic institutions. The Gary system itself became just another school system, its crusading days a memory; confronted with Negro migration, the Gary schools were just as segregated as schools without a reformist tradition.[20]

New York school reformers were badly battered by their experiences in the Gary school war. They had tied their reform program to the political fortunes of John Purroy Mitchel, not realizing that they ran the risk of losing everything. By acting imperiously and not building broad support for the plan, they failed to get public approval. By attacking teachers and supervisors while installing a new system, they sacrificed the staff's acceptance of the plan. By their tactics, they allowed the Gary plan to become a partisan issue. After a resounding defeat at the polls, they learned from bitter experience, to paraphrase Flexner, that education is far too complicated a process to be subjected to the distortions of campaign rhetoric. After the Gary school war, faced with a powerful Tammany Hall, reformers both inside and outside the school system took refuge in the cry, "Take the schools out of politics."

The editor of the *Journal of Education,* reflecting on the Gary school war, concluded that "the only good thing that may come out of the New York outrage is that good school men and women will not provide the excuse for political deviltry in any city for some years. . . . If education has gained in any respect, we would like to know how, when and where." [21]

JOHN HYLAN lost no time in letting the reformers know that they would have no influence in his administration. He appointed a Board of Education which did not include anyone of prominence in civic or educational affairs. The *Times* wondered, "In what secret recesses of personal or Tammany acquaintance did Mayor Hylan find his seven appointees to the new Board of Education? They are almost entirely unknown to the public, their names suggest nothing, there is no record by which their promises can be judged." Shortly after taking office, the new president of the Board of Education assured the mayor that "the duplicate programme has been eliminated almost completely from the lower grades." 22

The search for political dissenters and disloyal elements in the schools, unleashed by Mayor Mitchel, continued unabated under Mayor Hylan. Immediately after the mayoralty election, three teachers who had been involved in organizing faculty protests against a longer day were brought up on charges of "conduct unbecoming a teacher" and dismissed, not for their organizing activities, but for their political views. The state commissioner of education reviewed their case and concurred in their dismissal, stating that, "The same degree of loyalty is asked of a teacher as of a soldier." In the "intellectual Reign of Terror" (so named by the *American Teacher,* the Teachers Union publication) which followed the World War, only the PEA and the embattled Teachers Union, itself an object of persecution, defended academic freedom.23

During Hylan's two terms in office, the schools embarked on a massive construction program, as he had promised. Yet by the end of the 1920s, though nearly $300 million had been expended to create 475,000 new sittings, the accommodations problem remained acute. Because of population shifts and huge annual increases in pupil enrollment, there were still 50,000 pupils on part-time in 1930.

Throughout the Hylan administration, the Board of Education and the Superintendent of Schools, William Ettinger, warred with the mayor. Like Mitchel, Hylan and the Board of Estimate tried to control educational policy through the power of the purse. The city authorities even went so far as to take state school funds out of the public schools' account and transfer them to the city's General Fund for the Reduction of Taxation. Hylan used the Board of Estimate's financial power to create and fill positions at the Board of Education on a political basis. Associate superintendents and district superintendents were hired over Ettinger's veto, and friends and relatives of the Democratic boss in Brooklyn were awarded high positions in the school system. In 1923, the city administration sought an amendment to the Greater New York Charter which would have destroyed the Board of Examiners and vested total control of the Board of Education's employees in the municipal authorities; after a great outcry from the city about "spoils" legislation, the amendment was defeated in the state legislature. In 1924, when his term expired, Superintendent Ettinger, who had stubbornly resisted political domination, was not re-elected to his post.

The worst excesses of the Hylan regime in its relationship to the schools had their roots in the Mitchel administration. Mitchel's militant Americanism degenerated into a repressive ideological witch hunt after the First World War. And Mitchel's efforts to control the Board of Education set the precedent for Hylan's aggressive interference in educational affairs. When Mitchel tried to cut down on the independence of the educational authorities, it was in order to bring the schools into closer touch with the needs of the rest of the city. Hylan had no such honorable intentions; he expected his appointees to the Board to be subject to his instructions on matters of staffing and policy. He viewed the Board of Education like any other branch of city government; it was a rich source of patronage, except for the hindrance of the Board of Examiners. Despite his best efforts, Hylan never managed to eliminate that last obstacle.

BETWEEN
THE WARS

1920–1954

CHAPTER 21

Dividends of
the Depression

UNDER THE IMPACT of early twentieth-century progressive thinking, the public schools came to be thought of as the foremost agency for social betterment. Progressives believed that America could be saved and restored to the condition that existed before industrialization, before the emergence of the slum, before the arrival of alien masses glutted the melting pot and gave the upper hand to corrupt political machines. Whether idyllic conditions ever existed or whether they could ever be attained was doubtful, but what was certain was the progressive faith that American society could and must be improved, and that the school was the best available mechanism for pursuing social salvation.

In the aftermath of World War I, the search for normalcy became a reaction against progressive idealism. During the same period, however, progressivism in education entered its heyday. Prewar educational progressives, typified by John Dewey, had been preoccupied with uplifting society; postwar progressive education turned inwards and concentrated on freeing individual potential. Lawrence Cremin described the mood of the 1920s: "As the intellectual avant garde became fascinated with the arts in general and Freud in particular, social reformism was virtually eclipsed by the rhetoric of child-centered pedagogy." [1]

In New York, as in the rest of the country, progressivism was rebuffed and frustrated. The loss of the Gary school war and the resurgence of Tammany Hall were bitter blows to New York's progressives. They learned from the Gary experience that it was better to avoid political confrontations, to work instead on quietly implanting their assumptions and programs within the public school hierarchy. Their efforts were successful, in part because the public schools themselves had a rhetorical tradition not unlike that of the progressives. New York's public schools had been founded, after all, to improve society and to salvage the children of wayward parents. Even the Public School Society had seen a direct connection between the work of its schools and the elimination of crime, poverty, and other social disorders.

Educational progressivism, intent on expanding the scope of the school to "meet the needs of the whole child," was on the rise just as the public schools were under tremendous pressure to develop new programs for the children who were brought into school by the child labor laws and the compulsory-attendance law. Schoolmen recognized that the traditional, formal academic course was irrelevant and inappropriate for the many children who would soon join the work force, and that new solutions had to be found to reduce truancy and educational retardation, both of which were intensified by the compelled presence of unwilling students.

"Meeting the needs of the whole child" took different forms in private and public schools. In the postwar period, several private schools were founded which experimented with curricula drawn from children's life experiences; they emphasized the creative arts and otherwise encouraged individual self-expression. These schools were a breath of fresh air to a prosperous upper-middle class that was struggling to throw off the constraints of Victorian conventionalism.

But in the public schools, overflowing with the children of the immigrant poor, "meeting the needs of the whole child" meant something quite different. On one level, it meant school health examinations, vacation schools, libraries, and other specialized services on behalf of the child, his family, and his community. On another level, it meant different curricula for children who did poorly with straight academic fare. One of the first attempts to adjust the curriculum to the child was initiated by Julia Richman, the system's first Jewish female district superintendent, herself a progressive reformer; in 1898, she experimentally classified pupils in one school as bright, medium, and poor, then assigned the best teachers to the slowest classes.

As testing became more scientific, progressives thought it represented the key to reform, since testing made ability grouping possible, and ability grouping meant that each group would get the curriculum which best met its needs. Progressive educators believed that these policies would promote the individualization of schooling. To demonstrate the potential of "modern progressive school methods for public schools generally," the Public Education Association in 1922 sponsored an experimental school, called the "Little Red School House," within the public school system itself. Children were grouped according to their tested mental ability, though the curriculum and teaching methods were the same for all classes. Every effort was made to keep the classes informal and to relate learning to the children's direct experiences. Sol Cohen, historian of the PEA, has written that activities were "all-important: 'Here some two hundred small workers don overalls and go about their occupations,' building, hammering, painting, modeling, singing, dancing, and 'constantly creating and solving problems.' There was 'no retardation problem in Miss [Elizabeth] Irwin's classes, since no child is ever left behind.' Classroom discipline was maintained by 'popular agreement.' The aim of the program was to 'interest the children, and make their social relations easy and happy.'" [2] (Hostile school officials stopped funding the school in 1932, but the Little Red School House survived as a private progressive school.)

Public school officials seized upon intelligence testing and ability grouping, adapting them to the problems of mass schooling. Though they employed the rhetoric of individualization, schoolmen relied on the new scientific methods to help sort out their hundreds of thousands of pupils into groups which might be identified, labeled, and taught accordingly. A child who was identified by testing as a "slow" learner would be channeled into slow classes where he would be taught less than children in faster tracks. The label then became a self-fulfilling prophecy, which did more to meet the needs of the efficiency experts than those of the children.

William Ettinger, Maxwell's successor, revealed the public school officials' almost religious belief in the power of science to solve educational problems, as well as their appropriation of progressive rhetoric to different purposes:

> Perhaps the most characteristic advance in school administration during recent years has been the rejection of the assumption that all children are practically alike in physical and mental endowments. . . . Today progressive school administration requires that an earnest effort be made to sort out children on a scientific basis, so that group instruction may still be consistent with recognition of the fact that, as regards physical and mental traits, one group differs widely from another. Up to the present perhaps the greatest waste in education has been due to the crude classification of pupils. . . . In the last analysis, education must be wrought in terms of the individual child. [3]

Even a conservative official like Dr. William O'Shea, Ettinger's successor in 1924, found it useful to adorn the system's policies in progres-

sive trappings. O'Shea and the progressives in the PEA shared nothing
more than mutual revulsion, yet under O'Shea certain policies were instituted which had progressive roots. Vocational education was expanded; junior high schools were increased, in part to relieve high
school congestion, but also to provide a place for vast numbers of overage elementary school children. Standardized achievement and intelligence tests were prepared for the entire school system. Like progressives, Dr. O'Shea believed that the key to educational progress lay in
refined testing procedures, which would make it possible to discover the
education best fitted for each child. The difference, of course, was that
progressives sought individualization, while school officials sought classification techniques to make mass education more efficient. The rhetoric, however, was the same.

When O'Shea retired in 1934, he was succeeded by Dr. Harold G.
Campbell, an insider who had risen through the ranks of the bureaucracy. Campbell assumed the superintendency in the depths of the Depression.

The Depression's first impact on the schools was crippling, because
of a general curtailment of funds: teaching positions were eliminated,
programs were cut back or killed altogether; all school employees took
salary cuts; class sizes were increased; evening schools, summer schools,
athletic centers, and vacation playgrounds lost funds. And these cuts in
services occurred at a time when more students stayed in school longer
because there were no jobs for young people. But New Deal-inspired
Federal programs to combat the Depression injected new life into the
schools and made possible programs that reformers had dreamed of for
years: cultural and recreational programs, adult education courses,
free dental clinics and other health services, remedial educational
programs, new teaching materials, nursery schools, parent education
projects, a school lunch program. Not only were new programs created
and old ones rejuvenated, but Federal money also made possible the
construction of twenty-four schools, adding 60,000 new sittings.

The atmosphere of social tinkering and experimentation, embodied
by the New Deal in Washington and Fiorello La Guardia in City Hall,
reached into the office of the superintendent of schools of the city of
New York, much to the surprise of educational reformers. Harold Campbell had spent thirty-two years in the New York City school system before his designation as city superintendent; he was considered an educational conservative. Yet progressives found in Campbell an articulate
advocate of their fondest hopes. What they did not realize was that the
slogans of progressivism changed their meaning in the context of a mas-

sive school system. It sounded very progressive when Campbell told a civic conference that the old days, when "school success was possible only for an aristocracy of the intellect," were gone:

> In the newer conception a child is retarded if he fails to make progress in accordance with his individual interests, aptitudes and mental capacity. . . . Instead of considering a child "maladjusted" if he is unable to master the curriculum, particularly the fundamental knowledges and skills, he is to be considered maladjusted now if he is unable to get along happily in his surroundings, with persons as well as with books, and if he falls short in social sensitiveness and social responsiveness.[4]

The problem of educational retardation which had plagued the schools for decades was simply defined out of existence: no longer would children fail, because each would perform according to his own ability. This meant that a child in the slowest track could be promoted to the slowest track in the next grade, and the schools could boast of a 100 percent promotion policy.

In 1935, Superintendent Campbell, a Republican, a Presbyterian, and heretofore a conservative, allied himself with the new education by launching an experimental, child-centered "activity program" in seventy schools, involving 75,000 pupils. The experiment was based on the best features of progressive schools. Set up as a six-year trial of individualized and socialized education, the activity program shifted emphasis from subject matter to the child. The old, rigid curriculum was discarded for a program which grew out of children's interests. Formal recitations were modified by conferences, excursions, research, dramatization, and other participatory experiences. At its best, the activity program created classrooms comparable to those in the best private progressive schools. The State Education Department, which had initially encouraged the program, evaluated it in 1941 and recommended its extension throughout the city. Schoolmen wrongly assumed, as they had in the past and would again in the future, that a program which succeeded in a *selective* situation would also succeed in a *universal* situation.

The pride of the system was its provision for high-achieving pupils. Academically able pupils moved in fast tracks and enriched classes into academic high schools, and, if they passed the entrance examination, into the city's elite high schools. Competition for excellence was keen, and from these schools came a generation of remarkable professionals, scholars, writers, scientists, and intellectuals—many of them the children and grandchildren of the new immigration that ended in 1924.

The other end of the achievement spectrum continued to be the schools' knottiest problem. The low-achieving pupil was staying in school longer than ever because of the Depression, and the schools had to develop programs for him. Vocational education was one route; a diluted academic curriculum was another. Nonpromotion was frowned on, since it might impair a student's personality development.

The spirit of the late 1930s and early 1940s was well captured by Superintendent Campbell, who recognized that the public schools had a new clientele and that high schools were no longer the preserve of those with the highest ability:

> Experiments are needed now as at no other time, for we have in our schools, chiefly our secondary schools, vast numbers of children who have neither the taste nor the liking for the traditional school subjects. . . . Many have not the native ability to master these subjects; but the field of subjects by which we may educate children is practically unlimited. There are educative possibilities in almost every human activity when rightly directed. . . . Never let it be said that when called upon to solve the educational problem of any boy or girl that we threw up our hands and cried, "It can't be done." . . . There is no such child. Every child belongs in school and there is something we can do for every one of them. . . . Educate each one in the way in which he is capable of being educated.[5]

The language of progressive education and modern psychology had captured the bastion of the New York City school system. The schools fulfilled the progressives' demand that they take on the problem of "the whole child" by enlisting an army of psychologists and social workers. As the concepts of progressivism were adapted for mass education, they were bowdlerized beyond recognition. No longer did the school speak of failure or educational retardation, but of maladjustment, social problems, and personality difficulties.

The essence of a bureaucracy is standardized practices, rules, and regulations. One intended result of bureaucratic standardization is fair and equal treatment. An implied result of the same practices is impersonal, mechanistic relationships among the participants in the system. In the early nineteenth century, the Public School Society eliminated human caprice and error by prescribing every detail of permissible behavior on the part of the student and the teacher. The Board of Education, in its own fashion and in the language of modern social science, strived for much the same goal. The teacher was saddled with burdensome clerical detail and endless forms, census forms, health forms, dental forms, attendance records, test scores, and so on. There was unconscious irony, certainly, in the directives and circulars from central headquarters which mandated progressive attitudes and humanistic practices. The school system was attempting to standardize even individualization.

BY 1940 THERE WAS GOOD REASON to believe that the New York City school system had finally conquered the problems which traditionally had plagued it. The average class size was down to thirty-four, and fewer than 20,000 pupils were in short-time classes. The promotion rate was nearly 100 percent. The system's elite high schools were widely regarded as the best in the nation, and the achievement levels of the system as a whole were consistently above the national norms. Teachers competed to work in the school system; the staff included men and women of unusual caliber, who in better times might have become college professors. An atmosphere of experimentation, stimulated by the activity program, excited the admiration of educators from other cities.

Officials at the Board of Education and progressives at the Public Education Association were in agreement that the great advances of the era from the late 1920s through the mid-1940s came about because of the incorporation of the principles of social work and psychology into the schools. Though it was not apparent, the Depression was to a large extent responsible for the remarkable improvement in the schools. Large-scale Federal assistance had provided new schools and additional personnel. The scarcity of jobs made the teaching profession highly desirable. Meanwhile, elementary school enrollment was on the decline, easing the overcrowding problem. School enrollment dropped steadily in the decade from the mid-1930s to the mid-1940s, from more than 1.4 million to 827,000. New York City was in a lull between waves of immigration, a brief period of stabilization. By 1940, New York's schoolchildren were largely second- and third-generation Americans; the public schools were no longer coping with the overwhelming numbers and social problems of first-generation immigrant children. These demographic factors contributed significantly to the lessening of gross school problems.

But school officials, having struggled through the first four decades of the century to alleviate overcrowding and educational retardation, believed that the negative statistics had been reversed by the power of modern educational thinking, which they were proud to have introduced into the system. There was a sense of satisfaction, of complacency, because of the mighty struggles capped by mighty achievements.

And yet, at this very moment in time, a new strain of criticism began to be directed at the New York City schools: the huge and sprawling system, said a few isolated critics, was overcentralized. By its very bigness, it had lost the capacity for flexibility and adaptation to change. Researchers again noticed, like the authors of the Hanus Report in 1913, that teachers and principals were not sufficiently involved in shaping the curriculum; schools were not responsive to or reflective of their local community; relationships among teachers and administrators had become overly hierarchical; the rulebook was more important than it ought to be. The Board of Examiners, which controlled the appointment and promotion of school personnel, erected a maze of requirements that favored those who had graduated from city colleges and discouraged outsiders, regardless of their competence. The system which had been conceived as a revolt against formalism was charged with having become a school machine. While educational authorities noted the need for consolidation of small districts throughout the country, a small but growing number of educators came to believe that many big city systems required decentralization, in part to revive public interest, but also to encourage educational leadership among teachers and principals.[6]

So long as the system was perceived as successful, the reports of its critics were filed and forgotten. The Board of Education's newly won reputation for greatness was a shield against hostile scrutiny. No one then foresaw the massive demographic changes which were about to transform the character of the schools as well as the city. This comparatively halcyonic period ended after World War II, when two concurrent migrations, one of Southern Negroes, the other of Puerto Ricans, made New York City again a major port of entry for poor members of a different culture in search of a better life. Once more, the public schools were faced with spiraling enrollments, crowded and deteriorating buildings, academic retardation, vandalism, and a cultural conflict between teachers and pupils. In the 1890s, the reformers blamed the ward trustees and decentralization for inefficient administration and poor education; this time the reformers would castigate the professional staff and centralization.

CHAPTER 22

From Americanization to Integration

IN THE MIDDLE 1950s, a black migrant to New York told a reporter, "I'd rather be a lamppost in New York than the Mayor of a city in Alabama." Yet the *New York Times,* in its survey of Negro living conditions in the city in 1954 noted that New York was only "comparatively better" than the South for blacks. Although the city and the state had laws forbidding discrimination, Negroes still encountered gross discrimination in housing and employment, as well as de facto segregation in many public schools. Because Negroes were excluded from most private housing, they clustered in large numbers in Harlem and in sections of the Bronx, Brooklyn, and, to a much lesser extent, Queens. Where there were slums, there were slum schools, old (like the neighborhood) and overcrowded (like the neighborhood). Certain fields of employment were closed to Negroes, while in others, Negroes were concentrated in low-paying, low-status jobs.

Civil rights agitation had just begun in the South, and it was still possible for the leading official of the New York Urban League to say that, "Nobody is particularly mad at anybody else." [1]

Despite periodic bursts of energy over the years, reformers, with rare exceptions, had ignored the poverty and discrimination suffered by Negroes in New York. European immigrants did not become the object of reform zeal until their numbers made them a major social problem, and so it was with Negroes. After World War II, the size of the new immigration from the South and Puerto Rico qualified poor blacks and Puerto Ricans as a first-rank social problem. The Negro population of New York City had grown steadily since the beginning of the century, when it was only 2 percent. From 1920 to 1930, the black population doubled from 150,000 to 328,000; another sharp increase took place in the 1940s. By 1950, there were almost 750,000 Negroes in the city. Added to this new migration were large numbers of Puerto Ricans, especially after 1946. From 1940 to 1950, the Puerto Rican population more than tripled, to 250,000. Thus, by 1950, the new black and Puerto Rican immigration constituted one million people in a city which only ten years before imagined it had successfully absorbed and assimilated the last of the great migrations.

The movement to improve the lot of the black and Puerto Rican immigrants tended to focus on blacks, for a variety of reasons: first, because blacks were more numerous; second, because racial discrimination (which also affected dark-skinned Puerto Ricans) kept Negroes out of many residential areas and contributed to the spread of dense Negro ghettos; third, because efforts to eradicate bigotry were led by organizations formed specifically on behalf of Negroes, like the National Association for the Advancement of Colored People (NAACP), the Congress of Racial Equality (CORE), and the Urban League; fourth, because the Negro race, with its history of slavery and brutal mistreatment, had a special claim on the American conscience.

The newcomers arrived in New York as hapless as their European predecessors half a century before. Many crowded into slums inhabited previously by the poor of other nations; many moved into the low-paying, low-skill jobs held before them by poor Irish, poor Italians, and poor Jews. In some ways, poor blacks and poor Puerto Ricans had advantages that previous immigrants did not have; by the time of their arrival in great numbers, the city had a network of public and private welfare services. Though the resources usually were not sufficient to meet the demand, the city endeavored to supply low-rent public housing and free medical care, as well as a welfare system. Despite inadequate services, it was significant that the city accepted its responsibility to assist the needy, which it had refused to do in the nineteenth century.

Yet the immigrants from Europe had one great advantage over Negro and Puerto Rican migrants. They had arrived in New York at a time when the economy could absorb unskilled labor, even that of a child. Child labor was abhorrent to the social reformers, but it had been an important source of income to the poor immigrant family. The usual school-leaving age was fourteen, when many boys and girls found low-paying work in factories and shops. As child labor laws and compulsory-schooling laws were strengthened, industry replaced exploited children with machinery or moved away to find cheap labor elsewhere. By midcentury, society's demand for educational qualifications and credentials kept the unskilled and the unschooled at the bottom of the economic ladder, with little to hope for. The child who left school at age fourteen to go to work was called "industrious" in 1905; fifty years later, the same child would be labeled a social problem and a dropout, because the economy of the city could not legally use him, and in fact, would have difficulty finding a place for him at age seventeen or eighteen, when he was of age to work.

The Negro immigrants had another handicap: color. Those who arrived from the South were burdened by their experience of living at the bottom of a caste system where people like themselves could aspire for little and achieve little. Many came to New York with little education, without any special skills or a trade, ill-equipped to struggle and survive in a tough, competitive city. But whether from the South or born in New York, Negroes found a color line in New York, illegal, unwritten, but effective. They were, in W. E. B. Du Bois' apt phrase, "marginal" persons.

Racial barriers in jobs and housing threatened to confine blacks to permanent third-class citizenship. Consequently, efforts to improve the status of the Negro became demands for racial integration. For Negroes, integration was a necessary means to gaining equal opportunity and equal access to the advantages of American life. There had been civil rights activities in many areas of the city's life before 1954, but after the year that the United States Supreme Court struck down school segregation laws in the South, the drive to integrate the New York public schools consumed more attention and energy than any other aspect of the movement for integration.

Those who sought a way to break the vicious cycle of poverty and discrimination in which so many Negroes were trapped turned naturally to the school. They reasoned that if children of different races attended school together, racism would in time be eliminated. Even more important, education was traditionally considered the lever of economic opportunity. Presumably then, Negroes would overcome poverty by getting a good education, as European immigrants were supposed to have done. Thus, the public schools were assigned tasks which could have been more immediately approached through other areas of public policy. Attacking job discrimination, unemployment, and underemployment would have improved the economic status of Negroes faster than school integration; eliminating residential segregation would have at once ended school segregation. But neither of these approaches had the mythic appeal of the school.

The public school was the very symbol of the melting pot, the great cauldron of Americanization and assimilation. Once again, it was hoped, the school would be the agency which socialized and absorbed an alien culture. This hope was based on the popular myth that the public schools had *single-handedly* transformed immigrant children into achieving citizens. The myth began with the progressive faith that the schools could accomplish this miracle if they willed it, and grew with school officials' self-congratulatory justifications of continued public support for their programs. This view of the school as an institution capable of individual and group salvation took no account of the other factors that contributed to the assimilation of European immigrants or of the large numbers of immigrant children who were not successful in school. Early twentieth-century public schools performed no miracles for first- and second-generation children of European descent, but did facilitate enough mobility to succor the myth of omnipotence. The curriculum was rigid and irrelevant to children's lives, the classes were overcrowded, and the teachers (many of them second- or third-generation Irish or Germans) had no special affection for the immigrant children or for their parents' strange culture. But the ladder was there from "the gutter to the university," and for those stalwart enough to ascend it, the schools were a boon and a path out of poverty. The majority of immigrant children, who did not get to a university or even, in the first generation, through high school, owed as much or more to the nation's rapidly developing economy and to their own personal, familial, and cultural resources, as to the school. Difficulty in school was not uncommon for children from poor, rural backgrounds, whether they were white and southern European in 1910 or black and southern American in 1960.

THE QUESTION OF INTEGRATION never arose in relation to European immigrant children. "Assimilation" and "Americanization" of the Europeans were altogether different from integration. European immigrants lived clustered in slums and attended their neighborhood public school, where very often the majority of pupils did not speak English. Assimilation and Americanization included teaching them how to speak English; inspecting their heads for lice; lecturing them on cleanliness and hygiene; teaching them to salute the flag, to recite the "Pledge of Allegiance," to sing the national anthem, and to revere American heroes. Because so many were anxious to discard their European past and to become Americans, there was little resistance to the cultural evangelism of the school. And the neighborhood school, whatever its faults, was part of a reassuring community network.

The demand for integration which arose in the 1950s was a new stage in the schools' history. The schools of New York had always been neighborhood-based, and until the civil rights movement began to press for an end to de facto segregation, the neighborhood school was taken for granted, regardless of its ethnic homogeneity.

The drive for school integration had several sources. Some sought integration on pragmatic grounds, arguing that predominantly black schools did not get a fair share of the best personnel and resources. Others perceived school integration as a fulfillment of the American Dream, a society in which diversity is melded into a new and universal brotherhood. This perspective was rooted in Horace Mann's vision of the common school as a place where all segments of the community meet as equals and in John Dewey's conversion of Mann's common school into a miniature community which projects an improved social order. Ideally, this mixture of different social levels creates a democratic setting in which everyone learns to live and work together. In New York City, however, Mann's version of the common school never existed, because of the size of the city and the existence of neighborhoods that were not ethnically and economically mixed.

The school integration movement in New York City was mobilized by the United States Supreme Court's historic decision banning state-enforced school segregation. The decision was directed at seventeen southern and border states where dual school systems were maintained for the two races, but integrationists in the North saw in the language of the Court the basis for opposing *any* school which did not have a white majority; a predominance of nonwhites, in this view, was segregation. In the words of the Court:

> Segregation with the sanction of the law . . . has a tendency to retard the educational and mental development of Negro children and to deprive them of some of the benefits they would receive in a racially integrated school system. . . . To separate them from others of similar age and qualifications solely because of their race generates a feeling of inferiority as to their status in the community that may affect their hearts and minds in a way unlikely ever to be undone. [2]

The Court was addressing itself to states which had created a separate school system for black children in order to prevent integration; those states compelled segregation and in doing so degraded blacks *as blacks*.

Militant integrationists in the North, anxious to erase the stigma of racism that had permeated American society from its beginnings, read the decision to mean that any large clustering of blacks was evidence of racism. From this perspective came the proposition that democratic ideals required the dispersion of blacks in equal proportions, and always in a minority, throughout the New York school system. The assumption was made and stated openly by leaders of the integration movement and by school officials in both city and state that a school where blacks predominated was *by definition* incapable of providing equal educational opportunity. No matter what kinds of teachers or facilities or pupils a school had—even a school in which every pupil was the intellectual equivalent of Frederick Douglass or Stokely Carmichael or Roy Wilkins —a black school could never be a good school, in the eyes of militant integrationists. It was somehow never understood in the passion for school integration that it was racism and poverty that were stigmata, not blackness itself. Thus, the effort to eliminate all forms of racism, whether written or unwritten, turned into a well-intentioned but ill-fated demand that no school have an enrollment that was more than 50 percent black (or black and Puerto Rican).

The existence of schools in New York City which were overwhelmingly black was not the result of state law; in fact, any discrimination by race was forbidden by state law. Still, there were subtle administrative practices which were as racist in their effect as legal segregation: school-zoning lines drawn to keep minority group children out of white schools; districts gerrymandered into odd shapes to preserve the racial status quo in certain schools; junior high schools kept white by the arrangement of their elementary school feeders.

But not all de facto segregation was caused by racist machinations. Blacks and Puerto Ricans, like other groups before them, settled in distinct neighborhoods, a trend which was likely to have occurred even in the absence of racial discrimination; poor newcomers have historically moved as a group into slum areas, because of the ready supply of low-cost housing and the proximity of newcomers like themselves. Discrimination in housing was a negative reason for the clustering of ethnic groups, but the positive desire to live as part of a community—with its own social clubs, churches, groceries, restaurants, hometown associations, and sense of identity—was another reason. Some schools were black and Puerto Rican because they reflected the neighborhoods around them, just as some schools in the early part of the twentieth century were almost entirely Jewish or Italian. As the population movement of blacks and Puerto Ricans into New York City continued through the 1950s and 1960s, the integration movement discovered that its target had to be not just racism, but the neighborhood-school concept itself. The civil rights activists wanted New York to live up to the democratic promise of Horace Mann's vision. They were up against formidable obstacles: a population tide that steadily increased the numbers and proportions of blacks and Puerto Ricans; concentrated black and Puerto Rican residential areas that grew larger each year; a decreasing white population in the city, as whites moved to the suburbs; an exodus of white students to private and parochial schools; a strong tradition of neighborhood schools; white resistance to racial integration; middle-class hostility to mixing with the lower class; bureaucratic dislike for change; and the political and economic difficulties of massive pupil exchanges between white and black communities.

Racism was the key impediment to the demands of the school integration movement, but there would have been inhibiting factors even without the racial issue. Had it been decided at some earlier point in history that poor Italians or Irish or Poles should be transferred out of their slum schools and distributed evenly among other, better, schools, in order to expose them to middle-class children, there would very likely have been a two-sided protest: first, from middle-class parents who perceived the children of lower-class foreigners as ill-mannered, dirty, and educationally backward; and second, from the parents of the children being dispersed, who would have resented the imputation of their group inferiority and resisted sending their children to unfamiliar, perhaps hostile, neighborhoods. (In 1904, Lower East Side Jews refused to bus their children out of the neighborhood to less crowded schools. See p. 176.)

The course of the struggle for racial integration illustrates how the Board of Education tried to deal with conflicting pressures and failed; how that failure undermined public confidence in the Board of Education and led to its reorganization; how the impossibility of attaining city-wide racial balance caused many integrationists to abandon the principles of integration and common values, in favor of local control, even though local control had been the battle cry of those who fought integration; and how a battle which began with a vision of universal values ended as a movement to admit sectarianism—ideological, political, and racial—into the public schools, if it pleased the local community.

FOURTH SCHOOL WAR

Racism
and Reaction

CHAPTER 23

The Discovery of
Segregation and Scandals

THE ENLIGHTENED LIBERAL ATTITUDE of the early 1950s, arrived at in painful reaction to generations of racism, was to ignore racial differences. Liberal policymakers strived to be color-blind. Liberal organizations worked to stamp out racial stereotypes and to have racial identification removed from questionnaires and standard forms. This position was based on the conviction that all men are created equal and that therefore social policy should recognize no difference among races. The hopeful assumption was that if society emphasized merit and consciously disregarded racial and ethnic differences, a truly democratic order would one day prevail.

The public schools were in the forefront of liberal efforts to disregard color. There was no ethnic census of the system as a whole or of individual schools until 1957. Children were zoned to attend their neighborhood school, as always. Before the school-integration movement came to life in 1954, the only report on the problems of Negro pupils in the New York schools (whose academic retardation rate was then twice that of all pupils in the city) was prepared by the PEA in 1915.[1]

Therefore it came as a shock when Dr. Kenneth Clark, an associate professor of psychology at the City College of New York and the first Negro to receive a permanent appointment to the faculty of a city college, charged that the New York City school system was maintaining segregated schools where Negro children received inferior education. The challenge was laid down at a National Urban League symposium, only a few weeks after the Supreme Court's school decision in 1954. Clark's research on the harmful effects of segregated education had figured in the Supreme Court's decision. Born in the Panama Canal Zone,

Clark had been brought to New York at the age of five; he had attended Harlem public schools, Howard University, and Columbia University. Since 1946 he and his wife operated the Northside Center for Child Development, a treatment center for children with personality disorders. An intellectual and ardent integrationist, Clark commanded broad respect.

The president of the New York City Board of Education, Arthur Levitt, responded immediately by asking the PEA to conduct a full investigation of the public school education of Negro and Puerto Rican children. Levitt denied that any school had been intentionally segregated and agreed that all-Negro schools were bad for the children in them. In the fall of 1954, he said to an audience in Harlem that there was "no segregation in schools deliberately imposed by legislation. . . . Unfortunately we have schools populated 100 per cent by Negroes but that is because we have to build schools near the children and when the children live in a homogeneous community, you have segregated schools." These "segregated schools" were "not good educational policy" and their result was "a psychological scarring." Levitt's statement reflected the prevalent liberal view that an all-Negro school was educationally unsound and undesirable.[2]

The Board of Education took two steps at the end of 1954 to demonstrate its intention to comply with the spirit of the Supreme Court's decision. It appointed a top-level Commission on Integration, consisting of the nine Board members, twenty-three civic leaders (including Dr. Kenneth Clark), and five supervisors, and gave them a mandate to develop recommendations to further integration. Its cochairmen were Levitt, who had resigned as Board president to become state controller, and Charles Silver, a successful businessman and the new Board president.

Furthermore, the Board adopted a forceful policy statement, avowing its intention to eliminate de facto segregated schools:

We . . . interpret the May 17th decision of the United States Supreme Court as a legal and moral reaffirmation of our fundamental educational principles. We recognize it as a decision which applies not only to those cases in litigation, but also as a challenge to Boards throughout the nation, in Northern as well as Southern communities, to re-examine the racial composition of the schools within their respective systems in order to determine whether they conform to the standards stated clearly by that Court.

The Supreme Court of the United States reminds us that modern psychological knowledge indicates clearly that segregated, racially homogeneous schools damage the personality of minority group children. These schools decrease their motivation and thus impair their ability to learn. White children are also damaged.

Public education in a racially homogeneous setting is socially unrealistic and blocks the attainment of the goals of democratic education, whether this segregation occurs by law or by fact. In seeking to provide effective democratic education for all the children of this city, the members of the Board of Education of the City of New York are faced with many real obstacles in the form of complex social and community problems. In spite of these and other difficulties, the board is determined to accept the challenge implicit in the language and spirit of the decision. We will seek a solution to these problems and take action with dispatch, implementing the recommendations resulting from a systematic and objective study of the problem here presented.[3]

In the fall of 1955, the PEA study reported that there were forty-two elementary schools which were 90 percent or more Negro and Puerto Rican and nine junior high schools which were 85 percent or more Negro and Puerto Rican. Of 639 elementary and junior high schools in the city, not quite 8 percent fell into this category. The authors of the report absolved the Board of responsibility for intentionally segregating black and Puerto Rican children, but criticized it for not doing enough to promote integration. The report held that the makers of zoning policy attempted to be color-blind; zoning policy sought to minimize the distance from home to school, to avoid traffic hazards and topographical features, and to keep districts similar in size: "In general, the principles followed in zoning school districts ignore both the possibility of separation and integration of ethnic groups." Compared to schools which were more than 90 percent white, the predominantly Negro and Puerto Rican schools were older and less adequately maintained, had a higher rate of teacher-turnover, and had a smaller proportion of tenured teachers. The teacher-pupil ratios in the two groups of schools were not dissimilar, nor was there a significant difference in the expenditure per pupil. But segregated schools did exist, and these schools had fewer experienced teachers than did other schools.[4]

Now that the PEA had detailed the outlines of the problem, the Commission on Integration worked to make recommendations which would solve it. The commission held hearings during 1956 and 1957 and discovered that it was easier to issue a policy statement than to find a viable course of action, an experience that would be repeated often in the years ahead. Its cochairmen, Levitt and Silver, optimistically forecast that the Board would draft a "master plan" which would "set a pattern for desegregating schools that will serve as a model for the whole nation." However, the commission's two most important reports—on zoning and on teacher assignments—ran into intense opposition from the groups affected. Some white parents worried that rezoning for integration would lead to long-distance bussing and the compulsory assignment of white children to slum schools. Some teachers reacted with hostility to any suggestion of forced rotation or compulsory assignment of experienced teachers into slum schools; critics could not understand why the Board of Education did not affirm its power to assign its employees wherever they were needed. Spokesmen for both groups threatened that any compulsion by the Board would cause them to flee to the city's safe, white suburbs.[5]

Debates over these policies proceeded against a background of incipient crisis in the schools. Because of postwar population growth, the schools were again overcrowded, and again thousands of students were on short-time or double sessions. The Board could not get enough money both to renovate the scores of obsolete, nonfireproof buildings *and* to build new schools. Of even greater concern to the public were reports of juvenile delinquency, gangs, assaults, vandalism, and fire-setting in the schools, which were especially pronounced in slum districts. The more that middle-class parents read about the conditions in ghetto schools, the more frightened they were about any integration that might send their children into those schools. Some of the white fear was simply racist fantasy. But there was also a middle-class fear, common to both races, of sending their children to a school where lower-class behavior predominated, where there was a breakdown of discipline and adult authority, low academic standards, obscene language, and sporadic violence. Harrowing accounts of conditions in ghetto schools stirred the conscience of leading citizens who had no children in the public schools and heightened the anxiety of those who did.

In 1957, when the Commission on Integration's controversial reports on zoning and teacher assignments came before the Board of Education for approval, representatives of the city's leading civic, religious, and social-welfare organizations urged their immediate adoption, but counterpressure was strong enough to cause Board President Charles Silver to promise in advance of the vote that no long-distance bussing was contemplated. Public statements from the Board staff pledged that the reports would not disturb the principle of the neighborhood school. When at last the Board of Education unanimously approved the two reports, a qualifying statement was appended which made it impossible to know exactly what had been approved or what would be implemented. The

Board's qualifying statement said: "The Board of Education reserves to itself the privilege of interpreting the meaning to be attached to certain terms and phrases, of further studying and exploring the merits of specific recommendations and of resolving the administrative problems which necessarily arise in carrying out any policy for a school system as complex as that of New York City."

The *New York Times* was delighted with the Board's vote: "Refusing to be awed by a campaign of misrepresentation, selfish obstructionism and in some cases prejudice, the Board of Education has courageously approved in principle a program to equalize educational opportunity." If the Board had not done so, "how could the decent people of this city have held up their heads today . . . ? How could we have answered our conscience? What could we have said to our friends in the South?" 6

As the Board would discover on many future occasions, voting "yes" was easy compared to the difficulty of implementing new policy, for it was primarily the people affected who objected.

Superintendent of Schools William Jansen believed in the concept of the neighborhood school. He refused to compromise his point of view, while insisting that he was committed to integration.

When the first local integration controversy bubbled to the surface in 1956, Jansen refused to back down on his commitment to the neighborhood school. A newspaper report that the Board had quietly begun integrating some schools in the Bedford-Stuyvesant section of Brooklyn elicited a frontal attack on Jansen by a Brooklyn minister, the Reverend Milton A. Galamison. Galamison, the chairman of the education committee of the Brooklyn NAACP, accused Jansen of leaking false claims of success, while the recently completed Junior High School 258 remained segregated. Galamison called JHS 258 a concrete test of the Board's integration policy. But Jansen would not integrate JHS 258 because of its location in the Negro ghetto. He believed that white parents might cause violence if their children were forced to attend 258. Because of the impasse, Galamison sought and received the support of six other NAACP branches to form a city-wide coordinating committee to press for school integration. Galamison soon became a leader of the integration movement.

Superintendent Jansen was indifferent to the anxieties of Negro parents who knew that their schools had been branded inferior, but worked hard to quiet white parents' fear of integration. Jansen assured white parents in Queens that there would be few, if any, changes in their schools: "We have no intention whatsoever of long-distance bussing or bussing of children simply because of their color. If we bus children, it will be because there is room in one school and not in another, as we do now. We believe in the neighborhood school." [7]

Jansen's call for volunteers to teach in the city's difficult schools evoked twenty-five responses from the city's 40,000 teachers, when 1,000 teachers were needed. His "master plan" for zoning affirmed that elementary and junior high schools "are essentially neighborhood or community institutions which serve the children of families living within an area contiguous to the school building." He opposed permissive zoning (allowing children to transfer out of their neighborhood district) and bussing for integration. "It is impossible to establish any fixed ratio of children of different racial backgrounds for all schools. The ratio will vary among schools and will constantly change as the residential pattern shifts from one majority to another." The schools, wrote Jansen, cannot solve the problems created by residential segregation.[8]

It was a strange and unsatisfying situation. Defenders of the neighborhood school were alarmed because the Board of Education had issued bold statements of support for integration; some worried that the plans were being carried out surreptitiously. Blacks and their white allies in the civil rights movement were enraged because nothing was being done to implement the Board's integration policy, and the public statements of the superintendent of schools pledged that nothing more would be done in the future.

At this early stage of the integration struggle, the political legitimacy of the Board of Education began to crumble. The Board could formulate policy, but could not enact it. The mayor might have stepped in to give the Board the strength of his political authority, but Mayor Robert F. Wagner, elected for the first of three terms in 1953, did not believe in interfering in school affairs. Wagner, son of a famous liberal senator and himself a pillar of the city's liberal establishment, knew the workings of the city as well as any mayor before him. Civic groups like the PEA had said for years that the schools should be "above" politics, and Wagner agreed. He was a man who by temperament sought conciliation and shunned controversy. He fought to provide the schools with the funds they needed, but he studiously avoided being drawn into policy disputes.

The Board of Education, trapped between bitterly conflicting forces, was left to fend for itself. Lacking any clear political mandate from the people, the Board fell back on devices calculated to placate, to deceive, to buy time, or just to keep peace. It was a desperate strategy with no hope of victory; all it could hope to accomplish was the deferral of little crises, which would ultimately produce irretrievable situations.

IN THE TWO YEARS FOLLOWING the PEA report on the status of Negro and Puerto Rican children, the number of public schools which were 90 percent or more Negro and Puerto Rican increased from fifty-one to seventy-seven. The Negro and Puerto Rican population had grown; racial ghettos had become denser and bigger. Ghetto parents felt an increasing sense of desperation about their children's schools, which produced few high school graduates and had inexperienced teachers, decaying facilities, and low academic standards. Like other upward-aspiring immigrants, Negro and Puerto Rican parents believed that education was the key to their children's future. Negro leaders, pinning their hopes on integration as the best way to eliminate inferior schools, grew angry and impatient with the Board of Education's combination of high-flown rhetoric and callous inaction.

Renewed pressure for integration brought claims from Superintendent Jansen that integration had been advanced in more than fifty schools by the unpublicized transfer of minority-group students to previously all-white schools, which Jansen would not identify, to avoid stirring up white reaction. Integrationists did not believe it. Dr. Kenneth Clark openly spoke of "sabotage" of the Board's integration plans and accused Jansen of "dragging his feet." Other civil rights leaders charged that Jansen was paying lip service to integration while maintaining the status quo. Galamison, now the president of the Brooklyn NAACP, blamed the stalling of integration on recalcitrant officials: "It's a question of attitude. Many school officials simply do not want to reduce segregation. They offer many excuses as to why it can't be done, when basically they are content with the status quo." [9]

In 1958, Jansen reached the compulsory retirement age and was replaced by John Theobald, an administrator with a good political sense. Theobald had previously been president of Queens College and deputy mayor under Wagner. Theobald wanted to advance integration and understood that to do so required keeping the temperature low on all sides. He met privately on numerous occasions with civil rights leaders, so that they would understand that he was trying to promote integration without alarming white parents.

Theobald carefully avoided making public statements which would inflame frightened middle-class parents. At the same time, he tried quietly to push limited integration measures by calling them something else. His public pronouncements about pupil transfers were ambiguous. Pupils were shifted to utilize space better; if these shifts from crowded schools to underutilized schools also caused integration, that was incidental. Theobald declared his allegiance to the neighborhood-school concept, but, unlike Jansen, announced his intention to study the rezoning of high schools to facilitate integration.

While Theobald was moving cautiously, the Commission on Integration issued its final report, declaring unequivocally that segregated education is inferior education. In keeping with the temper of the times, the commission confused race with class; it might have been more accurate had the report suggested that slum education was likely to be inferior education, regardless of the race of the children. In not recognizing that distinction, the commission contributed to an unconsciously racist sentiment that the presence of blacks in large numbers was a threat to quality education. Operating on this premise, the commission report resoundingly endorsed integration as a goal that was within sight: "New York City, as it has so often in the past, is engaged in an enterprise of social pioneering of the greatest moment. The eyes of the country— indeed of the world—watch with great interest what happens here." The task of proceeding "with all deliberate speed toward integration of our schools is not an easy one. But the terrain has been surveyed, the route mapped, and without any question, the people of the City of New York want to travel that road to the end." [10]

Years later, critics of the Commission on Integration and its parent body, the Board of Education, would say that their optimistic statements were either incredibly cynical or incredibly naive. Yet, if the ethnic cast of the city had stabilized at that point, if the white exodus from the city had stopped, if the black and Puerto Rican proportion in the schools had stayed at 35 percent (as it was in early 1958), then perhaps the high hopes of the period would have had a chance of success.

DURING 1958 AND 1959, Superintendent of Schools Theobald assayed a delicate maneuver: he tried to advance integration without seeming to, while holding off restive ghetto parents. He was confronted with demonstrations *for* integration by Harlem parents and demonstrations *against* integration by Queens parents. Theobald damped down the Harlem protest by agreeing to transfer a group of boycotting students out of their neighborhood schools. The situation in Queens was not so easily resolved, for it involved the first implementing of the city's integration policy. In 1959, Theobald launched the first trial of permissive zoning: almost 400 black elementary pupils in Bedford-Stuyvesant were transferred to underutilized, all-white schools in the Ridgewood and Glendale sections of Queens. The move, he insisted, was necessitated by overcrowding and was made solely in the interest of effective school utilization.

The conservative Glendale and Ridgewood neighborhoods refused to have their fears calmed. Here finally was the confirmation of their anxieties, despite years of disclaimers from the superintendent of schools and numerous official bows to the concept of the neighborhood school. They feared that this move was preliminary to massive pupil exchanges —blacks into their schools, their children into slum schools. The middle-class families who lived in the neat frame houses of these communities wanted to preserve their neighborhoods and their property values, and they saw both threatened by integration. Their reaction was racist; but also, it was an expression of middle-class status anxiety. As an official of the Ridgewood-Glendale Taxpayers' Committee put it: "Good neighborhoods do not just happen. They are the result of hard and earnest work to establish and maintain a good and wholesome place in which to live and bring up their children by parents who accept their responsibility of seeing that their children have wholesome surroundings and companions. This is a fight for the principle of the neighborhood school." At stake was not just the quality of the neighborhoods, but the right of the communities involved to control their schools and to exclude outsiders who had not earned the privilege of residing in their neighborhoods.[11]

Theobald resisted the objections from Glendale and Ridgewood, where protest meetings had whipped up community groups to a fever pitch. He refused, however, to admit that his decision was related to integration: "I cannot, as Superintendent of Schools, sit by and have youngsters on part time and have extra seats easily accessible and not use them." Board President Silver echoed Theobald's insistence that the transfers from Bedford-Stuyvesant into Queens were not based on the integration program but were a temporary emergency measure which would last only until new schools were completed to ease overcrowding in Bedford-Stuyvesant.[12]

On the first day of school in September 1959, almost 50 percent of the white elementary students in Ridgewood and Glendale stayed home. The boycott ended after the first day; the system survived the first major test of permissive zoning. However, the school officials' insistence that the transfers had nothing to do with integration undercut any sense of victory for integration. Their refusal to own up to their intention to further integration was, of course, a transparent deception; no one was fooled. A public agency which relies on public support cannot continue to expect public trust when it habitually sacrifices candor for momentary political advantage. People who followed school affairs became aware that the actions and the words of the Board of Education were not necessarily the same.

IT HAD BECOME EVIDENT that the problem of inferior education would have to be confronted within the ghetto school, which was obviously not about to be eliminated by overnight integration; some civil rights advocates hoped that integration would move faster after ghetto schools were educationally up to parity with nonghetto schools. Equalizing ghetto education was easier demanded than done, but the Board of Education discovered in early 1959 that one of its experiments had the makings of an answer to the problem. The program, called the Demonstration Guidance Project, was an experiment in intensive compensatory education, based on recommendations of Dr. Kenneth Clark's subcommittee of the Commission on Integration. It began in 1956 at Junior High School 43 in upper Manhattan. The top half of each grade was saturated with special services, smaller classes, remedial instruction, cultural activities, extra counseling, and increased parental involvement. Academic achievement improved dramatically, and the dropout rate was halved.

Theobald and Silver, calling the project "the most significant breakthrough in any city toward equality in education," announced that the program would be rapidly expanded into thirty additional schools. Instead of limiting it to the upper half of each grade in junior high school, the new program, called Higher Horizons, included all children, beginning in the third grade. School officials ignored the dilution inherent in transforming a selective program into a universal one, and the fatal danger of adopting a new plan overnight without adequate planning, training, or resources.[13]

Higher Horizons quickly became a political plum which school officials used to buy off threats of demonstrations for integration. Later in 1959, when disgruntled ghetto parents announced their intention to boycott the schools, Theobald headed off the threat by agreeing to add eighteen Harlem schools to the Higher Horizons program. Each time Higher Horizons was expanded, its benefits were stretched thinner. Whereas the original Demonstration Guidance Project spent an additional $250 per pupil on the most promising pupils, Higher Horizons spent only $40 extra per pupil while encompassing entire schools. The conditions which produced a dramatic impact at JHS 43 were eliminated from the outset of Higher Horizons.

The rapid extension of Higher Horizons, despite its bright promise, was not sufficient to satisfy the civil rights movement, any more than the successful bussing of 400 black children into Queens had been. To those who believed that racially separate education is inherently unequal, compensatory education was but a variation of achieving separate-but-equal schools; and the victory in Queens was only a minuscule step toward integration, which was not even a precedent for further integration.

Theobald issued a "progress report" on integration in June 1960 that made clear just how difficult it was to gain ground. In his report, hopefully dubbed "Toward Greater Opportunity," Theobald pointed out that while blacks and Puerto Ricans were 40 percent of the city's total register, they were 75 percent of the elementary school enrollment in Manhattan. He described the massive outflow of whites from the city and their replacement in almost equal numbers by blacks and Puerto Ricans. To illustrate the frustrations of planning for integration, he gave the example of a new public school on Manhattan's upper West Side which was built to accommodate children from a low-income housing project and from a newly constructed private housing development. The public housing project was 75 percent Negro and Puerto Rican, while the private development across the street was almost completely white; both projects had more than 2,000 apartments. Only four children from the private apartments registered in the new public school; the rest enrolled in private and parochial schools.[14]

In the 1950s the white population of New York City had declined by more than 800,000, while the black and Puerto Rican population increased by over 700,000. This population trend, in combination with segregated housing patterns, contributed far more to the growth in the number of segregated black and Puerto Rican schools than any policies of the Board of Education.

Though Theobald could rightly claim that the number of all-white schools had declined, weary advocates of school integration saw his report as a catalogue of excuses. His unofficial advisory group of black and Puerto Rican leaders warned him that continued inaction might lead to a school boycott in September 1960.

The predicted offensive was led by the Reverend Milton Galamison, now the president of the newly formed Parents' Workshop for Equality, an organization whose headquarters were located in Galami-

son's church in Brooklyn. Born in Philadelphia in 1923, son of a post office clerk, Galamison graduated from Lincoln University in 1947 and
Princeton Theological Seminary in 1949; since then, he had been pastor
of Siloam Presbyterian Church and a civil rights activist. Charging that
the Board's ineffectualness had permitted the number of segregated
black and Puerto Rican schools to double since 1954, Galamison threatened a "sit-out" on opening day. Within days, Theobald announced the
adoption of Open Enrollment, a program which offered black and
Puerto Rican junior high school students the option of transferring out
of their present schools and into designated receiving schools which
were predominantly white. However, Galamison refused to call off his
planned protest unless the Board prepared "a plan and timetable for
school desegregation and immediate transfer for those children whose
parents want an integrated education." To stave off Galamison's threats,
the Board of Education extended Open Enrollment beyond junior high
schools to sixteen elementary schools. The enlargement of the program
was hastened by the discovery that very few of the eligibles at the junior
high level were exercising the option to transfer to predominantly white
schools. Of some 12,000 junior high school students eligible to transfer,
only 393 had applied. The reaction at the elementary level was similar:
of nearly 8,000 elementary pupils eligible, only 284 applied. Because
Open Enrollment was initiated as an abrupt capitulation to political
pressure, neither sending schools nor receiving schools were prepared
for the program; the lack of planning hindered the success of Open Enrollment, since some receiving schools did not welcome the transferees
or have adequate remedial services.[15]

When civil rights leaders complained about the cost of transportation, the Board agreed to provide free transportation, but there was no
significant upturn in the number of parents electing to transfer their
children away from their neighborhood school. In the ensuing years,
fewer than 5 percent of the eligibles chose to move to mostly white
schools. Critics of the Board of Education later charged that the failure
of Open Enrollment was the fault of racist principals and superintendents who wanted to sabotage integration. Some ghetto principals resented Open Enrollment because it drained away their best students
and most motivated parents; these principals worried that Open Enrollment lowered the morale of pupils, parents, and staff by labeling their
schools inferior simply because of their high proportions of black and
Puerto Rican pupils.

The low response of black and Puerto Rican parents to Open Enrollment raised other possibilities: that minority parents did not feel as
stigmatized by participating in neighborhood institutions as their lead-

ers thought they did, that black parents feared that their children might not be welcome in predominantly white schools, or that many parents might prefer to send their children to school near their home.

PUBLIC CONFIDENCE in the New York City school system sustained two major shocks in 1961. First came the revelation that the renowned system was slipping academically. The Board of Education released system-wide test scores which showed that in the previous year, third graders were reading a month below the national average, while sixth graders were generally a month or two ahead in both reading and mathematics. Only the year before, the same tests had shown all grades ahead of the national average, some by as much as four to six months. School officials were at a loss to explain the downturn; perhaps, they suggested, the rest of the country was catching up to New York City. They must have realized that the system was absorbing the impact of the new immigration, but either an habitual lack of candor or a stubborn belief in the mythical omnipotence of the school kept them from acknowledging it. Had they faced the facts squarely at this point and confronted the implications of a population shift in which thousands of middle-class pupils left the system to be replaced by thousands of new immigrant children, they might have been in a better position to prepare the system and the public for a resurgence of many of the same problems that stymied the schools in each previous era of immigration.

However, revelation of the biggest scandal in the history of the Board of Education diverted public attention from the schools' other dilemmas. In May 1961, a state investigation uncovered evidence of gross irregularities in the city's school-construction funds, involving payoffs, bribes, structural defects in school buildings, and safety hazards. Several middle-level Board officials were charged with direct complicity. While no member of the Board of Education was tied to the corruption, the enormity of the crimes undermined the viability of the entire Board. Less sensational, but equally damaging to the system's reputation, was the investigating committee's documentation of administrative chaos within the school-construction and repair divisions; red tape and interminable buck-passing among officials permitted safety hazards and structural defects to go uncorrected for years. A few weeks after the scandal broke, Theobald admitted that vocational-high-school students had built a fifteen-and-a-half-foot boat for him. He insisted that there had been no impropriety since he had paid for the materials and since students often built boats on order as part of their training.

Governor Nelson Rockefeller called the legislature into special ses-

sion in August 1961; special legislation was passed which removed the
Board of Education and directed Mayor Robert Wagner to name a new
nine-member Board of Education drawn from nominations made by a
select screening panel of civic and educational leaders. At the recommendation of State Commissioner of Education James Allen, Jr., the legislature also ordered a revival of local school boards, to provide an opportunity "for effective participation by the people of the city in the
government of their schools." Allen hoped that strengthening the local
boards would spur decentralization of the massive system, a course he
had publicly urged since 1958. Appointment of the new local boards was
taken away from the borough presidents and lodged in the Board of Education.[16]

The reform Board of Education was sworn into office in September
1961. The new Board did not ask Superintendent Theobald to resign,
but in May 1962 he announced his retirement. The reform Board was a
distinguished group, including Anna Rosenberg (a former assistant secretary of defense), Samuel Pierce (a young black judge), Clarence Se-
nior (an economist and sociologist), Lloyd Garrison (former dean of the
University of Wisconsin Law School), Morris Iushewitz (secretary of the
Central Labor Council), James Donovan (a prominent attorney), and
Max J. Rubin (also a prominent attorney). The new Board was evenly
divided among Catholics, Protestants, and Jews, a political tradition
of recent vintage.

Max J. Rubin, elected president of the Board, promised to speed decentralization and to eliminate bureaucratic inefficiency. As the first
steps toward reform, the Board divested its members of their customary
chauffeur-driven limousines and reduced their personal staff; an unintended effect of cutting back the Board members' own staff was to make
the members more dependent on the professional bureaucracy than before, since their sources of information and independence were diminished. Another reform was to reduce the number of closed executive sessions and to increase evening public meetings of the board. To revive
local school boards, the Board of Education decreased their number
from fifty-four to twenty-five, each with nine members appointed by the
Board from nominations made by local community screening panels.

The unsettled question of teacher unionism received the new
Board's immediate attention. In November 1960, the United Federation
of Teachers, a combination of several teachers' organizations, had conducted a one-day strike to demand a collective-bargaining election. The
Board had then polled teachers to determine whether they favored
collective bargaining. The teachers voted overwhelmingly "yes." After
the reform Board was installed in 1961, it approved an election to

choose the organization which would represent the teachers. The United Federation of Teachers, which was affiliated with the organized labor movement, competed against a newly created, less militant group called the Teachers Bargaining Organization, which was supported by the National Education Association; a third competitor in the contest was the Teachers' Union, which had been banned from the schools for the previous eleven years as a Communist-dominated union. On December 16, 1961, the United Federation of Teachers won easily, becoming the bargaining agent for the city's 43,500 teachers.

DURING THE 1950s AND EARLY 1960s, the schools were at the center of one crisis after another: juvenile delinquency, a soaring high school dropout rate, a perennial teacher shortage, union–management conflict, increasing racial segregation, the staggering cost of replacing old buildings and adding new ones. The stormy course of the integration battle, particularly in the early 1960s, contributed to the public sense that the Board of Education was incapable of providing forceful leadership and had lost touch with its public; scarcely a week went by without a demonstration against Board policies, by one side or the other.

Several studies in the late fifties and early sixties recommended major administrative changes in the system. These studies fed a spreading suspicion that the Board of Education as a system, rather than as nine individuals, was not coping effectively with its problems. The most critical report was prepared by Mark Schinnerer, retired superintendent of Cleveland, at the request of State Commissioner Allen after the 1961 construction scandals. Without staff, Schinnerer spent three months visiting schools and talking to teachers, principals, and students. In early 1962, Schinnerer issued a report which raked the school system over the coals. He urged drastic revision of teacher-procurement methods, a doubling of school expenditures within the next eight years, an immediate increase of the teachers' minimum salary, and abolition of the Board of Examiners. He held that the Board of Examiners' methods were devised at a time when there was a surplus of qualified teachers; because of lengthy delays in issuing teachers' licenses (anywhere from three months to a year), many applicants went to work in the suburbs. The system was hampered by "too much inbreeding" and a lack of independent thinking; it was dominated by committees and by a bureaucratic mentality that recited "the most frequently heard answer . . . this is the way it has always been done." He wanted to see the system administratively decentralized, giving each district superintendent authority for all the schools in his area, with no district having more than 30,000 pupils:

The purpose of decentralization of school administration is to place the responsibility and authority for decisions as close to the points of impact as possible. This means, consequently, transfer of authority, subject always to review if necessary, to lower and more diverse levels of the administrative hierarchy. . . . The decentralization of administration is not without hazards. If too much authority is dispersed to lower levels of responsibility, they may develop a chaotic condition resembling anarchy. . . . Decision-making should never be decentralized in areas which are beyond the expected competence of the ones who must make the decisions.

He did not favor greater participation by city officials in the school system; in fact, he thought the Board of Education should be fiscally independent, completely separate from the city administration. He felt that the Board of Education should be free to experiment with various forms of community involvement; to this end, he urged the repeal of the state law mandating local school boards.

He warned that unless changes were made quickly, the New York schools would become

not just second class, but even third class or fourth class. . . . There are a million girls and boys in the New York City public schools. They are going to make or break New York City in the future. They need a good education now. It cannot be put off until some later date. That is why nothing, but nothing, must be permitted to detract by one iota from the efficiency with which their educational program is carried out.[17]

The Schinnerer report was ignored by the Board of Examiners (which released its own proposals for improvements two weeks after the Schinnerer study appeared), and praised by Board President Max Rubin. Fred Hechinger, education editor of the *New York Times*, agreed substantially with Schinnerer, though he felt that the complete elimination of the Board of Examiners might be too radical a step. Hechinger noted that the great days of city schools predated the development of suburbia; the new great schools, he wrote, were now in the suburbs. While many suburban systems had seventy educators per 1,000 pupils, New York City had about forty-five. He concurred with Schinnerer that it made no sense to turn away teachers by a complicated examination system, when suburban districts were willing to accept them on the basis of interviews, records, and supervision.[18]

Schinnerer's recommendations went unheeded. In early 1962, there was an undercurrent of dissatisfaction with school affairs, but no clearly identifiable constituency for the basic changes Schinnerer was urging.

CHAPTER 24

Boycotts and Demonstrations

AFTER A NATIONWIDE SEARCH by a select committee, Dr. Calvin Gross, the young, articulate chief of the Pittsburgh public school system was selected to be superintendent of schools for New York City. It was the first time that an outsider had been chosen for the job. The press and the interested public cheered his designation and eagerly anticipated fresh initiatives.

Gross entered the New York system in early 1963, just as the integration battle entered a new phase. The realization had dawned that, although Open Enrollment reduced the number of predominantly white schools, the number of predominantly black and Puerto Rican schools continued to increase. Advocates of school integration were convinced that something more had to be done, that it was not enough to transport students out of the ghetto into white schools.

The City Commission on Human Rights, a group appointed by the mayor to promote integration and to improve race relations, had taken the lead in the fall of 1962 and forthrightly urged the Board to adopt a "two-way" bussing policy. The commission proposed that the Board give priority treatment in budget and services to ghetto schools: "The shifting of children in two directions—to and from Negro and Puerto Rican areas—should and could be brought about if one could be assured of an improved education by doing so." [1]

The commission's statement was the opening shot in a direct assault on the neighborhood school, the cornerstone of the city school system. Civil rights advocates accused school officials of paying lip service to integration, and the charge was true; it was also true, however, that school officials paid lip service to the concept of the neighborhood school. What officials tried to do was to give everyone what they wanted: by adopting a one-way bussing plan, whites could continue to have their neighborhood schools to which Negroes (who wanted to get away from their neighborhood schools) would be admitted. It was a precarious balancing act and did not keep the peace for long.

A few months after Calvin Gross' appointment, State Commissioner Allen dislodged the shaky status quo in New York. He ordered all school boards in the state to report what they had done to remove racial imbalance, and he declared that schools which were more than 50 percent Negro were racially imbalanced and were failing to provide equal educational opportunities. Only a school that was more than 50 percent mainland white met the state's definition of equal opportunity. The well-intentioned liberalism of Allen and the state's Board of Regents gave a useful prod to towns and villages where achieving integration was still merely a question of the will to do so. But in New York City, Allen's mandate revived demands for a goal which was steadily retreating, not simply because of bureaucratic foot-dragging or outright racism, but because of the ongoing shift in New York's population. By 1963, black and Puerto Rican students were 40 percent of the total enrollment; the number of elementary schools which had 90 percent or more minority-group pupils had risen from 42 in 1955 to 134 in 1963. In addition, the fact that black and Puerto Rican pupils represented 52 percent of all first graders indicated that within a few years these groups would become a majority of the entire enrollment. (In calculating racial balance, Puerto Ricans were counted together with blacks; thus a school that was predominantly Puerto Rican would be considered racially imbalanced.)

Whatever the beneficial impact of Allen's new policy on towns outside New York City, the result in New York was to label officially as inferior any school in which more than half the students were black and Puerto Rican, regardless of any objective educational conditions. Given the times, Allen's position was compassionate, humane, and liberal. But the underlying assumption was unexamined. Stating that predominance of a particular racial or ethnic group in a school makes that school inferior is, on the face of it, a libel against that group, whether they are black, Italian, Jewish, or whatever. Historically, schools in slum neighborhoods had always been educationally inferior schools; the reasons, having to do with slum conditions which spilled over from the neighborhood into the school, were remarkably similar no matter what the color or national origin of the children in the school.

These considerations were not apparent in 1963. The state's policy directive was announced at a time when the civil rights struggle was the major topic of the nation. The Federal Government had won a show-down with the segregationist governor of Alabama, George Wallace, forcing the integration of the University of Alabama. Nonviolent Ne-groes took to the streets in Birmingham to press for equality and were met with police violence. And the brutal assassination of the NAACP leader in Mississippi, Medgar Evers, shocked decent Americans of every race.

As the national civil rights movement concentrated its energies on a massive demonstration in Washington in late August 1963, the school in-tegration activists in New York City launched a new, militant thrust. Civil rights demonstrations were a daily occurrence. The national direc-tor of the Congress of Racial Equality (CORE), known for its direct-action techniques, declared that the integration of the New York schools would be CORE's major target. A new organization, the Harlem Parents Committee (HPC), emerged, which had little patience for promises and piecemeal compromises. HPC was led by Isaiah Robinson, a newcomer to the leadership echelon of the movement. In 1971 he would become the first black president of the New York City Board of Education.

That summer, civil rights leaders struck directly at the neighbor-hood-school concept. They threatened to boycott the schools unless the Board of Education prepared a timetable for integration by September 1, accompanying the city's report on racial imbalance to Commissioner Allen. To satisfy Allen's standard of racial balance, only massive and in-vountary pupil transfers would suffice. Open Enrollment and Higher Horizons, in this perspective, were tokenism. So, too, integration spokes-men denounced the construction of new schools in the ghetto to replace obsolete buildings or to relieve overcrowding as a "separate but equal" policy. (Very often, however, at hearings of city agencies and the Board of Education, parents from old or overcrowded schools testified *for* con-struction, preferring immediate relief of oppressive conditions in their neighborhood schools.)

To head off a September boycott, Gross began meeting with civil rights leaders. The chief spokesman of the school integration movement was Galamison, who, as president of the City-wide Coordinating Com-mittee for Integrated Schools, represented the NAACP, CORE, the New York Urban League, the Harlem Parents' Committee, and his own Par-ents' Workshop for Equality. Instead of a timetable for city-wide inte-gration, Gross offered the rights leaders a "free transfer" plan, which would permit any child in a racially imbalanced school to transfer to a school where space was available. Gross' proposal was flatly rejected.

Robinson walked out of the meeting, saying he had heard only "lots of tired phrases of gradualism and tokenism." Further meetings brought no progress. The discussions floundered over two issues: involuntary transfers and a specific school-by-school timetable for integration. Without a commitment to the former, the latter was clearly impossible. There could be no integration of ghetto schools unless white children were bussed into the ghetto, and Gross refused to agree to involuntary bussing. Civil rights leaders stepped up their plans for a school boycott in September.[2]

When Gross announced the Board of Education's response to Commissioner Allen's directive on racial imbalance, civil rights leaders were unimpressed. The Board expanded Open Enrollment into a "free choice transfer policy," increasing the number of students eligible to transfer out of their neighborhood schools. Its report described rezoning to spur integration; monthly meetings between Superintendent Gross and rights leaders; intensified recruiting of Negro and Puerto Rican teachers; improved education in ghetto schools; increased use of curriculum materials dealing with contributions of Negroes and Puerto Ricans to American history; and possible introduction of school pairing along the lines of the Princeton Plan for September 1964.° Superintendent Gross called New York's plan "the most comprehensive effort to achieve maximum integration . . . of any city school system in the country." It represented "all possible steps we have been able to devise, short of the compulsory interchange of Negro and white students between distant communities." State Commissioner Allen, who doubted the desirability and feasibility of massive pupil transfers, called the city's response "an imaginative and constructive program of action." [3]

School integration leaders considered the Board's program too little, too late, and continued to plan a large school boycott in September. Concurrent with these events in New York during the summer of 1963, the civil rights movement was strengthened and unified by its March on Washington on August 28, attracting over a quarter of a million blacks and whites. The movement at this point was dedicated to nonviolence and interracial cooperation; its theme song, "We Shall Overcome," characterized the spirit of expectancy and righteousness which gave the movement moral force. In this atmosphere, New York's rights leaders were not inclined to make a political or pragmatic settlement of what they held to be a moral issue.

° The Princeton Plan paired two schools, one which was predominantly white with one which was predominantly black; all children from kindergarten to third grade attended one school, and all children from fourth through sixth grade attended the other.

Given the emotional momentum of the rights movement, school officials feared that the threatened boycott would become a reality. One rights group had been expelled by police from City Hall after a forty-four-day sit-in; another group chained themselves to cranes in Queens to demand construction jobs for blacks and Puerto Ricans. Within days of the scheduled opening of school, the rights leaders agreed to cancel their boycott, after a meeting with school officials arranged by Stanley Lowell, chairman of the City Commission on Human Rights. Exactly what the Board of Education agreed to was later cast into doubt, but the rights groups and Lowell thought that the Board agreed to prepare a specific timetable for integration of every school district and to make this plan available by December 1.

The boycott, for the moment at least, was off. But when the white reaction to the agreement set in, the level of conflict around the issue escalated dramatically.

The very day after the pact was announced, a ruling from the New York State Supreme Court cast doubt on the legality of any forced integration. White parents in Brooklyn had gone to court to prevent their children from being zoned out of one school and into another to promote integration; they contended that their children were being used to satisfy a racial quota and were discriminated against as whites. The court held that zoning a school for racial reasons was a violation of the State Education Law, which stated that "no person shall be refused admission into or excluded from any public school in the state of New York on account of race, creed, color or national origin." [4]

Organizations began to spring up around the city, especially in Brooklyn and Queens, to resist any forced transfers and to fight for the neighborhood-school concept. Most of the groups called themselves "Parents and Taxpayers" (PAT), and a city-wide council was formed, which claimed to represent 300,000 members.

At the same time, a new rights group called EQUAL was formed. It was small, predominantly white and committed to total integration. EQUAL attacked the free choice transfer policy, because it put the burden of integration on parents instead of the Board of Education. EQUAL eventually surpassed even some black organizations in its militant advocacy of total racial balance.

THE DYNAMICS OF PROTEST POLITICS makes compromise appear as weakness. Leadership either moves to an extreme position or is replaced by new leaders who are more militant and less willing to compromise. Those who accept compromise are made to appear compromised themselves and are quickly eclipsed in media coverage and in reality by others. The presence of the mass media is critical to this process, because the media have the power to christen a man a "spokesman" or "a leader," whether he has a following or not; the interest of the media in providing good copy or good entertainment causes reporters to encourage conflict and to seek out provocative spokesmen. Old-style leadership was based on winning an elective position or heading an organization with mass membership; protest leadership by-passed the old sources of legitimacy and went instead to those with the appropriate media image.

This phenomenon had important consequences for the politics of the public schools. Any integration proposals from the Board which offered less than a city-wide timetable for racial balance were unacceptable. The leader who compromised on a limited plan might have looked like an "Uncle Tom" and opened an opportunity for more militant leadership. But at this moment in time, to continue to press for total racial balance on the premise that any school which was less than 50 percent mainland white was inferior required a determined avoidance of reality. For the reality was that the New York public schools were very soon to be majority black and Puerto Rican; in 1966, the minority enrollment would pass the 50 percent mark. After that date, it would be impossible to have a white majority in every school. Yet the leadership of the integration movement continued to demand the impossible, to reject compromise, and to insist that school officials make promises which could not be fulfilled.

Dr. Kenneth Clark stayed aloof from the competition for media-ordained leadership. While other civil rights leaders waited hopefully, if skeptically, for the timetable for city-wide integration which the Board had agreed to deliver by December 1, Clark spoke out against forced transfers. Though Clark never ceased to believe in the superiority of integrated education, he recognized the political dynamite inherent in the bussing issue. He predicted that middle-class parents, white and black, would not permit their children to be bussed into ghetto schools because they were "clearly and woefully inferior." He warned that efforts to force long-distance bussing would occur "only under conditions of intense and prolonged protest" which would "clearly lead to a disruption of the educational process and will not affect positively the education of any child." He called such suggestions "unrealistic, irrelevant, emotional and diversionary." In Clark's view, later expressed in his widely read *Dark Ghetto*, the central problem in the ghetto schools was the teachers' assumption that black and Puerto Rican children cannot learn because of their slum background. He saw this as a self-fulfilling prophecy, which led teachers to incorporate their low expectations into their classroom behavior. He called for a program "which makes effective teaching the exclusive concern of these schools" and for a "thorough reorganization of these schools in terms of personnel, curriculum, teaching methods, supervision and accountability." Teaching conditions in Harlem, he maintained, reflected a "normative, consistent pattern of criminal neglect." [5]

The promised report from the Board of Education was released on December 9, 1963. It described the start of the free-transfer plan and listed zoning changes for the following fall. There was no timetable for city-wide integration. Superintendent Gross called the record of integration in the New York schools "one of trail-blazing achievement." He conceded that the report offered no panaceas, but was nonetheless an "honest and practical attempt" to deal with the problem.[6]

The rights leaders were outraged. They had called off their boycott in exchange for what they thought was a firm commitment to produce a specific timetable for full integration; they received instead a progress report on a program they had rejected three months before. The paramount issue again was the Board's failure to order involuntary transfers into the ghetto. Gross acknowledged that "the things they apparently want can only be achieved by the involuntary bussing of children over a long distance. I don't think any parent, Negro or white, would stand for this. It would lead to chaos." Galamison said that while bussing was only one way of achieving integration, "anyone who talks about integration and is against bussing is not serious about the matter." He announced that the boycott was on again and would be held in February 1964.[7]

As an atmosphere of crisis intensified, Board President Max Rubin resigned and was replaced by James B. Donovan, a tough and outspoken lawyer who had been vice-president of the Board. Donovan's blunt statements moved all sides farther apart. He opposed forced bussing and favored the neighborhood school. He ridiculed the rights organizations for making impossible demands. He called them "so-called civil rights groups" and said that the Board was doing more for integration than those who were "screaming 'freedom now' and 'integration now.' . . . These are just jingoistic slogans that so far as I'm concerned haven't aided any program. I again point out to you that you cannot get from these people a constructive, practical plan. . . . You simply cannot put one million children on wheels and send them all over the city of New York."

Galamison recognized that forced integration was politically unwelcome in white areas, but he held that it was the responsibility of the

Board of Education, not the rights groups, to develop a workable plan. He pledged that the boycott would last until the black schools were desegregated by compulsory attendance of white children: "We realize that every school can't be desegregated right now, but we want a timetable. It's my personal opinion that three years is a reasonable period of time in which to do it. There's no reason why junior high schools couldn't be integrated by next September. The following year they could integrate the elementary schools. Finally, they could finish up the whole school system." [8]

The trouble with Galamison's proposal, which seemed reasonable on the surface, was that there were not enough white pupils to go around. Within two years, non-Puerto Rican whites would be a minority in the system. Besides, residential patterns made total integration a logistical monstrosity. If whites were brought into ghetto schools from surrounding areas, then the fringe-area schools would need to bring in whites from outlying districts; and to make room in crowded ghetto schools, black and Puerto Rican students would have to be transported, not to the fringes (which were already balanced or in danger of "tipping"), but to the outlying districts. Theoretically, all "segregated" schools (i.e., schools that were more than 50 percent black and Puerto Rican) could have been integrated with a 50 percent white enrollment, but only by cross-borough, long-distance bussing *and* a retention of the white pupil population. The first requisite was politically unpopular; there was no way of assuring or controlling the second.

Galamison invited Bayard Rustin, organizer of the successful March on Washington, to organize the school boycott. Under the aegis of Galamison's City-wide Committee for Integrated Schools, which included all the major rights groups, Rustin won the support of many churches and community groups, particularly in ghetto areas.

The Board of Education attempted to head off the boycott. A month after Gross termed integration in the New York schools a "history of achievement," members of the Board of Education let it be known that they were disappointed in Gross' integration plan because it did not go far enough. Gross, hospitalized with pneumonia at the time, was incommunicado. At a public hearing Board President James Donovan stressed that the plan which was under attack by rights groups was Gross' plan, not the Board's. The Board, he promised, was drafting its own proposal, which would be ready by February 1 and would include "a specific timetable for every specific item where it is possible." By disowning Gross' plan, the Board revealed its loss of confidence in his leadership and diminished his future effectiveness.

Fred Hechinger, of the *New York Times,* warned that the Board's plan might promise more than Gross could deliver; or alternately, if it failed to promise enough, might further infuriate the boycotters. Hechinger wrote that, despite slogans to the contrary, "almost every private conversation with educational experts on city and state levels reveals the generally felt, but never publicly stated, belief that integration of the schools in such large areas as Harlem and the Bedford-Stuyvesant section of Brooklyn is impossible, either now or on any future timetable."

The Board's plan was revealed on television by Board President James Donovan and Deputy Superintendent Bernard Donovan. (Gross was out of the state, recuperating from his recent pneumonia.) It was a three-year plan to improve the racial balance at 30 of the city's 165 predominantly Negro-Puerto Rican schools, end part-time schooling, and upgrade education in ghetto schools. The plan specifically rejected the demand to produce racial balance in all the city's schools as "not feasible. . . . The mandatory transportation of great numbers of children over wide areas would inevitably result in educational chaos." The thirty ghetto schools affected by the plan included twenty elementary schools, which were to be paired with predominantly white schools, and ten junior high schools, whose elementary-school feeders would be shifted.[9]

On paper, the plan looked like a timid gesture; it was certainly not a city-wide solution. In practice, even this limited goal proved impossible to implement, since blacks rejected it as too little and defenders of the neighborhood school rejected it as too much.

The immediate purpose of the plan was to cancel the boycott, but it was too late. Its momentum was too strong to be halted by anything less than an unequivocal commitment to balance racially every school in the city.

The white liberal leadership, which had always been sympathetic to the rights organizations, was in a quandary. The *Times* called the boycott "tragically misguided" and "pointless." PEA criticized the boycott and its major goal, cross-bussing, which it believed would drive whites out of the system. Mayor Wagner tried to remain uninvolved and issued an ambiguous, carefully worded statement: "No child should be offered as a sacrifice on the altar of either discrimination or antidiscrimination. We must not sacrifice the learning of one so another may learn; but no child has an inborn right or claim to advantage. . . . We must give every assistance to the weak without handicapping or holding back the strong. There is need for an exquisite balance, a balance which cannot be finally achieved by zone lines or numbers, but only by endless zeal and effort, and by deep understanding." Gross, still out of town, remained silent.[10]

No matter what the trepidations of the white liberal community, the boycott was on. On February 3, 1964, pickets marched at 300 of the city's 860 public schools; 44.8 percent (460,000) of the city's pupils did not attend school, compared to the usual absentee rate of 10 percent (100,000). Ghetto schools were most affected by the boycott. About 3,500 demonstrators, including many children, marched to the headquarters of the Board of Education, where they demanded instant integration and the ouster of Board President James Donovan.

James Donovan belittled the boycott as "a fizzle. . . . All these people proved, is how easy it is to get children to take a holiday instead of going to school. They also showed that parents could be frightened into keeping their children at home by a campaign of intimidation and threats of possible violence." He refused the City Commission on Human Rights' invitation to meet with civil rights leaders, unless they were prepared to present "specific, constructive proposals." The citywide PAT organization put the Board on notice that it would sue to prevent forced pupil transfers and to preserve neighborhood schools.[11]

The Board, despairing of developing a feasible plan which could satisfy Galamison's group, asked State Commissioner Allen to review its integration plan and to make recommendations which would correct the growing racial imbalance in the schools. Allen agreed and turned the problems over to his Advisory Commission on Human Relations and Community Tensions, composed of Dr. Kenneth Clark, Rabbi Judah Cahn, and Teachers College President John Fischer.

Repercussions from the boycott were still being sorted out when Galamison announced that a second boycott would be held on a date selected to "penalize the Board of Education economically" (state aid payments were based on periodic reports of pupil attendance). Galamison had said before the first boycott that he would rather see the school system "destroyed—maybe it has run its course anyway," than to permit it to perpetuate racial segregation.[12]

The boycott and the threat of another one caused several relationships to come unglued. The New York Times urged James Donovan not to seek re-election as president of the Board. Though the Board reaffirmed its confidence in both Donovan and the still silent Calvin Gross, Board members privately critized both and acknowledged that their days in office were numbered. A crisis also occurred in the traditional alliance between Negro integrationists and white liberals. Black leaders were angered by the white liberals' ambivalence about the boycott. On the other side, many whites who had previously backed Negro demands did not agree with the City-wide Committee's campaign for integrating every school in the city by massive pupil exchanges.

Galamison's announcement of a second boycott led to a deep split within the rights movement itself. Some groups which had supported the first boycott withdrew from the City-wide Committee, rejecting both the second boycott and Galamison's leadership. Most of those who pulled out were affiliated with national civil rights organizations, whose doubts reflected their hardheaded assessment of the national political situation. National civil rights leaders were working with President Lyndon Johnson to get a civil rights bill passed by the same Congress which had failed to pass it for President John Kennedy; and they were looking ahead to the 1964 party conventions. They feared that pushing too hard would not only cost them white allies but stir up the conservative white backlash that Alabama's Governor George Wallace was already exploiting in Democratic presidential primaries.

Rustin, the organizer of the first boycott, refused to participate in the second. He believed that a second boycott would be a tactical error; he knew that it was bound to be less successful than the first, and that diminishing returns might wipe out the psychological advantages already attained. After the March on Washington, he felt that the movement had carried protest to its logical conclusion and that blacks now had to enter a new phase, the acquisition of political and economic power. In hopes of re-creating the liberal coalition of the New Deal era, Rustin and labor leader A. Philip Randolph formed the A. Philip Randolph Institute to promote coordination among blacks, unions, intellectuals, and socially progressive businessmen.

THE SITUATION in the city continued to deteriorate as Galamison, on the one side, and PAT, on the other, stepped up their activities. Galamison declared that not only a second, but a third boycott would be held, though only the Harlem Parents Committee, a few chapters of CORE, and his own Parents' Workshop for Equality supported a second boycott. At the same time, the Board's proposal to pair twenty white elementary schools with twenty ghetto schools filled the PAT with fighting spirit. PAT members picketed neighborhood schools, organized local protest rallies, and planned a mass rally at City Hall.

The intensity of the opposition to bussing and pairings endangered even the limited plan which the Board had agreed to implement. Board President Donovan believed that there was nothing left to concede to Galamison, nothing the Board could do to avoid the second boycott. Donovan feared that Galamison's stance gave fuel to PAT. He went to

Bedford-Stuyvesant to plead with leaders of local parents' associations. He told them that the schools were trying to "lead the way," but that integration of the schools depended on integrated housing and better job opportunities. Trying to dissuade them from supporting the new boycott, he said: "You have to help us. Do not play into the hands of those who have started 32 law suits to stop our integration program." He was hooted and jeered by members of Galamison's organization.[13]

On the eve of the PAT mass rally and the second school boycott, the Appellate Division of the State Supreme Court reversed a lower court decision of the previous September which had denied the Board the right to rezone schools for the purpose of integration. The court simultaneously upheld the right of pupils to attend schools nearest their home. The court ruled that the Supreme Court's 1954 decision forbade segregation by law, but did not require integration:

> It does not mean that, regardless of school zones, or the residence of Negro or white students, there is an absolute, affirmative duty on the part of every board of education to integrate the races so as to bring about as nearly as possible racial balance in each of the schools under its supervision. . . . It does not mean, for example, that white children who live in non-contiguous or outlying areas must be "bussed" into a Negro area in order to desegregate a Negro school. . . . . Each child has the right to attend only the public school in the zone or district in which he resides; and, in our opinion, this means the school nearest to his home.[14]

Both sides hailed the decision. Galamison called it "very exciting news," because it overturned a decision which made zoning for integration impossible. PAT was pleased by the court's reaffirmation of the principle of the neighborhood school. So long as this decision stood, the only elementary and junior high school integration which was possible was in fringe areas, which were rapidly retreating as ghettos expanded.

A few days before the second rights boycott, PAT held its mass rally at City Hall to demonstrate support for the neighborhood school. About 15,000 people turned out to march. They carried signs such as: "Can a bus bring a sick child home?"; "Have child—Won't travel"; "Princeton Plan in the garbage can."

On March 15, 1964, the second boycott took place. It was half the size of the first one: 26 percent of the register was absent, 168,000 more than usual. Galamison told a crowd of 2,500, mostly teenagers, assembled at Board headquarters: "If we have to come down here again, we are going to take this building." [15]

Galamison was delighted with the results of the boycott, accomplished without the support of any of the major civil rights organizations. It established him as a leader on his own. The large and powerful organizations lost their ability to control him and, in the bargain, ended up looking as though they had been bought off by a puny, piecemeal program. In fact, the school integration movement was moving steadily closer to a complete collapse, having exhausted any possibility of settlement which did not include forced transfers. Galamison's tactics made partial successes and small compromises unacceptable by ridiculing anything less than total integration.

The Board's integration plan was left with few friends. It was assailed by the PAT and ridiculed by Galamison. Even committed integrationists on the Board lost faith in pairing after a delegation from the Board visited paired schools in Greenburgh, New York, and discovered that Negro students were plagued by the same learning handicaps as in New York City. They came to believe that pairing was just a "gimmick" which had no impact on the educational problems of Negro and Puerto Rican children.[16]

The Board began to backtrack under pressure from white parent groups and the overwhelmingly white organization of school supervisors. School officials quietly agreed not to try to push through pairings where the local school board, district superintendent, and parents' associations opposed them. Eventually, of the twenty pairings publicly announced by the Board in January 1964, only four pairings actually took place.

While the Board's integration plan was being eroded during the spring of 1964, Galamison and the militant Brooklyn CORE (which had been expelled by the national organization for its excessive zeal) engaged in a series of attention-getting demonstrations, blocking truck deliveries to the World's Fair, trying to stall traffic on opening day of the Fair, threatening to call for water-wasting during a drought. They managed to raise the level of racial tensions and to demonstrate that, so far as the integration of the public schools was concerned, a dismal impasse had been reached.

CHAPTER 25

The Allen Report

Just when the school integration movement seemed to be most dispirited and confused about the future, the Allen report was made public. The movement got an immediate revivifying shot in the arm, but the medicine proved to be hallucinogenic.

The Allen report, named for State Commissioner of Education James Allen, Jr., was prepared at the request of the Board of Education after the first civil rights boycott in February. Authored by Allen's Advisory Commission on Human Relations and Community Tensions, the report was released in May 1964. It excoriated the Board of Education for its ineffectual attempts at integrating the schools and held that desegregation could be advanced by reorganizing the school system.[1]

The Allen report was at once hailed as a battle-plan for the civil rights movement. An NAACP aide said, "It really is an excellent report. No one has ever attempted so comprehensive a plan for desegregation and good schools. It is a model for every Northern city." Galamison was delighted by the report and called it "a forceful, extremely imaginative approach, a giant step in the right direction. It is far and away the best thing we have had from this kind of quarter." Galamison hoped that the report would become a rallying point which would pull the civil rights movement back together again. Gross announced he would meet with civil rights leaders and "move rapidly" to implement the Allen report.[2]

Those rights activists who lavished praise on the Allen report were responding above all to the report's harsh criticism of the Board of Education. The report reviewed the Board's past integration efforts as well as its current proposals and found all of them insufficient. It dismissed Open Enrollment and the Free Choice Transfer policy (too dependent on voluntary choice by Negro and Puerto Rican parents), zoning (too limited in impact), the school building program (too many schools built in the ghetto), junior high school feeder changes (too limited in impact), and pairing of schools (even if all the proposed pairings were adopted, segregation would be reduced by only 1 percent). While the report granted that New York's public schools were more integrated than those of Detroit, Philadelphia, and Chicago, and that the increase in segregated schools "cannot be attributed to intentional policies of the Board of Education in New York City," it castigated the Board for failing to hold the line against increasing segregation: "We must conclude that nothing undertaken by the New York City Board of Education since 1954, and nothing proposed since 1963, has contributed or will contribute in any meaningful degree to desegregating the public schools of the city. Each past effort, each current plan, and each projected proposal is either not aimed at reducing segregation or is developed in too limited a fashion to stimulate even slight progress toward desegregation."

But the Allen report was not equal to the encomiums heaped on it. Its very title, "Desegregating the Public Schools of New York City," contributed to the widespread but erroneous conviction that it was a plan to desegregate the city's schools. It was not. In fact, while stating forcefully the educational and moral virtues of integrated schools, the Allen report vividly detailed the demographic and sociological factors which made city-wide integration unattainable.

Its statistical analyses projected that by 1980 the Negro and Puerto Rican enrollment in the city schools would reach 70 to 75 percent of the total. It conceded, "most painfully, overcrowding in the residential ghettos makes many efforts to desegregate futile, for distance and population concentration work against nearly all forms of redistribution. Many features of this population process are beyond the control of either the Board of Education or any other agency of city government." The commission recognized that "total desegregation of all schools . . . is simply not attainable in the foreseeable future and neither planning nor pressure can change that fact." It noted, too, that the relative increase in the number of Negro and Puerto Rican pupils "helps to *precipitate* movement out of the public schools and out of the city" by white pupils.[3]

Having acknowledged the Board's powerlessness to reverse these segregating trends, the commissioners nonetheless stated that they were "persuaded by the fact, however, that *there are steps which could be taken by the Board to desegregate the public schools.*" Lest anyone jump to the conclusion that the report intended to advocate forced transfers, the report immediately added:

We believe that part of the educational desirability of any plan at this time must be its *mutual* acceptance by both minority groups and whites.

It should be obvious, but does not always appear to be, that integration is impossible without white pupils. No plan can be acceptable, therefore, which increased the movement of white pupils out of the public schools. Neither is it acceptable, however, unless it contributes to desegregation.[4]

This was a large order indeed: to devise an integration plan which was (1) acceptable to both minority groups and whites; and (2) effective in desegregating a system in which Negroes and Puerto Ricans would soon be the majority.

The report did not even mention two courses of action which would have increased the number of white pupils available: either the public schools' absorption of private and parochial schools in the city, which had about 400,000 pupils (less than 10 percent of them black), or the creation of a metropolitan school district, merging the New York City school system with overwhelmingly white suburban school districts. Politically, neither of these proposals seemed feasible, but they were at least capable of producing integration, which the Allen plan itself was not.

The Allen plan, in essence, was a reorganization of the grades from 6 years (elementary)–3 years (junior high school)–3 years (high school) to 4–4–4. The first four years in the primary school would be devoted to achieving quality education; many of the primary schools, it was conceded, would be racially segregated, since they would be "in the neighborhood and as close as possible to the homes of the children." This recommendation implicitly rejected the assumption of the civil rights movement and Commissioner Allen that a segregated school is by definition an inferior school.

The greatest movement toward integration was supposed to come in the new middle schools. The report urged the prompt elimination of all twenty-five segregated Negro and Puerto Rican junior high schools.* "Some" middle schools would still be segregated, but the report predicted that "their ethnic mixtures, especially in the ghetto fringe areas, can approach desegregation" because their students would be drawn from both predominantly white and predominantly Negro and Puerto Rican primary schools.

At the high school level, the report called on the Board of Education to "introduce a large scale program of construction and develop-

---

* In 1963 there were one hundred and thirty-five junior high schools; twenty-five were 90 percent or more Negro and Puerto Rican, twenty-nine were 90 percent or more white, and eighty-one were middle-range, having at least 10 percent of both groups.

ment of *four year comprehensive high schools.*" It urged that most of the new high schools be built in predominantly white sections of the Bronx, Brooklyn, and Queens, which would assure their desegregated status in the next decade. The comprehensive high school, combining academic and vocational programs under one roof, was advocated as an educational reform, aimed at eliminating the socially stigmatized vocational high school. Since the vocational high schools then had a bare majority of white students and the academic high schools were more than 75 percent white, combining the two also promoted integration.[5]

The report offered an array of other recommendations, including the organization of educational clusters (a middle school and its primary feeders) under a single administrator with a high degree of autonomy; the development of educational parks, where several schools are built on a single site, drawing from a large and heterogeneous pupil population; preprimary programs; improved recruitment and advancement of minority teachers; and an increase by one-third of the system's expenditure per pupil.

Though there were highly placed supervisors at the Board of Education who believed that a grand scheme was being dropped on their laps with little regard for the critical details of implementation, the Board directed Gross to draw up a plan for adoption of the report.

Only Fred Hechinger of the *Times* viewed the report skeptically. He believed that the commission had tried to appease both the rights groups and the PAT groups by using the Board as a scapegoat: "The Board's integration proposals are made to appear ridiculous; even the policy of permitting Negro pupils to transfer voluntarily out of 'ghetto' schools—a policy often hailed in the past as revolutionary—is harshly dismissed in the report." He called the report a package deal which included everybody's favorite educational experiment, as well as some distant visions, which could be debated for years without any expenditure of dollars. Hechinger saw the Allen report not as a revolutionary step forward, but as a demonstration of passing the buck. The report's call for an additional $250 million for the school system came at a time when the mayor had just cut the Board of Education's budget, the governor had just turned down a request from Commissioner Allen for $10 million for the city's integration program, and federal money for education was still nonexistent. The net result was that expectations had been raised by an agency which had no responsibility to implement its proposals, and that the problems of implementation could heighten frustration and conflict.[6]

Many integrationists seized on the 4–4–4 scheme as the yardstick against which all subsequent plans of the Board would be measured. In fact, the Allen report undermined support for most of the programs that integrationists had been fighting for. While accepting the neighborhood primary school, the report rejected bussing, enforced transfers, pairings, or any program that was not *mutually* acceptable to both minority groups and whites. But under the mistaken impression that the Allen report was a blueprint for truly integrating the school system, rights groups picketed the homes of Board members with signs reading "All the way with Mr. A."

The Allen plan became the backdrop for a series of highly emotional confrontations, based largely on misunderstandings of its implications. Since the Allen panel had said that the schools could be integrated, PAT forces threw themselves into an all-out campaign against long-distance bussing and compulsory transfers. Instead of being placated by the report's endorsement of the neighborhood primary school, PAT leaders focused only on its recommendation to transfer fifth and sixth graders away from their neighborhoods to middle schools; any compulsory transfer was viewed by them as the foot-in-the-door to wholesale bussing, no matter how many assurances were offered to the contrary.

The PAT groups saw in the Allen report only what they feared; the rights groups saw only what they wanted. Galamison promised that he would back the Allen plan and added his own recommendations: that no middle school be segregated and that the neighborhood primaries be paired for integration wherever possible. He disregarded the fact that the Allen plan rejected pairing or any other plan that was not mutually acceptable. Neither the report nor any of the rights leaders tried to explain how integration could be secured in a school system whose Negro and Puerto Rican enrollment was expected to reach 70 to 75 percent by 1980.

Soon after the Allen plan was released, Gross presented a plan for "quality integrated education" which involved the transfer of 40,000 students to promote integration. The plan was approved by the Board of Education, but attacked by both the PAT and civil rights groups. Rosemary Gunning, leader of the city-wide PAT, said: "We are shocked by this plan. It will not only destroy the rights of many children to go to their traditional neighborhood schools in September, but obviously envisions wholesale involuntary transfers in the future." Rights leaders called the plan "a colossal deception" and a "fraud" which was "diametrically opposed to the spirit and intent of the Allen recommendation." They had wanted the Board to take dramatic action, like closing down all junior high schools which could not be integrated. Within the system there was also resistance. High school principals opposed the shift of ninth graders to their already overcrowded schools; junior high school principals objected to the elimination of the junior high school in favor of middle schools.[7]

Faced with opposition from within and without, Gross tried to win over allies for his stewardship of the schools. While the storm was breaking over the integration plan, he pledged his support to a project for improving slum schools; the proposal was prepared by school professionals and backed by the UFT. It embodied the latest and best educational thinking: small classes, prekindergarten classes, team teaching, educational television, nongraded classes, intensive professional support services, and special programs for weekends and summers. The program —called More Effective Schools (MES)—started in the fall in fewer than a dozen elementary schools.

Simultaneously, Gross began meeting with rights leaders. They agreed to suspend demonstrations and to work together to plan for broad-scale integration in September 1965. In mid-June 1964, the rights groups switched their position on the integration plan and decided to support it.

The leaders of PAT called a boycott to demonstrate against long-distance bussing at the opening of the fall term in 1964. Even though such bussing had not been proposed either in the Allen report or in the Gross integration plan, PAT's fears were not to be quelled. On the first day of the new term, 27 percent of the total enrollment was absent. City officials managed to avoid making any concessions to PAT, but PAT pressure was strong enough to evoke statements of opposition to long-distance bussing from both Senator Kenneth Keating and his election opponent, Robert F. Kennedy.

That fall, the brief *détente* between Gross and the rights leaders crumbled. He could not satisfy their demands. Among other things, he could not or would not stop the construction of new schools in the ghetto. Galamison warned that he would engage in civil disobedience to step up the pace of integration. In the first three months of 1965, Galamison led sporadic boycotts at segregated junior high schools and at special schools for maladjusted students; the boycott, opposed in several instances by parents, lost momentum and dwindled out.[8]

In March 1965, the Board of Education asked for Gross' resignation. Most observers believed that the top staff at the Board of Education resented the importation of an outsider and failed to cooperate with him. In any event, the Board made a scapegoat of Gross, just as others had made a scapegoat of the Board, for not providing strong leadership on integration and administrative decentralization.

The Board of Education appointed Bernard E. Donovan, Gross' right-hand man and deputy, to succeed him. Donovan, a member of the system since 1930, promptly released the integration plans prepared by Gross for September 1965. It became apparent why Gross lost the support of the civil rights groups: the plan included no compulsory assignment of whites to Negro schools, no new pairings, and no junior high school zoning changes. Based on the recommendations of the Allen plan, it endorsed the four-year comprehensive high school and authorized large-scale shifts of sixth graders into junior high schools and ninth graders into high schools.

The plan was a notable step towards implementing the recommendations of the Allen report; but it did little for integration. Frustrated rights spokesmen complained that, "out of the shreds and tatters and misinterpretations of the Allen report, the Board has come up with a program calculated to produce a maximum of confusion of activity and a minimum of action. Never have so many changes been contemplated to affect the integration of so few." The same criticism might well have been directed at the 4–4–4 plan itself.[9]

The Board then proclaimed its intention to implement the Allen recommendations. It would achieve maximum integration by converting the system to 4–4–4. All 138 junior high schools would be abolished by 1973; the changeover to the intermediate school would begin in September 1966. This new forcefulness may have pleased civil rights leaders, but there was a fantasylike quality to the Board's new sense of determination in meeting the pressures for integration. For, total desegregation of the New York City public schools, in the words of the Allen report, "is simply not attainable in the foreseeable future." Unhappily, the Board of Education found it more comfortable to pretend that it was.

CHAPTER 26

Enter Lindsay

DURING 1965, lethargy settled over the civil rights movement, nationally as well as locally. After the lifting of a self-imposed ban on demonstrations during the 1964 presidential campaign, the civil rights movement never regained its momentum and élan. Something important had changed. Working closely with Negro leaders, President Lyndon B. Johnson had won passage of major legislation to benefit minority groups: a sweeping civil rights act, an economic opportunity bill to attack poverty, a voting rights act which would cause the registration of hundreds of thousands of disenfranchised Southern blacks. Despite these accomplishments, discontent in ghetto communities was on the rise. The spirit of universal brotherhood which bound the civil rights movement together was palpably waning. Perhaps the revolution of rising expectations had overtaken events with demands that could not be easily satisfied. Perhaps the accomplishments of a decade of bloody struggle were too abstract; urban ghetto riots in the summers of 1964 and 1965 shocked the nation into the realization that poor blacks were angry and dissatisfied with the pace of visible change.

The new mood found form in the many organizations which sprang up to wage the "war on poverty" started by the Johnson administration in 1964. The community-action component of the poverty program gave money to organizations which were developed and operated with the "maximum feasible participation" of the poor. The motivating principle of the program, that a major cause of poverty is powerlessness, was developed by Richard Cloward of the Columbia University School of Social Work, who was a founder of New York's pioneer community-action program, Mobilization for Youth. Acting on this assumption, the Office of Economic Opportunity funded community organizations controlled by neighborhood residents; community-action programs gave the poor a voice in shaping the decisions that affected their lives, as well as control of a payroll. It was expected that the poor would know which problems oppressed them most and would devise programs which were most needed in their communities. It was a good idea which ran into problems immediately. Where there were elections for the controlling board, participation was low, usually less than 5 percent of those eligible to vote; in some instances, social-work professionals manipulated unsophisticated boards; payrolls sometimes became patronage; some communities were split by political battles for control of the program; and in many mixed communities, ethnic groups vied for power.

Community-action programs became for poor blacks and Puerto Ricans what Tammany Hall once had been for poor Irish and Italians. Tammany Hall acted on behalf of the immigrant to help him where the government would not; the local poverty program acted on behalf of the poor black and Puerto Rican, in place of a government which had become too remote to reach him and too bureaucratized to care about him. Because the program controlled funds, it became a source of power and patronage. For the ambitious young person, it was the locus of power in the community, a path to leadership not strewn with civil service rules and regulations. And at the same time, it had a vast potential for good works where the need was great.

While the civil rights movement had been constantly urged to act responsibly (lest their actions reflect badly on the Negro race), incentives were built into the poverty program to do the opposite. The more trouble that swirled around a community, the more funds there were channeled into its community-action program. Since one of the functions of the poverty program was to identify problems and draw attention to them, the tempo of demonstrations picked up as the program gathered momentum. New Yorkers had lived through years of civil rights demonstrations without trepidation, since they were usually characterized by nonviolence and assertions of goodwill. But demonstrations organized by antipoverty workers took place with a different tone, for in the background was the warning of Watts, California, scene of angry rioting and looting. New Yorkers began to experience a sense of foreboding and malaise and to wonder whether New York was doing enough to stave off trouble in its ghettos.

The time-for-a-change theme played a significant role in breaking the Democrats' hold on the mayoralty in 1965. The Democrats, in power continuously since 1942, nominated City Controller Abraham Beame as their mayoral candidate. The Republican and Liberal parties combined behind the dashing, handsome forty-one-year-old congressman from the Silk Stocking district, John Vliet Lindsay. A graduate of the St. Paul's

School in Concord, New Hampshire, Yale College, and Yale Law School, Lindsay was a Protestant and a Republican in a city that was neither. New York was traditionally Democratic, but Lindsay had a well-financed campaign and strong newspaper support. Helped along by a third-party candidate and defections from the Democractic column, Lindsay narrowly defeated Beame.

Lindsay came into office with the attitudes typical of a Republican, upper-class reformer in a Democratic city. He had little direct experience with municipal government. He assumed that the dilemma of the city was caused by a failure of imagination and a lack of character. He assumed that those who ran the city before him were time-servers and petty politicians, who resolved conflicts by making under-the-table deals. Lindsay vowed that he, unlike his predecessors, would bring the best men to government service. He promised to find new solutions which would "disturb the tired, self-perpetuating ways of the past. . . . Jobs will change. Deals will be canceled. Fat and easy lives will be disrupted." By dint of superior character and intelligence, he expected to bring efficiency, nonpartisanship, and good government to New York.[1]

What Lindsay did best was to dramatize his concern for the problems of the city. He used the media brilliantly to communicate with the citizenry. He required that a top-level commissioner be on duty twenty-four hours a day, in case of crisis; he tried to set up "little city halls" throughout the city, but was blocked by the borough presidents, who saw them as incipient political clubhouses; he initiated a regular television program; he took "walking tours" through troubled sections of the city, to show his accessibility.

Lindsay's passion for institutional reform and new approaches inevitably was directed at the educational system of the city. His campaign statements on education gave no hint of the active role that he would later play in reshaping the system. In his position paper on education, candidate Lindsay carefully avoided controversy, endorsing educational parks (which pleased integration leaders and educational reformers) and opposing "the bussing of young children out of their neighborhoods in order to achieve a better ethnic balance" (which pleased the neighborhood-school forces).[2]

Lindsay inherited a Board of Education that had been appointed by Mayor Wagner. Its president, Lloyd Garrison, was one of the most distinguished men ever to hold the office. Garrison succeeded the outspoken James Donovan as president of the Board in 1965. Garrison, a descendant of the great abolitionist William Lloyd Garrison, had been dean of the University of Wisconsin Law School, a past president of the

Urban League, and a founder of the Democratic party reform movement in New York City; his stature and liberal credentials were unassailable.

But despite the qualities of individual members of the Board, an aura of ineffectiveness and indecision hung about the school system like a fog. After its years of being the center of conflict and the object of critical studies, little was left of the system's public reputation. Six months before Lindsay took office, Martin Mayer, a well-known journalist and local school board chairman, attacked the school system in devastating terms: "No other sizable city in America spends as much as 70 per cent of what New York City spends per pupil, and, despite all the propaganda to the contrary, only a few suburbs spend more. . . . Public confidence in the school system is fearfully low and dropping: white children are leaving the city public schools at a rate of 40,000 a year." Mayer pointed out that few members of the Board or top school officials had children in the public schools. "Even worse, the Negro middle class has almost entirely disappeared. Of the Negro leaders of the integration drive . . . not one has or has had a child in a New York public school." The good teachers who entered the system during Depression days, when teaching was a highly desirable and secure job, were nearing retirement, and young teachers were fleeing to the suburbs. The staff lived, he wrote, with an expectation of failure; to the outside world, the system turned a face of "vanity, arrogance, self-absorption. . . . All change is resisted because it implies a criticism of the present—and feared, because it will be made without consulting the teachers who will have to live with the results. The hope for leadership has been disappointed so often that people have turned in upon themselves, learned to live with meaningless and fantastically detailed rule books, lost any sense of the possibilities outside the narrow structure of the hierarchy of jobs."

Mayer wrote that no one was willing "to say publicly what everyone close to the situation knew was true—that relatively little could, in fact, be done in terms of meaningful and useful integration for the majority of the city's Negro and Puerto Rican children. . . . The central fact, which everyone knows but nobody seems to wish to discuss, is that by the age of 12 the average Negro or Puerto Rican white child in the New York schools is more than two full years behind the average mainland white child in academic accomplishment." He concluded that the necessary revolution in education could not occur until the schools and teachers were liberated "from the suffocating discipline of central control." [3]

Two weeks after Mayer's article appeared, Superintendent Bernard

Donovan promulgated an administrative decentralization plan for the system. Beginning in September 1965, there would be thirty locally administered school districts, each headed by a responsible district superintendent. Central headquarters would be phased out and turned into a service agency. But when the new administrative setup took effect, there was no visible change in the way the schools operated; it led to no greater public participation, no noticeable improvement in the system's efficiency, and no real shift of authority from the central headquarters to the field.

From the perspective of the Board of Education, its administrative decentralization was a major internal reform. From the perspective of the public and the new mayor, nothing had changed. When Lindsay took office in 1966, he complained that New York spent more per pupil than any other major city without having a "demonstrably better" school system. He warned that he intended to scrutinize school spending "to insure that every dollar is spent with maximum effectiveness." At the outset of Lindsay's first term as mayor, the school system was embroiled in the complex changeover from a 6–3–3 organization to the Allen plan's 4–4–4. While school officials talked hopefully about a six- or seven-year target date for completion of the reorganization, there were pressures building in ghetto communities for some tangible evidence that a decade of protest had not been wasted.[4]

CHAPTER 27

IS 201:
An End and a Beginning

1966 WAS A PIVOTAL YEAR for the New York City schools, not in the sense of conscious decisions made, but because of the confluence of events and tides. It was the first year of the mayoralty of John Lindsay. This was significant for the Board of Education because Lindsay had not appointed the Board, and therefore took no responsibility for its actions. This circumstance meant that the Board was adrift politically and would have no anchor or strength in dealing with controversy. 1966 was also the year in which black-power ideology became a national issue and spurred a growth in black separatism and militancy. It was the year in which the black and Puerto Rican enrollment in the New York public schools at last passed the 50 percent mark, in a trend that was certain to continue in the foreseeable future. And it was the year in which the Board of Education, pursuant to the Allen plan, opened its first intermediate schools; among them was IS 201, located at 127th Street and Madison Avenue in Harlem.

WHAT HAPPENED IN THE SUMMER and early fall of 1966 at IS 201 cannot be understood without some account of the ideas and career of Preston R. Wilcox. Preston Wilcox, a charismatic black social worker, played a seminal role in the course of events at IS 201.

In the early 1960s, Wilcox was the director of the East Harlem Project, the community-action arm of two neighborhood settlement houses. He became deeply involved in the life of the East Harlem community, since the project provided staff assistance to local community groups and worked on problems of schools, tenant groups, urban renewal, voter

292

registration, and community planning. He had a passionate concern for the poor, for those whose lives were dependent on public agencies. As a community organizer, he worked to involve the poor in community action and to transmit organizational skills. He found that the poor had internalized society's low estimate of them and had little incentive to demand what was rightfully theirs, for fear of being considered troublemakers. As individuals, they were scorned and powerless; united, they might achieve political power as well as the dignity to which they were entitled.

In 1961, Wilcox prepared a report for the New York City Commission on Human Rights on the bussing of Negro children from East Harlem into schools in Yorkville, a white community to the south of East Harlem. While Negro leaders were unsuccessfully fighting to reduce the number of de facto segregated schools, Wilcox quietly raised the questions that would survive the demise of the drive for school integration. Wilcox pointed out that the schools of East Harlem were overutilized, racially segregated, and likely to remain so. He noted, "One can reasonably suspect that the schools in East Harlem wear a stigma of being inferior to schools in all-white or mixed neighborhoods." This sense of stigma was increased by integration programs like bussing, rezoning, and Open Enrollment; these policies, though well-intentioned, had the effect of dampening "student motivation, parental interest, teacher investment and community practices." He worried that the Board's integration programs drew off the "cream"—the most highly motivated children of the most highly motivated parents—from the East Harlem schools, removing leadership potential from schools that could ill afford the loss. Wilcox was convinced that, "If one can believe that a predominantly *de facto* segregated' white school can be a 'good school', then, one must believe that a *de facto* segregated' and predominantly Negro and Puerto Rican school can also be a 'good school.' If one can believe that one's potential has no ethnic dimensions—then one must behave in this manner." [1]

Wilcox strongly believed that one of the conditions for a successful school was the expectation that success was not only possible, but highly probable. The integration program, on the contrary, was explicitly based on the tenet that a predominantly black and Puerto Rican school could *never* be a good school. A leaflet distributed by Galamison's City-wide Committee for Integrated Schools stated this position succinctly: "*Can segregated schools be made equal?* Throughout history segregated public schools have been separate but *never equal*. In the Harlems throughout the country, separate minority-group public schools have always resulted in inferior education. It is on this basis that the Supreme Court handed down its historic 1954 school decision." [2] Wilcox understood that the implication of this view was that there was something inherent in the minority group which was inferior, and he could not accept this.

Wilcox joined the faculty of the Columbia School of Social Work in 1963, but continued to be involved in East Harlem antipoverty politics. In 1964, shortly after the Office of Economic Opportunity was approved by Congress, Wilcox wrote a program proposal on behalf of a network of East Harlem community organizations, requesting recognition and funds for MEND (Massive Economic Neighborhood Development) as the area's accredited antipoverty program. MEND embodied the philosophy of Richard Cloward, Wilcox's colleague, which held that powerlessness was itself a major cause of poverty. Community organizers believed that the process of organizing and participating would help to overcome the neighborhood's sense of powerlessness.

In early 1966, the impending opening of Intermediate School 201 gave Preston Wilcox the opportunity to merge his philosophy of community involvement with the educational concerns of the East Harlem community.

THE BOARD OF EDUCATION had decided in 1958 to build a new junior high school in Harlem to relieve severe overcrowding in the area. However, opposition from the school integration movement to new construction in Harlem caused a postponement of the project. By 1960, with overcrowding unabated, the plans were revived. In 1962, architectural drawings for a windowless, air-conditioned building were approved by city officials. Though the integration movement still objected, some parent-association leaders from East Harlem appeared before the City Planning Commission to urge the speedy completion of the school.[3]

Construction of the new school began in 1964. In 1965, Superintendent Donovan announced that 201 would be one of the city's first intermediate schools. Donovan and the local district superintendent were pressed by community organizations and school parent leaders for assurance that the new school would be integrated. School officials insisted that the intermediate school was part of the overall integration program outlined by the Allen report; they held that its proximity to the Triborough Bridge would make the school accessible to white students from Queens and the Bronx, who would be attracted to 201 by its superior educational program and outstanding facilities.

The local district superintendent met repeatedly with the local school board and community leaders, who sought a commitment to integrate the new school. Little came of these meetings, because school officials would not admit in advance that the school would not be integrated. There was a growing feeling among those active in the community that they should participate in planning IS 201, but a realization that their involvement was not welcome. The naming of the school was symbolic of their suspiciousness and of the Board's secretive manner, one reinforcing the other. Parent and community leaders thought that they ought to have the right to choose a name for the school. But without consultation, the Board designated the school the Arthur Schomburg School. Community leaders were outraged, not only because they had not been involved, but because they thought that 201 had been named for a German Jew instead of a black person. In fact, Schomburg was an African of Puerto Rican descent who had amassed a major collection of black history—now maintained by the New York Public Library—while working as a bank clerk.

A group of community leaders, most of them affiliated with antipoverty organizations, particularly MEND, formed an Ad Hoc Parent Council. Notable in this group were Alice Kornegay, leader of the East Harlem Triangle Community Association; Babette Edwards, a MEND worker; Helen Testamark, a worker for the United Block Association and president of the District Council of Parent Associations; Isaiah Robinson, chairman of the Harlem Parents Council and president of the parent association of JHS 139; Berlin Kelly, a MEND staff worker with a master's degree in sociology; the Reverend Robert Nichol, a young white minister from the East Harlem Protestant Parish; Dorothy Jones, who worked for the Office of Church and Race of the Protestant Council and as a consultant to the City Commission on Human Rights; and Preston Wilcox.

In January 1966, the Ad Hoc Parent Council warned Superintendent Donovan that IS 201 would not be considered just another segregated school. Isaiah Robinson wrote him:

> We have knowledge that your office with deliberate calculation intends to use Puerto Rican students on a 50–50 basis to give the impression that IS 201 is racially integrated. This tactic, sir, will attract the strongest, most militant protest from this organization and others allied with us in the struggle for real racial integration of New York City schools. This in itself will turn IS 201 into a battleground.[4]

A few weeks later, confirmation from the district superintendent that the school would open with an enrollment that was equally divided between Negroes and Puerto Ricans set off a flurry of activity. Organizations participating in the Ad Hoc Parent Council circulated a petition calling on the Board to keep 201 closed "until such time our community is satisfied that the education to be offered by the school meets the critical needs of our children." [5]

Preston Wilcox saw in the response of the Ad Hoc Parents Council an unusual opportunity; he recognized the emergence of an issue around which the community might be mobilized. He circulated a position paper called "To Be Black and To Be Successful" among the 201 activists, which proposed that 201 become an experimental school with "the responsibility for educational and administrative policy in the hands of the local community." He called for the establishment of a "School-Community Committee" selected by parents of the children in the school. This committee would have broad powers, including the authority to hire the principal and the top administrative staff. It would review reports from the school staff to the Board of Education; administer after-school and weekend programs; hold open meetings for parents and teachers; and work to bring the community into the school ("How many 'fish fries' have been held in the schools in Harlem? How about 'house-rent' parties, weddings, funerals, birthday parties and club affairs?").[6]

When the Ad Hoc Parent Council met with Superintendent Donovan in March 1966 and presented the Wilcox plan, he rejected it as unlawful. A month later, the local school board held a public meeting about IS 201, attended by Mayor Lindsay and Superintendent Donovan. Speaker after speaker denounced the public school system and IS 201. At this meeting, Preston Wilcox presented an expanded version of his proposal. He held that if the school system "can do no more than it is already doing, then the communities of the poor must be prepared to act for themselves . . . just as they must become involved in the direction of all programs set up to serve their needs." He argued that the community committee should select the principal, to ensure that he was accountable to the community and would understand the values and expectations of the community. Under these circumstances, "one can expect the school in the ghetto to become what schools in more privileged areas already are, a reflection of local interests and resources, instead of a subtle rejection of them. For the operating philosophy of the existing system is too often manifested in a conscious or unconscious belittling of the values and life styles of much of its clientele." He predicted: "This effort will be wrought with controversy and conflict, but it must be made." [7]

Wilcox's position was strikingly similar to that of Bishop Hughes over a century earlier. Contending that the interests and values of blacks were so divergent from those of the Board of Education that only the local community could administer the schooling of its children, Wilcox proposed that the school be operated with public funds by the community. Hughes had had the same objections to the Public School Society and had sought the same remedy. Hughes did not complain about the miseducation of Catholic children, for few of them attended schools of the Public School Society. It was the Society's values he objected to, and it is this point on which he and Wilcox concurred. The educational trappings of the school were not nearly as important to Hughes or Wilcox as its value system. Hughes wanted the schools to reinforce Catholicism; Wilcox wanted IS 201 to affirm the strengths of black culture. The outlook of each was based on the premise that his group had to free itself from the assimilationism of the common school concept and to control publicly funded schools of its own, for its own purposes.

When the Wilcox proposal was published in July 1966, it was accompanied by a skeptical critique by Robert Dentler of the federally funded Center for Urban Education. How would the community committee be selected? Are the objectives of parents necessarily the objectives that the school should adopt? Can a principal be "accountable to the parents" and at the same time use his independent judgment, which might lead him to oppose the parents' wishes? If the experiment is appropriate to 201, would Wilcox "seriously propose that New York City's public schools be monitored, regulated and programmed by more than 875 'selected' school committees?" Does a good relationship between school professionals and parents necessarily produce better education? Dentler pointed out that the limited research available indicated that quality of education was primarily related to the level of expenditures and the social class of students and faculty.[8]

However, the summer of 1966 was not a time for doubts about the new idea. It had a momentum of its own and was not to be waylaid.

DURING THE SPRING AND SUMMER OF 1966, the Ad Hoc Parent Council made the Wilcox plan the basis of its demands and added to it a demand that the new principal of IS 201 be black or Puerto Rican to enhance the children's self-image. At the same time, the group continued to insist that the Board carry through its promise to integrate the school with white children from other boroughs.

School officials, hoping to placate community feeling, maintained that IS 201 would be a model school with unsurpassed facilities and program. But leaders of the Ad Hoc Parent Council would not be assuaged by promises of quality education. They determined to take their stand at IS 201 for integration or for a new kind of school governance.

Though the community group adopted a public stance of "Integration or Community Control," few if any of those involved believed that integration was a remote possibility. There was growing doubt that integration was even desirable. Livingston Wingate, a member of the inner group and executive director of HARYOU-ACT, Harlem's largest antipoverty program, told a conference of school superintendents that summer: "We must no longer pursue the myth that integrated education is equated with quality education." He called for an end to the debate about bussing and integration, and for immediate action to improve the quality of black schools.[9]

That same summer the Board of Education announced the demise of Higher Horizons, its major compensatory education program: it had failed to have a measurable impact on pupil achievement. It had been expanded too fast, and the original scheme diluted to ineffectiveness. The pilot program seemed to show that more money and better teaching could make a difference; Higher Horizons showed that a little more money made no difference. It also appeared to demonstrate that the Board of Education had no viable program to improve ghetto schools.

Thus, the members of the Ad Hoc Parent Council were convinced that the Board of Education could deliver neither integration nor quality education. Their distrust of the system was so deep that they wanted neither involvement nor participation, but a role sufficiently powerful to guarantee that their voice would not be ignored. There were two obstacles: first, getting the Board of Education to authorize a community committee, and second, replacing the system's designated principal with a black or Puerto Rican of their choosing. The strategy of the parent and community activists was to keep IS 201 closed until their demands had been won.

The Board of Education, following its usual procedures, selected a principal for IS 201 from the Board of Examiners' competitive lists. He was Stanley Lisser, white, Jewish, a Harlem principal with a good reputation. Lisser hand-picked a teaching staff for IS 201 which was almost equally divided between whites and blacks. Lisser was unacceptable to the Ad Hoc Parent Council for obvious reasons. He responded to the rules and regulations of the Board of Education, while the local leaders wanted a principal whom *they* handpicked, to respond to their requirements. When Lisser routinely sent out a notice that IS 201 would open for a summer remedial reading program, he was immediately denounced by the activists. They accused him of a calculated effort to undermine their negotiating stance by opening the school. Lisser soon capitulated, canceled the summer session, and put off the opening of IS 201 until the fall.

Getting the Board of Education to accede to the demand for a community committee was no easy task. The activists were told directly by Board President Lloyd Garrison and Superintendent Donovan that the proposal was contrary to the State Education Law, which did not permit the Board to delegate its authority. The parent and community group spent the summer of 1966 trying to bring outside pressure on the Board of Education to settle with them. They met with Mayor Lindsay, who told them that forced integration was out of the question; they filed a complaint with the City Commission on Human Rights and got a wait-and-see response. State Commissioner Allen would not see them, because it was his policy not to intervene in local disputes unless invited in by local officials. A New York congressman arranged an appointment for them with Harold Howe, the United States Commissioner of Education, who told them that there was no way he could help them.

As the opening of school neared, the IS 201 activists prepared to boycott the school. The situation came to the attention of the press in early September, and the 201 spokesmen had a chance to reach out for public support. The designated leader of the group was Mrs. Helen Testamark, wife of a carpenter and mother of seven. Mrs. Testamark told the press and television crews: "Either they bring white children in to integrate 201 or they let the community run the school—let us pick the principal and the teachers, let us set the educational standards and make sure they are met." Another parent said to reporters: "I don't want any more teachers who make excuses for not teaching, who act as if they're afraid of a seven-year old child because his color is different. I don't want to be told that my daughter can't learn because she comes from a fatherless home or because she had corn flakes for breakfast instead of eggs." [10]

Up until this point, nothing had been said about the teachers; not even the original Wilcox proposal had accorded the community committee the right to hire and fire teachers, yet this function was a logical component of total community control. The teachers who had been brought to IS 201 by Lisser did not seem to feel threatened during the preboycott period. They attended special training sessions during the summer and had no contact with the brewing political dispute. A few days before the fall term was to open, they organized themselves as a chapter of the UFT. A UFT representative told them that it was up to them to decide whether they wanted to support the parents' boycott and promised them the protection of the union against reprisals by the Board of Education in the event that they supported the boycott.

Under the threat of a parent boycott, Garrison and Donovan tried again to work out a compromise. With the opening of school only a few days away, the two top officials met in Harlem with the activists and offered to create a representative community council to work with the school staff in planning programs; the council could help screen candidates for teaching and administrative positions, but would not have the power to hire or fire. The 201 spokesmen rejected the proposal as insufficient. They would accept nothing but real, not advisory, power.

For the moment, at least, the Board stood its ground. At their next meeting, Garrison read a prepared statement to the 201 negotiators and their entourage of observers:

> The Board has reluctantly reached the conclusion that it would be a waste of time to work out the details on an over-all program of this sort. So long as this group maintains the position it has taken up to now—that it is not interested in any plan which would fail to give the community the power to hire and fire teachers and supervisors, to determine the curriculum and to direct how the funds allocated to 201 should be spent. The Board cannot lawfully do this and it cannot do by indirection what it cannot do directly.[11]

Garrison's statement precipitated a mass walkout and threats from the forty or so parent and community representatives to boycott not only 201, but every school in Harlem. Within minutes, an automobile with a loudspeaker moved through the 201 area, urging parents to keep their children home on the first day of school. Handbills demanding total community control were widely circulated, with a pledge to "keep 201 closed by whatever means are necessary until these demands are met."[12]

As the activists had hoped, pressure mounted on the Board of Education to resolve the conflict. During the summer of 1966, disorders and riots had broken out in Atlanta, Chicago, and Cleveland. The tension around 201 seemed to offer tinder for an explosion in Harlem, and the 201 negotiators kept the implicit threat of violence in their arsenal. SNCC, the Black Panthers, CORE, and other militant groups stood by on the fringes, disdainful of integration but prepared to lend their support to demands for total community control. The *New York Times* blamed the Board for being insensitive to the parents' point of view, for promising integration when it was impossible, and for leaving the dispute unsettled until the eleventh hour. The *Times* feared that a boycott might be taken over by "extremist elements" and urged the Board to reach a "dignified truce" by working out "the ground rules for an effective but legally acceptable community council. . . . It is, of course, out of the question that the Board give the community 'total control' over the administration of this or any other school." [13]

Donovan and Garrison, trying to avert a boycott, invited the 201 group to join them in round-the-clock discussions and agreed to keep 201 closed during the first week of school. The parent and community group elected a ten-man negotiating team, as well as sixteen observers and two advisors. Wilcox and Wingate were the advisors.

During the first day of negotiations about sixty pickets marched in front of the empty 201. One picket shouted, for the benefit of reporters, "We got too many teachers and principals named Ginzberg and Rosenberg in Harlem. This is a black community. We want black men in our schools." These kinds of comments, some antiwhite, some anti-Semitic, appeared on television and in the newspapers; they contributed to changing the public image of the 201 activists from aggrieved parents to that of aggressive militants.

To show that he recognized the legitimacy of the negotiating group, Superintendent Donovan agreed to send out a special delivery letter to parents of IS 201 children, explaining why the school's opening had been postponed and temporarily reassigning their children to their previous elementary schools; Donovan cosigned the letter with Mrs. Testamark, chairman of the IS 201 Negotiating Committee. He also agreed that all statements to the press would be issued jointly by him and Mrs. Testamark. Donovan came prepared with a detailed proposal for a community council with numerous specific functions. Members of the 201 negotiating team had two major objections: first, that the council's functions were only advisory; and second, that agreement on a black or Puerto Rican principal was imperative before the school could open.

As the week wore on, Donovan improved on his original offer. He expanded the powers of the community education council, permitting it to work with the school staff in screening and recommending candidates for jobs. More importantly, no one would be hired "if there were sound and serious objection to his assignment by either the Council or the staff."

While negotiations proceeded and 201 was closed, the staff and children of 201 had been reporting to elementary schools which had neither room nor educational programs for them. About half the children failed to report at all. The parents of these children were becoming restive, and the negotiating team prevailed on Donovan to move the 201 children and teachers to PS 103, a closed-down, antiquated school in the neighborhood. Donovan directed the children and staff to report to PS 103 on Monday of the second week of school.

At the end of the first week of negotiations, only one issue was unresolved: the principal. The negotiating team told Donovan that Lisser was unacceptable and that "the community wanted a Negro or Puerto Rican male heading the school." According to an account written by a member of the negotiating team,

> Donovan requested that the committee prepare a list of people who would be acceptable to them for the Monday negotiations, and if Lisser's transfer were acceptable, they could immediately begin screening possible candidates for the post. He promised to bring the situation before the Board of Education. The clear inference was left with all present that Donovan was suggesting that Lisser would be replaced with a Negro or Puerto Rican principal acceptable to the community.[14]

On the same Saturday that this meeting took place, the *New York Times* published the full text of the community council proposal, which was supposed to be secret during the negotiations. Donovan's willingness to grant the community a veto for "sound and serious" but undefined reasons appeared to give the community council the chance to exclude professionals on a racial or religious basis. The same account also reported the negotiating team's demand that the white Jewish principal of 201 be replaced by a Negro or Puerto Rican. Publication of these demands hurt the cause of the 201 group in the eyes of the public, for it appeared that Donovan was capitulating to black extremists.

While public opposition to Donovan's concessions began to mount, the 201 teachers met to decide whether to teach at PS 103. They were concerned that the school was "educationally unsound," since its facilities and supplies were makeshift and the building itself was decrepit. They agreed to meet at PS 103 early Monday morning and decide then. The UFT leadership made no effort to influence the IS 201 staff on whether or not to teach in PS 103. The union's president, Albert Shanker, stated that the union did not want to become involved in the negotiations: "We feel that the involvement of parents in a community, their greater involvement in the educational process, if done in a constructive way, could result in great educational improvement." He had no objection to the controversial screening process, so long as it did not apply to teachers already assigned to 201: "We assume a sound reason would not mean color of a person's skin or somebody just not liking a person." [15]

Monday, the beginning of the second week of school, was a decisive and tumultuous day. The 201 teachers met outside PS 103; a delegation inspected the building and emerged with a negative report. The staff voted, 44–10, not to teach in 103, but to report instead to 201, to show that they were prepared to teach under "proper circumstances." The teachers had had no contact with the IS 201 activists; they did not consider the impact their decision would have on the negotiating team's efforts to keep 201 closed. They made their decision on professional grounds, unaware of its larger political consequences. Sometime during that morning the Board of Education decided to open IS 201 the following day.

Superintendent Donovan called an emergency meeting with the 201 negotiating team to inform them that 201 would open the next day, with Lisser still the principal. They were furious and promised a full-scale boycott. The negotiating committee realized that its position could be destroyed as soon as 201 was open; the threat of a boycott is stronger than a boycott itself, since the boycott might not succeed. Donovan, anxious to avert a boycott, told them not to make any decisions and promised to meet with them later that same day.

Within an hour, Lisser asked to be transferred. When this information was relayed to the teachers, still meeting at IS 201, they concluded that his sudden request was not voluntary and agreed not to report for work unless he were reinstated. They were, after all, a staff especially selected by Stanley Lisser. The staff adopted a resolution: "We stand in opposition to any group threatening violence. We will stand together behind Mr. Lisser for quality education and community cooperation. We ask the Board of Education not to honor Mr. Lisser's resignation." [16]

Donovan met with the negotiating team that Monday afternoon to put the final touches on their agreement. The negotiators agreed not to boycott IS 201, and Donovan, speaking for the Board of Education, agreed to accept the proposal for a community council, as well as to honor Lisser's request for a transfer out of 201. The negotiating team expected the screening process to produce a new principal immediately.

The reaction to Monday's events was immediate. Tuesday morning the staff of IS 201 set up two picket lines, one at 110 Livingston Street, the other at the school. Their placards carried the same message: "All of us or none of us." Thirty-five Harlem principals signed a statement objecting both to Lisser's "so-called voluntary transfer" and to the agreement to set up a community screening committee for professional staff. The Council of Supervisory Associations threatened to sue the Board of Education to prevent it from recognizing the 201 community council. The *New York Times* criticized Donovan for making a truce out of fear of extremist groups and guilt for past injustices, though ten days before the *Times* had urged a truce for the same reasons.

Before noon Tuesday, the Board of Education abrogated the agreement Donovan had made with the IS 201 negotiating team and announced that Lisser had withdrawn his request for reassignment, as a result of the support shown him by his staff. However, the Board stood behind the proposal for a representative community council.

The parent group learned of Lisser's reinstatement on the radio and sent off telegrams to Lindsay, Donovan, and the Board of Education:

> We of the Harlem community are shocked at the irresponsibility shown by the Board of Education towards the Harlem community by breaking a public agreement between the Board and the parents. Once again the Board has shown contempt for the Harlem community by appeasing those forces who wish to deny equal education to our ghetto children. The Board of Education's action has pitted white against black and we will hold them responsible for any actions the community may take. The conditions for further negotiations will be the removal of Stanley Lisser as principal of IS 201.[17]

Without formally declaring a boycott, the negotiating team asked parents to keep their children home. Keeping the boycott informal was a face-saving device in case it was not successful.

On Wednesday morning, September 21, IS 201 finally opened. Stanley Lisser arrived surrounded by police escorts. Five pickets were arrested as they tried to block his path. Fewer than fifty pickets marched in the rain, and many of them were whites from EQUAL, the militant integrationist organization.

More than two-thirds of the school's 600 students reported for school, despite the rain and the controversy. Months later, the parent-community group would remember only that the teachers' refusal to work in PS 103 had "broken" their boycott, when in fact it had reduced their ability to keep IS 201 closed with *threats* of a boycott. Once 201 opened, the boycott was broken by the parents who sent their children to school.

A week later, the picketing ended and attendance was normal. A ceasefire was in effect, but there was no sense of conclusion for either side.

The collapse of the 201 boycott did not discredit the 201 negotiating team. The negotiators may have been selected by a self-appointed group of community spokesmen, and they may have embraced community control more quickly than the parent body of the school, but the ideas they represented attracted wide support because of the objective situation. Ghetto schools were unquestionably failing to bring their pupils up to a par with nonghetto students; this was proven by every indicator of academic distress—truancy, dropouts, reading levels, and so on.

Integration, which many people had relied on to equalize education, was no longer numerically or politically possible; besides, the Board of Education had no plan to upgrade ghetto schools dramatically. Past performance made it unlikely that any Board of Education plan would be anything more than promises and paper. Thus, with the problem of ghetto education admittedly acute, and with no prospect of a solution from the Board of Education, the idea of community control had undeniable appeal. After all, parents had a right to demand better education for their children, and in this instance, parents were asking for a chance to succeed where the educational experts had failed. Some public officials were glad to deal with a ghetto group that, for once, was not demanding integration. Community control appeared to be a way out of the schools' dilemma. If the parents assumed control, they would have only themselves and their appointees to blame for failure.

The idea of black control of black schools appealed to a surprising cross-section of whites. Some school officials were attracted to the idea because parents would learn firsthand how difficult it was to solve problems. Some in the foundations and universities saw community control as a constructive way to engage the energies of black militants by ceding to them a part of the system at no sacrifice to anyone outside the ghetto. Sympathetic whites saw it as the best hope for rescuing ghetto schools through the power of aroused and committed parents (who had always been criticized for being apathetic about their children's education). Conservative whites recognized that black control of black schools implied white control of white schools, which they could comfortably support, for it guaranteed that black problems, black dissidence, and black pupils would be safely contained within the ghetto.

THE AFTERMATH OF THE CONTROVERSY at IS 201 left the city's political and educational authorities deeply antagonized and gave momentum to the feeling that basic changes in the school system were overdue. Lindsay complained that the Board of Education was isolated from the community, and that he intended to take a more active role in school matters.

A week after the departure of the last picket, the Borough President of Manhattan, Percy Sutton, warned that the issue had to be resolved, "because of the explosive nature of the situation. There's a chance of a very dangerous conflagration at IS 201 unless something is done." Lindsay concurred: "I know Mr. Garrison is well aware that the problem of 201 is not solved. The school is functioning, it's a very handsome school, excellent teachers, first-class principal, but the problem is not solved."

In early October, a new plan was put forward, authored by Dr. Kenneth Clark and supported by the 201 parent/community group. The plan proposed to place IS 201 and its elementary feeder schools under a board jointly operated by university personnel and parent representatives. It aimed at disengaging IS 201 from the control of the Board of Education. The Clark plan was endorsed by Mayor Lindsay, State Commissioner Allen, the *New York Times*, CORE, SNCC, the NAACP, and the New York Urban League.[18]

The UFT, however, attacked the Clark plan as an attempt to remove those schools and their faculties from the body of law and contractual agreements under which they presently operated. Union president Albert Shanker saw it as a precedent which could ultimately lead to an abrogation of the teachers' rights and their protection against punitive transfers. The UFT's stand brought into the open the resentment which the 201 activists had harbored against the UFT since the teachers refused to work in PS 103. The 201 negotiating committee wrote to Shanker that they discerned "a consistent picture of UFT cynicism under the banner of professionalism and trade unionism." The UFT was seeking, they charged,

> to retain the license to continue to miseducate and destroy Black and Puerto Rican children of New York City without being accountable to anyone. . . . If a clearly established and reasonable role for parent and community involvement in the schools cannot be established, other steps will be taken to assure that teachers in ghetto schools teach. We intend to use whatever means necessary to stop teachers White or Black from coming into the ghetto to cripple our children's lives.[19]

The language of the militant protest groups took on an importance of its own, for words, no less than actions, have consequences. Implicit threats of violence ("whatever means necessary") and accusations of monstrousness ("educational genocide") contributed to an atmosphere in which flamboyant hyperbole replaced rational discourse. Where rational communication is impossible, so also are settlements which satisfy both sides. How can anyone compromise with someone who is "destroying" his children? If politics and negotiation are scorned, the only alternatives left are total victory or total defeat. At later stages in the school war, this antipolitical strategy caused militant groups to snatch defeat from the jaws of victory.

The Board of Education rejected the Clark plan, citing a legal opinion of the State Education Department that under the existing law the

Board could not delegate its authority to any outside agency. The Board offered instead to establish an "advisory panel" of representative parents, teachers, and community members. David Spencer, newly elected chairman of the 201 group and a staff member of MEND, accused the Board of showing "complete contempt for the Harlem community." Isaiah Robinson called the Board's counter-proposal "a glorified parent-teachers association. . . . The children are being slaughtered through educational genocide—and everyone seems to be so proud." [20]

Stalled once again, the Board tried to appoint a city-wide task force of educators, public officials, and community leaders to develop ways to improve ghetto education. McGeorge Bundy, president of the Ford Foundation for less than a year, was invited to chair this task force. Bundy came to New York after six years at the pinnacle of national government, having served both Presidents John Kennedy and Lyndon Johnson as special assistant for national security. Before his government service, he had been dean of the faculty of arts and sciences at Harvard University. While deciding whether to accept the Board's invitation, Bundy despatched one of Ford's education specialists, Mario Fantini, to Harlem to review the situation. Fantini discovered that the IS 201 group adamantly opposed the task force, considering it just another delaying tactic of the board, a ploy to reduce the level of ferment without changing the status quo. Those who had been active in the 201 controversy refused to serve on the task force. They told Fantini that they wanted recognition and legitimacy, not the introduction of a third force. Bundy decided not to become involved in the task force. No prominent black would put himself in the position of appearing to oppose or compromise the interests of the 201 group; and no white of stature would serve on a committee boycotted by every leading black. The task force idea was dead.

The 201 group determined to embarrass the Board whenever possible, in order to show that the Board was incapable of managing conflict within the system. All the members of the local school board in Harlem resigned to protest the Board of Education's failure to reopen negotiations with the 201 activists. Statements of the negotiating committee repeatedly referred to the Board of Education as the "Board of Genocide."

The 201 group linked up with dissidents from other parts of the city in a bold three-day takeover of the Hall of the Board of Education in December 1966. At a public hearing on the board's budget, a Negro mother from the Brownsville section of Brooklyn demanded to speak out of turn. Supporting shouts from the audience disrupted the meeting to the point where the Board members called an early adjournment and walked out. The leather chairs of Board members were promptly filled by several members of the audience who declared themselves the "People's Board of Education" and designated the protestor from Brownsville as the superintendent of schools. The group then continued the hearings, which consisted of attacks on the regular Board of Education. Among the members of the People's Board of Education were the Reverend Robert Nichol, David Spencer, and Babette Edwards from IS 201.

The Reverend Milton Galamison was elected president of the People's Board of Education and soon arrived to take his place. Other educational activists who joined the takeover of the Board's hall were Evelina Antonetty, the Puerto Rican head of United Bronx Parents; Rosalie Stutz and Ellen Lurie, white leaders of EQUAL, which had followed the lead of militant blacks in switching from a strident advocacy of integration to a strident advocacy of community control; and Father John Powis, a white priest and a member of what was then called the Brownsville-Ocean Hill Independent Local School Board. Galamison declared that the goal of their demonstration was "obviously . . . to get rid of the Board of Education. There's no other solution." [21]

This highly publicized sit-in made the Board look ridiculous as well as ineffectual. It had been pushed out of its seat of power, literally and figuratively, for a three-day period. More and more, the Board had the appearance of a paper tiger, a giant apparatus incapable of asserting power. The mayor would not defend the Board; it wasn't his Board, its policies were not his policies. The more the Board tried to be reasonable, the more absurd it looked, for its adversaries wanted not to deal with the Board but to destroy it.

The coming together of the People's Board had an immediate effect on the participants. Dissidents from different parts of the city met each other and synchronized their political consciousness to a common program: community control of the schools. When the Board of Education spoke of decentralization, it meant *administrative* decentralization, increasing the power of district superintendents and school principals. When the People's Board of Education used the same term, it meant that "the school system must be decentralized to give local parents and community groups effective control over the education of their children." [22]

Within a brief period of time the opponents of the Board of Education had totally reversed their aims. In 1964, the City-wide Committee for Integrated Schools, under Reverend Milton Galamison, wanted to see the school system destroyed because of its unwillingness to break the neighborhood school pattern. Little more than two years later, Galamison wanted to see it destroyed in order to achieve total local control. By the end of 1966, many of those who had most passionately advocated integration and overriding of local parent organizations had been converted to the cause of community control of the schools, even though it meant abandonment of school integration. Integration, which had been described as a moral imperative, was quickly discarded as the feasibility of total racial balance receded. To those who were zealous by nature, local control became the new moral imperative. As in every previous school war, the idea of the school as an omnipotent shaper of values and destinies inspired moral dogmatism among school reformers; and once again, the rhetoric of the battle distorted legitimate educational concerns into inflexible political postures. Once again, school reformers discovered a political slogan which had the potent appeal of a panacea: community control. Community control, it was hoped, would lead to the ouster of recalcitrant bureaucrats and racist teachers, increase the number of black and Puerto Rican professionals, improve the self-image of minority children, and restore control of ghetto education to local parents, who would safeguard the education of their children. Local control of neighborhood schools, held to be "legally and morally wrong" in the days of the integration movement, was suddenly perceived as a natural right, irrespective of whether it produced schools which were racially segregated or ideologically exclusive.

THE DAY AFTER THE ARREST of the People's Board of Education, the Board of Education released school-by-school reading scores for the entire city for the first time. A fifth of the city's pupils were two years behind their grade level, and most reading failure was concentrated in ghetto schools. The lowest reading scores were in poor areas, the highest reading scores in well-to-do neighborhoods.

Many professional educators, particularly those in the school system, saw the economic variation in the scores as proof of what they had been saying for years: that poor children do not do as well in school as middle-class children because of the handicaps of a poverty environment. Regardless of race, poor children had problems which were not shared by their middle-class peers. The children of the poor tended to live in overcrowded, rat-infested housing; without proper medical care; with a poorly educated parent or parents; in a home without books or other inducements to learn what was taught in school; in a neighborhood where crime had greater immediacy than textbook studies; in a community where families moved two, three, or four times in a year because of pressure from landlords or government agencies.

However, the publication of the school system's reading scores made an altogether different impression on militant critics of the schools. It provided ammunition for their charges of the system's incompetence, and worse, malevolence. They saw the low reading scores in ghetto schools as evidence that the long parade of compensatory education programs had been an expensive hoax: it was proof that white children were taught to read, but black and Puerto Rican children were not. The theory of cultural deprivation was rejected by school critics as the professionals' excuse for poor teaching and inferior schools. Dr. Kenneth Clark viewed the "cult of cultural deprivation" as a sophisticated form of racism, a new way of saying that black children can't learn. He believed that this line of thinking offered alibis for "the fact that these children, by and large, do not learn because they are not being taught effectively and they are not being taught because those who are charged with the responsibility of teaching them do not believe that they can learn." He questioned whether Negro children were innately less able to learn (which he clearly did not believe), or "has inferior education been systematically imposed on Negroes in the nation's ghettos in such a way as to compel poor performance from Negro children?" His speculation lent credence to the black nationalists' contention that American institutions were so deeply and insidiously racist that they were programmed to keep black children undereducated and subservient. This was the basis of the phrase "educational genocide." [23]

This perspective was the educational justification for the advocates of total community control. Once they agreed that the system consciously conspired not to educate their children, it became imperative to wrest away control of the schools. The corollary to this view was the inference that black children *would* learn in a black-controlled school, where there was no clash between the culture of the teachers and the pupils.

The New York public schools were pinioned by their own legend, which they and their critics subscribed to. In the legend, the public schools had successfully educated the children of poor immigrants; in the legend, some children fared better than others, but the difference was individual merit, rather than any variation among national groups. The poor record of the New York schools in educating black and Puerto Ricans seemed to indicate either that those groups had unprecedented cultural deficiencies or that the schools were racist and/or educationally bankrupt in comparison to their greatness of former days. The legend of success supported this analysis, but the legend itself was not true. It was forgotten in midcentury that the rate of failure among first-generation European immigrant children had been high; in some years, as much as

40 percent of the schools' enrollment was academically retarded. Furthermore, the rate of school success and failure had been different among national and cultural groups. Jews, with their tradition of learning, had done very well as a group in the public schools (though there were also Jews who dropped out or failed). Other groups, such as Italians, had performed poorly. Some educational researchers in the 1920s and 1930s wondered whether inferior genetic endowment kept Italian children from matching the academic levels of other groups.[24]

To those who believed the legend of former greatness, it seemed that the school system either could not or would not educate black and Puerto Rican children as it had once educated other poor children. The flawed analysis set the stage: IS 201 was a signal battle in what would finally become an all-out school war over the issue of community control of the schools.

CHAPTER 28

Decentralization Emerges

JOHN V. LINDSAY was an activist mayor. He did not like to sit by and watch problems fester. Ordinarily, there was nothing he could do about the Board of Education other than to wait for members to retire and, in time, appoint a new Board. However, during his first year in office, he discovered a way to become involved. The path to change was provided by a report of the city's Temporary Commission on City Finances in the summer of 1966. It suggested division of the system into five borough school boards and urged the mayor to take a more active role in restructuring the system. More importantly, it informed Mayor Lindsay that the city school system was shortchanged by the state. Even though the system covered five counties, it was considered for purposes of state aid to be a single unit. Because state aid was based on real estate valuations, Manhattan's unusually high valuation caused the city as a whole to receive $100 million less in state aid than it would receive if each borough were considered individually.[1]

Lindsay, ever anxious to find new sources of revenue, asked the legislature on the first day of its 1967 session to consider the city school system as if it were five separate districts, one in each borough, for the purpose of dispensing state aid. The legislature agreed to do so, on the condition that the mayor submit a decentralization plan by December 1, 1967. Legislative leaders wanted to avoid the possibility that other cities might divide themselves up into artificial districts in order to qualify for more state aid.

A bill directing the mayor to prepare a decentralization plan was signed into law on April 24, 1967. The *New York Times* urged Lindsay to ask the Board of Education to write the new decentralization plan. But Lindsay had no intention of putting this opportunity for major surgery into the hands of the patient. On April 30, he appointed an Advisory Panel on Decentralization of New York City Schools and named McGeorge Bundy, president of the Ford Foundation, as its chairman. It was henceforth known as the Bundy panel, and Mario Fantini of the Ford Foundation was selected to head the panel's staff.

Fantini was already deeply involved in advancing decentralization on another front. Since late 1966, he had played a key role in forging a surprising *entente cordiale* between the UFT and the IS 201 negotiating committee, far from the glare of publicity. He became interested in IS 201 when Bundy asked him to investigate the Board of Education's ill-fated task force on ghetto education. Fantini began meeting with the 201 leaders and with Shanker and his chief aides. Soon all three parties met together in East Harlem, with Fantini's presence appearing to bless any pact with the promise of Ford Foundation largesse. Superintendent Donovan approved of the efforts made by the community group and the teachers to develop a cooperative plan. Two Yeshiva University professors, Sol Gordon and Harry Gottesfeld, worked out a plan which offered something to both sides. Gottesfeld was also a research associate on the staff of MEND, which indicated that the plan had the political approval of the MEND–201 power structure. Fantini stayed in close touch with both sides, and meetings were occasionally held at Ford Foundation offices.

The plan that gained the approval of the union and the 201 group was called "Academic Excellence: Community and Teachers Assume Responsibility for the Education of the Ghetto Child." The Yeshiva plan had two parts: it created a governing board for IS 201 and its elementary school feeders; and it spelled out an educational improvement program, based on the More Effective Schools model.[2]

The proposed governing board consisted of four community leaders, eight parents, four teachers, and one supervisor. Any conflict between groups on the governing board would be resolved by "consensus." It was agreed that "if an impasse developed, any question would be referred to a mediating body composed of a teacher, a community representative and an impartial party acceptable to both the others." This provision was crucial to the teachers, for it assured them that they would have an equal role in decision-making. Job rights of teachers were carefully defined and protected.°

° The Yeshiva plan, without any reference to home environment or cultural deprivation, acknowledged the failure of education in the ghetto and established a common ground on which parents and teachers could unite:

Teachers are at one with the parents of the city's school children in their concern with lack of achievement and reading retardation in the schools. . . . The school system cannot continue as an autonomous bureaucracy. Parents and community leaders must fulfill their right to exercise influ-

The plan aimed to enable parents and teachers to eliminate the distant hierarchical authority of the Board of Education which separated them. Both sides gained, but both sides had to compromise. The 201 group had to modify its demand for total community control and admit the professionals as partners in power, but it gained recognition and legitimacy; furthermore, by joining forces with the UFT, the 201 group had a better chance of getting approval of the plan.

The UFT recognized both benefits and risks in the plan. Any substantial improvement in ghetto schools was in the interest of the teachers who worked there; and too, the success of the plan would recoup the union's image of being in the vanguard of social action, an image which was important as the schools' population became majority black and Puerto Rican. Furthermore, the union's participation guaranteed its ability to write protection of its membership into the program. On the other hand, the union took calculated risks: the participants were removing themselves from the protection of the Board of Education, and, in the eyes of some of the membership, taking a chance that militant blacks would respect teachers' rights. What made the gamble worthwhile to the union was the expectation that the Board of Education would subsidize the academic portion of the plan.

---

ence in educational policy. . . . The teacher . . . must be allowed to take responsibility for exercising independent action and making expert judgment while performing his work.

The governing board was to have the following functions: selection of an administrator for the program; setting educational standards; recruitment and selection of staff "through an approved, objective method which does not lower standards"; determination of curricula; the power to contract for the assistance of a university; the right to allocate its monies.

On the teachers' job rights:

The governing board . . . must establish for itself a procedure through which unfit and/or unsatisfactory professional personnel may be removed. . . . The board should employ or appoint an independent, expert group [to evaluate] the total school program, including personnel performance. Should this committee . . . find it necessary to terminate the services of a professional, it shall have the right to seek such termination. Reasons for termination must be sound and serious, objective, substantiated, and subject to impartial review with provision for teacher defense.

The educational side of the program started from the base of an MES program (small classes, team teaching, additional professional services) and added to it a concentrated reading program; health services; a training program for community workers; in-service training for teachers (as well as an orientation course for professionals given by community leaders); and measures aimed at enhancing the students' self-image, e.g., emphasis on "pride in one's ethnic background."

In April 1967, as Lindsay prepared to name the members of the Bundy panel, the Board of Education approved a new administrative decentralization policy. The new policy increased the operating authority of the thirty district superintendents, while requiring them to consult with their local school boards. At the same time the Board announced that Superintendent Donovan had been asked to formulate specific proposals for "demonstration projects promoting greater community involvement." This statement referred to the projects that Donovan, the Ford Foundation, the UFT, and several community groups had been discussing since the beginning of the year.[3]

In these discussions, the role of Mario Fantini of the Ford Foundation was pivotal. His presence and the availability of his office as a neutral meeting ground seemed to assure the ultimate commitment of large sums of money by the Ford Foundation. All the other parties to the planning—the UFT, the militant community groups, Donovan—were to some degree distrustful of each other. Fantini was in a unique position to bridge the gaps among the other participants. As a former teacher and administrator, he appeared to understand the concerns of the teachers' union; as the representative of an education-minded philanthropy with an endowment of $3 billion, he commanded the respect of the superintendent of schools; in his writings and speeches, he was completely sympathetic with the aspirations of those who were in revolt against the present system.

Fantini had come to the Ford Foundation in search of ways to promote his chief interest in education: institutional change. He was convinced that large, bureaucratic systems were by their nature hostile to change, and that restructuring these institutions took precedence over any other reforms. In trying to improve education for the disadvantaged, he had written, the problem "is not with the learner. It is with the institution." In exploring methods of introducing institutional change in a large school system, Fantini became especially interested in the idea of a "model sub-system." The model sub-system was a small decentralized unit operating within a large system, "a laboratory which can become a vehicle for changing the major school system . . . free to depart from the established practices, e.g. hire from the outside, new curricula, etc." It could either "feed out ideas to the larger system or it could gradually expand and replace the larger one." [4]

In the spring of 1967, Fantini had an unparalleled opportunity to put his theories into practice in the nation's largest school system. Not only was he directing the staff of the Bundy panel, which had a mandate to draft a decentralization plan for the entire system, but he was instrumental in creating three model sub-systems that would pave the way for the citywide plan.

In May, the Ford Foundation informed the Board of Education that it would make planning funds available for three projects that had received the mutual consent of the community group and the union. In early July, the Board of Education announced the creation of three small experimental projects, in the IS 201 district in East Harlem, the Ocean Hill-Brownsville district in Brooklyn, and the Two Bridges district in the Lower East Side of Manhattan; each project was carved out of a regular district.

The Board stressed that it had approved only the "concepts," and that the proposal developed in each district would be subject to subsequent review and approval by the Board. The Ford Foundation made a planning grant to a community organization in each district: $51,000 to the Community Association of the East Harlem Triangle, Inc.; $44,000 to Our Lady of Presentation Church in Brooklyn; and $40,000 to Two Bridges Neighborhood Council Inc. Each organization was to act as fiscal agent on behalf of the planning group; the planning group in each district was composed of those parents, teachers, and community leaders who had participated in creating the proposals.

At the Ford Foundation's press conference announcing the grants, the disparity between what the board had authorized and what the community groups were expecting was succinctly described by David Spencer of 201: "The Board of Education will call it decentralization. We like to carry it further. We call it community control." [5]

By THE TIME the projects were launched, the initial enthusiasm of the UFT had waned. To the union's dismay, Superintendent Donovan informed everyone involved that the Board would not give the demonstration districts any additional funds. Donovan said that the purpose of the experiment was to see whether community participation would make a difference in achievement; if the demonstration schools got more money, the significance of the community's role would be obscured. This meant that the entire educational improvement portion of the program would be cut out.

The union thought that the community groups should join with them and refuse to initiate the projects until the Board agreed to restore the program funds. Fantini thought the community groups should take the governing structure while they could and work for program funds later. The community groups were sorry to lose the extra funds, but saw several advantages in not delaying. First, any delay might provide the Board of Education an excuse to kill the projects; second, they thought that once the projects were established, they would have sufficient political power to demand extra funds; third, those who were fervent about community control were convinced that control in and of itself was a major breakthrough, and that the programs cut out were just compensatory education; fourth, some felt that true community control required that the governing boards start fresh and develop their program requests on their own, subject to no prearrangements.

Consequently, what had begun as a cooperative coalition between the community and the teachers turned into something quite different.

While the carefully negotiated agreement between the UFT and the militant community leadership was dissolving, the UFT was under pressure to protect its membership in several ghetto communities. Especially in the Bedford-Stuyvesant and Brownsville sections of Brooklyn, teachers complained of harassment by militant blacks. Parent groups and community groups, usually connected with an antipoverty organization, singled out teachers and principals for removal; in one school, teachers' mailboxes were stuffed with anonymous hate mail. Robert "Sonny" Carson, head of Brooklyn CORE, demanded that all principals in Bedford-Stuyvesant submit to him written plans for bringing their classes up to grade level and threatened to remove those principals who failed to do so: "If Donovan thinks we are kidding, he had better wait until September and see what happens when those teachers, those principals try to come back to our community." [6]

Ghetto teachers, long under attack for the failings of their schools, felt directly threatened by the hostile atmosphere stimulated by militant organizations like Brooklyn CORE. In their fear and insecurity, teachers sought protection from their union. The UFT leadership was demonstrably more liberal than its membership. Rank-and-file conservatism within the union surfaced in 1966 on a different issue, the question of supporting Mayor Lindsay's proposed civilian review board to hear complaints against the police. Though the union's executive committee endorsed the board 30–2, the delegate assembly (representing schools all over the city) passed it by only 486–375.

The president of the UFT, Albert Shanker, was a tough and combative unionist who believed that the interests of the union were consonant with liberal causes. His family had been poor and Jewish; his politics were socialist. Born in Long Island City in 1928, he attended New York public schools, the University of Illinois, and Columbia University. He began teaching junior high school mathematics in 1952 and became active in the New York Teachers Guild in 1953; the Guild was then only one of many teachers' organizations and represented less than 5 percent of the teachers. Shanker was elected president of the UFT in 1964.

Most UFT members, like Shanker, were from Jewish families which had immigrated from Eastern Europe and formed a cultural milieu where unionism, socialism, anarchism, and radicalism were avidly discussed. Many had grown up in poor neighborhoods in New York, gone to public schools, graduated from free city colleges, and then become teachers in the city school system. While teachers in other cities might consider it "unprofessional" or undignified to affiliate with organized labor, the unusual concentration of East European Jews among New York's teachers predisposed them to favor association with the labor movement, especially because the teachers' union was historically identified with social progress. The recognition in 1961 of the UFT as the teachers' collective bargaining agent capped forty-five tumultuous years of struggle. At the time that the union was formed in 1916, and its first card issued to John Dewey, trade unionism was a minority movement among teachers. Over the years, the union often fought alone for free speech; equal pay and equal rights for women teachers; decent wages and decent working conditions; academic freedom; and protection of teachers from harassment by the Board of Education. When the union was taken over by a Communist faction in 1935, its leadership withdrew to form a new organization, the Teachers Guild, which continued to espouse the social idealism of the original union.

The Teachers Guild, and then the UFT, was criticized by its liberal allies during the integration battle for resisting forced rotation of experienced teachers into ghetto schools. Advocates of community control attacked the UFT because the union contract made it difficult to dismiss a tenured teacher who was incompetent or ineffective; principals would shift a teacher to another school rather than go through the time-consuming dismissal procedure, which involved formal charges, substantiated evidence, and professional witnesses. Intended to protect teachers against arbitrary supervisors and punitive transfers or dismissals, the union rules protected not only the able, but also those who were mediocre. To those activists who wanted to improve the quality of teaching in ghetto schools; and to those blacks who wanted to get rid of teachers who were insensitive to ghetto children; and to those separatists who wanted black teachers in black schools: the UFT blocked the way. The UFT was especially vulnerable to attack from blacks because more than 90 percent of the city's teachers were white; if New York, like Chicago, Detroit and Newark, had had substantial numbers of black teachers, the gulf between teachers and pupils might have been understood as a class difference, rather than a racial difference. But the small proportion of black and Puerto Rican teachers and supervisors gave credence to charges of racism in the Board of Examiners' procedures, in the Board of Education's teacher recruitment efforts, and in teacher attitudes in the classroom.

In the spring of 1967, with hostile acts against ghetto teachers on the rise, the UFT entered its biennial contract negotiation with the Board of Education. The union had two demands aimed at easing the situation of its members in ghetto schools: the UFT asked for an expansion of the MES program (which doubled the expenditure per pupil in schools so designated), and requested that teachers be permitted to suspend "disruptive" students from class. When the union announced these demands in the spring, the leading militant community groups, then hammering out a working arrangement with the union, were not critical; by the end of the summer, as the relationship soured, the union's demand for power to oust disruptive pupils was interpreted as racism.

In other years, the UFT's demands would have been seen as items on the bargaining table, subject to compromise; in 1967, they put the UFT at odds with newly emerging participants in New York City school politics.

CHAPTER 29

The Making
of a Power Struggle

DURING THE SUMMER OF 1967, as contract negotiations moved along at a snail's pace, and the Bundy panel staff worked on a city-wide decentralization plan, the planning groups in each of the three demonstration districts plunged ahead into uncharted territory.

The demonstration district that moved fastest to set up its program was Ocean Hill-Brownsville. At first glance, this was surprising, because Ocean Hill had few recognized leaders and established community organizations. But the paucity of local leadership facilitated the creation of a unified, new organization; by contrast, the 201 district was plagued with internal fights for leadership.

Ocean Hill-Brownsville had never been thought of as a community before; it was an area that lay between two recognized communities, Brownsville and Bedford-Stuyvesant. Like its neighbors, Ocean Hill-Brownsville was predominantly poor and black. Unlike its neighbors, Ocean Hill-Brownsville had no indigenous antipoverty organization, nor did it have any other kind of organizing force that would create a sense of community. The absence of countervailing powers in the area gave additional power to the group which received the Ford Foundation's grant to plan a demonstration school district.

Ocean Hill-Brownsville's designation as an experimental district grew out of the protest activities of two different groups. One group, the Brownsville-Ocean Hill (later reversed) Independent School Board had formed in September 1966 to protest the fact that the area had no representative on the local school board for District 17. The leader of the Independent School Board was Father John Powis, a white priest from Our Lady of Presentation Church. The formation of this group coincided with the eruption of the community control crisis at IS 201. The Independent School Board wanted a voice for the local community in the running of the schools. One of its goals, a reflection of the underorganized state of the neighborhood, was "to stimulate an almost inactive Parents Association in the schools in this section." Father Powis was a member of the militant People's Board of Education that took over the Hall of the Board of Education in December 1966.[1]

Another group in the Ocean Hill-Brownsville area originated with a newspaper story which described serious problems in an unnamed ghetto junior high school in Brooklyn. Recognizing their school, several teachers called UFT headquarters to ask for help in getting rid of the principal. One of the UFT's field representatives, Sandra Feldman, met with the school's UFT chapter. She encouraged the teachers to work with the parents. The teachers and parents picketed together at the district superintendent's office and won a commitment that a new principal would be chosen for the new Intermediate School 55, which would soon replace their school. They also won additional services for their school, as well as some relief of chronic overcrowding. The teachers and parents who had picketed together began to realize that the new IS 55 offered them an opportunity to work together, to plan new programs and perhaps to participate in the designation of the new administration.

In January 1967, the two groups merged, because of their mutual interest in IS 55. They designated their new organization "The Steering Committee for IS 55" and drew up a proposal to give the community a strong voice in the selection of the new principal and the new school's program. In April, Father Powis met with Mario Fantini and began to discuss making an experimental unit of eight schools, including IS 55, in the Ocean Hill-Brownsville area.

The Steering Committee, which included parents and teachers, formulated a proposal, drawing heavily on the Gottesfeld-Gordon plan for 201. Like the 201 proposal, it had two basic sections: provision for an elected governing board and provision for educational improvements. This proposal, submitted to the Board of Education on May 27, 1967, was the basis of the Ocean Hill-Brownsville project. The proposed governing board had limited functions, such as setting educational standards, developing curriculum, selecting principals and supervisors; the proposal did not include the power to hire and fire teachers. It was a plan for community participation, not a plan for community control.

Teachers' rights were protected in a section stating that evaluation of the school program would be conducted by "an outside, independent group of experts. . . . If, after all help and appropriate supervision, a teacher is found by the principal . . . to be incompetent or unfit for service, a procedure for removal shall be followed. This procedure shall call for impartial review of sound and serious, substantiated charges, and shall insure all judicial rights of the accused." The impartial review was to be carried out by a board—consisting of a college representative, a teacher, and a community person—appointed by the governing board.[2]

Immediately after the Board of Education approved the proposal "in concept," and the Ford Foundation made a grant of $44,000 to Father Powis' church, a planning council was picked to plan for the operation of the demonstration project. The planning council for Ocean Hill-Brownsville consisted of parent and community leaders who had participated in the Steering Committee for IS 55; the teacher members were elected at each school to serve for the summer only. The teachers' understanding was that a plan would be devised during the summer and that, at the opening of the fall term, the faculty of each school would vote on whether to take part in the project.

The planning council appointed Rhody McCoy to be its acting unit administrator. McCoy, the acting principal of a "600" school in Manhattan and a veteran of eighteen years in the school system, came to Ocean Hill-Brownsville to be interviewed by the planning council for the job of principal of IS 55, but was chosen for the top administrative position in the district. Like Galamison, McCoy was born in 1923, the son of a postal employee, and started his career in New York City in 1949. He grew up in Washington, D.C., in a black section and earned degrees at Howard University and New York University. While Galamison was making his mark as a crusading minister, McCoy, a soft-spoken man, had risen doggedly in the public school system, mastering the inner workings of the giant bureaucracy, but repressing his accumulated rage at the instances of racism which he perceived. He felt that he had been passed over several times for promotion in favor of less competent administrators who were white and Jewish. Like many of the critics outside the system, McCoy felt that the system benefitted its staff more than its pupils; he believed that the high rate of failure among minority children was not accidental. In his post at Ocean Hill-Brownsville, he would be able both to use his experience and to vent his inner rage.[3]

One of the first acts of the planning council was to announce that an election for the governing board's parent representatives would be held twenty-six days later, on August 3, 1967. Nominating petitions were circulated, and sixty-one candidates were nominated. Some 1,800 prospective voters were registered in a door-to-door campaign by nuns from

Father Powis' church, poverty workers, and paid parent-canvassers. Of twenty paid canvassers, thirteen were also candidates in the election; among the thirteen were the district's parent-association presidents.

The Board of Education appointed the Niemeyer committee (headed by Dr. John H. Niemeyer, president of Bank Street College of Education) to evaluate the system's overall decentralization program. The committee later reported that the Board of Education, which had expected the entire summer to be devoted to the planning process, was "suddenly . . . confronted early in August 1967 with the fact that an election was in full swing" (the planning proposal did not state that an election would be held quickly, but did say that the governing board would take office in "early August 1967," a phrase which no one at Board of Education headquarters seemed to have noticed). The election itself, wrote the Neimeyer committee, was "unorthodox." Voting lasted for three days; canvassers went from door-to-door to obtain votes; preregistration was not necessary to vote; the votes themselves were kept in unlocked boxes and drawers; the election was not supervised by any accredited agency like the Honest Ballot Association. Nonetheless, the Niemeyer report concluded that there was "an honest effort to get votes from all the parents." [4]

Between 1,000 and 1,100 people voted in the election. Seven women won places on the governing board. Six of the seven winners had been paid canvassers, and five of them were also parent-association presidents. The election winners met on August 7 and elected five community representatives to the governing board. The five included four members of the planning council (which had hired the election canvassers); the fifth new member, who had not been a member of the planning council, was the Reverend C. Herbert Oliver, pastor of the Westminster Bethany United Presbyterian Church, who became the board's chairman. One of the community representatives selected was a paid canvasser who had lost her race for parent representative. Parent members of the governing board continued to receive weekly payments as election consultants for several weeks after the election, and in September a budget category was created to assure them a weekly salary from Ford funds. [5]

Later, when the Ocean Hill-Brownsville district became the center of a bitter, city-wide teachers' strike, supporters of the teachers' union directed numerous criticisms at the absence of election procedures, at the haste with which the election was conducted, at the planning council's failure to bring its proposals to parent-association meetings, and at the fact that some paid canvassers were also candidates themselves. The essence of these criticisms was that the election was a deliberate vehicle

to transform the self-appointed planning council into the elected governing board. The district defended itself by asserting that since there were
no election guidelines, there was nothing to prevent paid canvassers
from running for the board.[6]

Well before the election, relations between the teacher representatives and the other members of the planning council had begun to deteriorate. The teachers complained that they had not participated in the
choice of the acting unit administrator. They believed that the final plan
for the district would be worked out by the consensus method, with
nothing adopted over the teachers' objections (as in IS 201's plan); they
soon discovered that the planning council was not informing them of all
its meetings. They had expected to help rewrite the planning proposal
into the final plan. Instead they concluded that the plan was being re-
worked without their participation; suggestions for changes were presented to them in writing. They later claimed that their objections were
ignored, "or, when we objected enough, the topic under discussion was
dropped rather than revised. We were told by Mr. McCoy that the Governing Board itself, when it took power, would make its own rules and
expand on the plan. Without our knowledge or presence part of the plan
was rewritten and submitted to Dr. Donovan."

The teachers felt growing hostility from the other participants on
the council. They recalled "a constant stream of remarks to teachers
which stated that teachers were bigoted, incompetent, uninterested, ob-
structive and were attempting to sabotage the plan." When they discovered that the other members of the planning council had met with Donovan, they sent a telegram to him to protest their exclusion. In the final
plan, the concept of "consensus" decision-making was dropped. The
teachers felt that their role had been reduced to that of an impotent minority.[7]

During July, the teachers and the rest of the planning council disputed the section in the original proposal which protected teachers'
rights. McCoy noted in his progress report for July that "The issues that
appear to be of major concern to the teachers are: observations, evaluations and terminations," that is, who is to observe and evaluate teachers,
and how their jobs are to be terminated. The original plan permitted the
governing board to select the entire board which reviewed charges
against teachers, but teachers now feared the concentration of too much
power in the hands of the governing board. It was a section which
might have worked in an atmosphere of mutual trust, but not in an aura
of hostility and suspicion. McCoy later stated that this clause "was modified at the request of the teachers though not to their satisfaction." The

language which was finally substituted was: "The Board shall make provisions for periodic evaluations of the total program. Such evaluations
will include the Unit Administrator, principals, teachers, community
workers, etc. This is not to be construed as meaning the Board will do
the evaluating. Existing Board of Education procedures for evaluating
teachers will remain intact." [8]

Teachers unsuccessfully sought to vote with the parent representatives in choosing the community representatives. McCoy felt that "the
issue emanated from the determination of teachers to see that no militants or black-power advocates were selected. This to us was an attempt
to exclude a vital segment of the community and to deny the exercise of
free choice." [9]

McCoy's progress report on the plan in July concluded with the
statement that "it is understood that no part of this plan shall abrogate
or abridge the UFT Contract, the Board of Education By-Laws and the
Civil Service Laws of New York City and State." The final plan, however, did not include this statement, saying instead that the governing
board would be "responsible and answerable to the New York City Superintendent of Schools and the State Commissioner of Education in all
matters pertaining to the schools of this district." [10]

The final plan was ready in late August. One of its three pages was
a strident preamble, averring that "men are capable of putting an end to
what they find intolerable without recourse to politics. As history has so
frequently recorded, the ending of oppression and the beginning of a
new day has often become a reality only after people have resorted to
violent means." Ocean Hill-Brownsville, it warned, was "at such a point
of desperation" when the experiment was authorized. The decentralization district represented "the last threads of the community's faith in the
school system's purposes and abilities." [11]

The plan itself was a minimal description of the governing board
and the unit administrator. The specific powers claimed for the governing board were: the hiring of a unit administrator, a business manager,
community workers, and principals. In nonspecific terms, the proposal
declared that the governing board would " determine budgetary needs"
and determine policy "in the areas of curriculum, program and professional personnel." Nothing was mentioned about the power to hire, fire
or transfer professionals. Nor was there any educational proposal, as
there had been in the original plan approved in the spring, nor were
there any bylaws for the new board.

McCoy's nomination of Herman Ferguson to be a principal of the
new IS 55 underscored his disinterest in maintaining or restoring good

relations with the district's teachers. Ferguson, an assistant principal at a Queens elementary school, had been indicted two months earlier for conspiring as part of a black extremist organization to murder moderate civil rights leaders Roy Wilkins and Whitney Young. (He was convicted in 1968 and jumped bail in 1970, after the United States Supreme Court declined to review his conviction.) At his last assignment, Ferguson had been an outspoken opponent of his school's MES program, which made him unpopular with the UFT. The selection of Ferguson confirmed for some teachers their fears about the militancy of the unit administrator as well as their fears for themselves as white teachers in a black district. A member of the governing board told reporters that Ferguson "is not afraid of changes from old methods of teaching, and he's the person to keep teachers in line." The teachers did not think that the governing board should be looking for ways to keep them "in line," and they liked even less the choice of Ferguson as the man to do it.[12]

McCoy and Reverend Oliver met on August 30 with representatives of Donovan, the State Education Department, and the Niemeyer committee. The first item on the agenda was McCoy's choices for principalships; he was asked to withdraw the name of Herman Ferguson. He later informed the governing board:

> This I refused and was supported by Reverend Oliver. There was much discussion by all present from reasons to legality. Reverend Oliver stated, "No one had the right to judge or determine the guilt or innocence of the candidate. Since he was still a Board of Education employee, we were within our rights to consider him." This was agreed but it was requested his name still be removed because of the publicity and the feelings of Dr. Donovan and the Board (rumor). Reverend Oliver stated that too often a minority member was accused, tried and sentenced by the press or public opinion before the trial and in most cases damaging to the person who may have been found innocent.[13]

Later the same day McCoy met again with the representatives of Donovan and the state, who had since conferred with the Board of Education. They informed McCoy that all his nominations but Ferguson were acceptable to the Board. On learning this, McCoy abruptly terminated the meeting.

He recommended to the governing board members that they either accept or reject his candidates for principalships; that they inform the Board of Education of their action; that they request a meeting with the president of the Board of Education; and that they send telegrams to State Commissioner Allen and Mayor Lindsay, requesting clarification of the problem.

The governing board members were now on notice that to abandon Herman Ferguson's nomination would embarrass their spokesmen as well as encourage a racist vendetta against a black man. On August 31, when McCoy's nominations were put to a vote, the teachers on the board argued vociferously against Ferguson. They warned that his selection would bring the project bad publicity; they urged at least a delay, since IS 55 would not be ready until later in the fall. The other members of the governing board voted unanimously for Ferguson; the teachers abstained. Their abstention infuriated the other members of the governing board, who saw it as a refusal to cooperate.

When the newly elected governing board voted to retain Rhody McCoy as its unit administrator, the teachers obstinately nominated the white principal of JHS 271 for the job. The other members of the governing board bitterly resented the teachers' action as a symbolic gesture of "no confidence" in their unit administrator's leadership.

By summer's end, a city-wide teachers' strike loomed, since the Board of Education and the UFT had not reached a settlement on the union contract. The union's demand for stricter controls over disruptive children offended civil rights organizations and black militants, since most suspended children were black and Puerto Rican. From the perspective of the demonstration districts, the union was striking for the right to put black and Puerto Rican children out of class more easily. The demonstration projects determined in advance of the 1967 strike to keep their schools open. UFT officials sought support for the strike in these districts and were rejected. District officials thought the UFT was offering them a deal: go along with the strike and the UFT will go along with the demonstration projects. It may be that those relationships were too far gone to allow cooperation during the strike, or that the project leaders saw themselves as employers and wanted to exercise their new role. Besides, some board members did not want their first action of the new semester to be the closing of their schools.

Yet something else had begun to emerge by the end of the summer of planning: the boards felt themselves to be in a power struggle with the union; they had begun to see that one of the chief hindrances to their control of the schools was the union. Just as they wanted their unit administrator to be loyal to the local board rather than the Board of Education, they wanted their teachers under local control. The impending strike heightened their sense that UFT teachers had another master. The boards wanted to make decisions on their own; they wanted legitimacy and authority. They did not want to have to deal with the union, except on an employer–employee basis.

This conflict of interest and power was not apparent in the spring of

1967, when the projects were planned. At that time, the Ford Founda-tion and the Board of Education had agreed that no demonstration dis-trict would be approved which did not have good working relations be-tween the community and the UFT. None of the preliminary proposals asked for the power to hire and fire teachers. But the 1967 teachers' strike made the leadership at 201 and Ocean Hill-Brownsville aware that this power was a necessary precondition to community control of the schools.

When the teachers struck during the first two weeks of the fall se-mester, schools all over the city were closed, but the schools in the dem-onstration districts were kept open. In Brooklyn, Herman Ferguson and Albert Vann, leader of the African-American Teachers Association, con-ducted workshops to equip parents to teach during the strike. Picketing teachers in Ocean Hill-Brownsville complained of harassment and threats by people connected with the governing board. McCoy com-plained that nonstriking supervisors covertly refused to cooperate with him during the strike.

In the IS 201 district, a member of the planning board announced that all striking teachers had resigned and would not be allowed to re-turn. The planning board adopted a policy that striking teachers would have to face a "screening panel" before they could teach again. When the strike ended, screening plans were revealed to the press by Herman Ferguson, who had been appointed to the staff of the planning board. After vigorous protest by the UFT, the deputy superintendent of schools informed the IS 201 planning board that it had no power to prevent striking teachers from returning to work.

The breach between the UFT and the demonstration projects was complete when the UFT joined in the supervisory union's court attempt to prevent the districts from picking their principals outside of the regu-lar civil service channels. The UFT's participation in the suit was retri-bution for the organized strike-breaking in the demonstration districts. The districts showed the union during the strike that they were not afraid to stand up to the union; the union showed the districts that they were not afraid to stand up to pressures to get out. It became a matter of time until both sides stood their ground at the same time and same place.

CHAPTER 30

The Bundy Report

During the middle 1960s, criticism of public schooling was unusually intense. Critics of the schools, like Paul Goodman, Herbert Kohl, Edgar Friedenberg, Nat Hentoff, John Holt, and Jonathan Kozol, questioned the means and the ends of the public schools in books, articles, and reviews. Some saw the school as a mechanism for repressing children's individuality; some saw it as an oppressive instrument for imposing the mores of the white middle class on the poor; others criticized the school as Dewey had for not taking on the task of social change; still others thought that the school should be the center of its community, providing cohesion, and, in poor communities, jobs and political leadership. Dissidents in the ghetto and dissidents on campuses articulated discontent with the status quo, the social and economic arrangements that had come to characterize modern American life. They looked at American society and saw a series of oligarchic systems which appeared to stand in the way of sweeping social change. Student radicals spoke of "participatory democracy," black militants spoke of community control; both urged new forms of participation in governmental affairs.

Because it is part of the stabilizing mechanism of the American political system to absorb dissidence by amelioration, government responded to demands for greater participation by creating new programs. At the federal level, local participation was the hallmark of not only the poverty program, but also the preschool Headstart program and the Model Cities program. Among municipalities, none reacted more earnestly to demands for citizen participation than the administration of Mayor John Lindsay.

Lindsay, from the start of his mayoralty, sought to establish his concern for blacks and Puerto Ricans. When Lindsay was directed by the legislature to prepare a decentralization plan for the city's schools, his own inclinations clearly favored the construction of a system which would satisfy demands for participation by ghetto community groups. The fact that the Bundy panel, which prepared the decentralization plan, was selected by Lindsay was significant to the direction that the report took. The panel and its staff were attuned to the views of the system's critics.

The trend of the report was further indicated by the choice of Mario Fantini as staff director. Fantini's opposition to centralized bureaucracy was well known. In addition, he had worked closely with the most militant community groups in planning the demonstration districts; during this period, he shared their view that a prolongation of the status quo would cause blood to flow in the streets of the ghettos.

The chief consultant to the Bundy panel was Marilyn Gittell. Gittell was a prolific critic who had developed a considerable *oeuvre* on the New York City school system. In 1966, she authored Staff Paper 9 of the Temporary Commission on City Finances, which proposed decentralization of the system into five borough districts and an increase in the mayor's participation in school affairs. Later that same year, she was a consultant to the militant People's Board of Education. In 1967 and 1968, she advised Harlem CORE in its efforts to get foundation funds and legislative approval for an independent board of education for Harlem. Staff Paper 9 provided the rationale with which Mayor Lindsay had gotten more state aid for the schools in return for a promise of a decentralization plan; indirectly, the Bundy panel owed its existence to Staff Paper 9.

Early in 1967, Gittell, an assistant professor of political science at Queens College, extended the analysis of Staff Paper 9 into a monograph called *Participants and Participation*, which was widely circulated. Gittell's basic conclusion was that the school system of New York was rigid, hostile to change, paralyzed by problems, and locked in a "precipitous downhill trend." [1]

The Gittell study advanced two related contentions: one, that the school system was characterized by a "persistent lack of fundamental change"; and two, that the system's failure to adapt and innovate was due to the monopolization of power by the central supervisory staff. In other words, the system was failing, and it was failing because participation in policy making was closed to all but a handful of inbred, professional bureaucrats. These factors were responsible, Gittell maintained, for the ineffectiveness of the Board of Education's compensatory education program and its integration policy.

Her prescription for change was to diminish the bureaucracy's power by opening the school system to more participants—the mayor, local school boards, civic groups, and the general public. She especially urged that the mayor play a more forceful role in the formulation of school policy, reversing what she called the "depoliticalization process." Overall, she proposed decentralization of the schools in order to stimulate local involvement, either by strengthening local school boards, by dividing the city into fifteen or more school districts, or by creating five independent borough school boards.

Participants and Participation was particularly important because it appeared at the same time that black and Puerto Rican children had become a majority in the school system, causing many of the people who wanted to reform the schools to abandon all hope of integration and to seek a new strategy for change. Decentralization, it appeared, was synonymous with educational reform, as well as with progressive social policy.

Yet while Gittell ably demonstrated the dysfunctional aspects of overcentralization, her prescription for change was not appropriately related to the two conspicuous policy failures, integration and compensatory education. Marilyn Gittell (and David Rogers in *110 Livingston Street*) blamed the collapse of these programs on the bureaucracy. But both areas of educational policy can be seen as instances where the Board of Education undermined its own high intentions by giving in to outside, public demands.

The Board of Education's unequivocal commitment to eliminate racially homogeneous schools had the support of major civic organizations and civil rights groups in the late 1950s and early 1960s. "It is possible," wrote Gittell, "that if implementation had been immediate and firm, the strong civic group support for the early recommendations would have been sufficient to carry forth the plan, and opposition might have been minimized. But delay and postponement allowed time for opposition coalitions to develop." In the Ridgewood-Glendale integration dispute, she noted, thirty-four citywide civic groups and the *New York Times* favored bussing, but "delay and indecision in the inner circle of educational decision-makers" compromised implementation because of the "protests of a small group of parents." [2] In sum, if the Board of Education had acknowledged the participation of only that part of the public that agreed with its policies, it might have been able to succeed. It is hard to see how things would have turned out differently if there had been decentralization, for the more public participation there was, the more difficult it was to implement integration. If the issue had been left to the local school board in Ridgewood-Glendale, it is unlikely that there would have been any bussing of black children into those communities.

Although civic groups backed the Board of Education's policy, public resistance to forced integration was so widespread that neither Mayor Wagner, Mayor Lindsay, nor any other major politician was willing to endorse it; therefore, it seems unlikely that mayoral intervention would have changed the outcome. With a handful of rare exceptions, American cities have not integrated in response to public demand, but in compliance with court orders. In New York City, as elsewhere, it was political pressure that blocked involuntary pupil transfers. The failure to achieve racial balance was due to the Board of Education's inability to impose a politically unpopular program as well as to its powerlessness to reverse a shifting ethnic tide.

Similarly, the overextension and dilution of compensatory programs can be attributed to the effect of outside political pressure on the school system. Both the Board of Education and the superintendent of schools responded to demands for quick results by expanding compensatory programs far beyond their capacity to produce quality education, without planning, without sufficient funds, and without sufficient training of the staff. The capitulation of the Board of Education can be traced to its lack of political authority, either the authority of having been elected or the authority of a mayor who was willing to stand behind controversial policies. Strong school officials might have refused, for instance, to bow to political pressure to expand prematurely a successful program like the Demonstration Guidance Project; they might have insisted on extending the program only as adequate funds were made available and they might have urged parents to work on behalf of increased appropriations. Instead, school officials acquiesced to public demand, exchanging an effective program in one school for an ineffective program (Higher Horizons) in several dozen schools.

Nonetheless, the Gittell analysis, which identified decentralization with educational reform, was the framework for the report of the Bundy panel, entitled *Reconnection for Learning, A Community School System for New York City*. Its recommendations for decentralization were a radical departure from the mild administrative decentralization which the Board of Education had pursued since 1961. The Bundy report envisaged a redistribution of political power of the system and an enlarged role in school affairs for the mayor. The panel recommended the reorganization of the system into a federation of thirty to sixty largely autonomous local school districts under the general supervision of a small central educational agency. The local boards would be partly elected by parents of school children, partly selected by the mayor. The plan gave each community board the power to hire a community superintendent of schools, to hire personnel, and to grant tenure; existing tenure rights were to be protected, and collective bargaining was to remain centralized. The Board of Examiners and the city-wide system of examination and licensing would be abolished. The community school districts would have authority over all regular elementary and secondary schooling within their boundaries, subject to state educational standards. The panel proposed draft legislation and urged its passage during the 1968 session of the state legislature.[3]

The premise of the Bundy report was that the once great New York City school system was "caught in a spiral of decline." (Gittell had called it a "precipitous downhill trend.") As evidence, the report cited the large numbers of children who were failing to achieve at grade level. It also attributed the exodus of white families to the suburbs and to private and parochial schools to loss of confidence in the public schools. The report held that "even in prosperous neighborhoods, parents' confidence in the public school system is diminishing."

However, this assertion that dissatisfaction was universal was never demonstrated. In trying to show the need for an overall revision of the system, the report failed to emphasize that academic retardation was not spread randomly through the system. Low achievement on standardized tests in reading and arithmetic was in fact concentrated in schools with large numbers of children from low-income families. Schools in middle-class areas had reading scores a year or more above the national average. The report insisted, though, that the entire system suffered from malaise; the report did not confront the persistence of low academic achievement by disproportionate numbers of lower-class children, a problem which was not only national, but historical.

A central inconsistency in the report was its attitude toward the potency of decentralization. On one hand, decentralization was described as a mechanism to reverse the educational deterioration of the system; on the other, the report stressed that decentralization alone would not improve the education of disadvantaged children. This discrepancy reflected an apparent difference of opinion between the staff, which did believe that decentralization, in and of itself, would have a revitalizing impact on the schools, and the panel, which cautiously qualified its view of decentralization:

> It is important to emphasize that the best possible reorganization of the New York City schools can be no more than an enabling act. It will not do the job by itself. Reorganization will not give New York the additional funds it needs to improve schools in all parts of the city. It will not wipe out the generations of deprivation with which hundreds of thousands of children enter the schools. It will not meet the great deficits in health and

welfare services that beset many families. It will certainly not wipe out the poverty and physical squalor to which too many children return when they leave school every afternoon. It will not wipe out the shortage of qualified, imaginative, and sensitive teachers and supervisors. It will not automatically provide insights into the uncharted terrain of the basic mechanisms of learning and teaching.

The panel further stated that, "Decentralization is not attractive to us merely as an end in itself; if we believed that a tightly centralized school system could work well in New York today, we would favor it." But the top staff of the panel felt otherwise. Fantini, Gittell and Magat, in their book, *Community Control and the Urban School*, stressed their conviction that community control itself was a vital ingredient in the radical improvement of education.[4]

Interestingly, the Bundy report did not cite in its bibliography any precentralization works on the New York school system; the authors lacked an awareness of the problems of the ward system or of the reasons that led to its abolition. Had they known the public schools' history, they would have known that small districts guarantee neither community participation nor educational reform; they do bring the schools closer to the parents and make them more reflective of local interests. This, however, can have results which do not necessarily please educational reformers, for the interests of the parents may be out of step with the latest educational thinking and may actually serve to insulate the schools against reform.

The publication of the Bundy report set off a heated public debate; reactions to the report were based not on its careful qualifications, but on its advocacy of decentralization as an educational reform strategy. Alfred Giardino, president of the Board of Education and a member of the Bundy panel, dissented from the report: he warned against attempting to recast "in one quick stroke, the largest educational system in the world." All the major organizations of educational professionals in the system attacked the report. Predictably, the most controversial section was the proposal to abolish the Board of Examiners. The panel, adopting the view of past reports critical of the Examiners, felt that its elimination would increase the pool of available teachers and raise the ratio of black and Puerto Rican staff (then less than 10 percent). But professional supervisors saw this proposal as an end to the merit system and the return of pork-barrel politics to the schools.

The UFT was willing to go along with the abolition of the Board of Examiners, but insisted that it be replaced by the national teachers' examination, to retain some objective standard for admission to teaching. The union strenuously opposed the report's provisions on tenure:

Tenure is a precious right. Tenure gives teachers the security they need to teach honestly, free from community pressures. Under the tenure concept a teacher can be dismissed only for cause after a hearing on the basis of charges brought against him *by other professionals who are competent to evaluate professional performance. Under the Bundy report, charges could be brought against a tenured faculty member by a community board of laymen with no professional expertise.* This proposal is anti-professional. It would encourage local vigilantes to constantly harass teachers. No teacher with professional integrity could teach in such a district.[5]

The teachers' union warned that the Bundy plan would "irreparably harm the educational system. The Bundy model is based upon a glorification of the old-time rural school structure. . . . (It) is not decentralization, it is Balkanization."

The UFT called instead for decentralization into no more than fifteen districts with school boards elected by the public (not just parents, as Bundy suggested). The union supported decentralization because it seemed inevitable, not because it seemed desirable. The union's position was that decentralization was educationally irrelevant: "*The basic shortcomings of our school system are not due to the fact that there are three districts or thirty, but to decades of financial starvation.*" Albert Shanker pledged that the union would spend "whatever was necessary" to defeat the Bundy plan.

The Bundy plan was attacked by those who refused to give up all hope for integration. It was criticized strongly by the Puerto Rican Borough President of the Bronx, Herman Badillo, who feared that it would encourage ethnic frictions as groups competed for control of school boards. The New York Board of Rabbis called it "a potential breeder of local apartheid." Several staff members at the Center for Urban Education published their rejection of the plan, warning parents to "beware of easy answers. . . . Nothing in the Bundy plan convinces us that there is any connection between the type of administrative changes it proposes and the improvement of education in our public schools." [6]

Though the plan received the endorsement of the Metropolitan Council of the American Jewish Congress, it was a curious statement of support, which acknowledged the report's worst potential features while hailing it as

a far-reaching and imaginative approach. . . . It is possible that by giving the power to hire and fire to local boards, we may open the door to personal and politically motivated appointments and increase racial and religious interference in the selection of staff. . . . In the long run, however, we believe that the very real desire of the parents in the community for the best education possible for their children will act as the most powerful deterrent against ill-made appointments.[7]

On January 1, 1968, Mayor Lindsay submitted the Bundy plan, with modifications, to the legislature. The chief change was that the high schools were withdrawn from the purview of the local boards and placed under the central board. Now known as the Bundy-Lindsay plan, it began to wend its way through the legislative process.

Lindsay's initial enthusiasm had cooled by the time the plan became a bill, in part because of his own assessment of the opposition aroused by the plan, but also because of critical staff reports from his Budget Bureau, which cautioned that the proposal was an idea rather than a workable plan. Lindsay performed certain perfunctory gestures, like sending officials to testify on its behalf, but he did not campaign for its passage.

Early in March, the Board of Education sent its own moderate decentralization plan to the state legislature. It proposed making the thirty existing local school boards operational instead of advisory, with the power to hire a district superintendent on a contract basis (that is, without tenure), with the power to grant or deny permanent tenure to teachers on the district superintendent's recommendation, and with the power to appoint elementary school principals on the basis of a qualifying rather than a competitive examination (so that principals could be selected who were not at the top of the civil service list). Their legislation also would have vested the Board of Education with authority to delegate any of its powers to local school boards.

The announcement of the Board's cautious plan mobilized supporters of the Bundy-Lindsay plan into action. Up to this point, decentralization lacked an organized lobby. A few days after the Board's plan was presented, a meeting of decentralization advocates was convened in the home of Mrs. Joseph E. Clark, Jr. Mrs. Clark, a member of a socially prominent family, was a member of the Citizens Committee for Children and other civic organizations.

At the meeting, a statement was drafted: "On Saturday, March 9, 1968, fifteen people speaking as individuals or for organizations but representative of a broad spectrum of the interests and concerns of the people of this city agreed to join together to indicate to their fellow citizens and to the state legislature the necessity for a meaningful plan for the decentralization of the New York City school system." At the next meeting of this group, a representative from the mayor's staff was present; an observer at the meeting noted that "Mrs. Clark has an appointment with the Mayor tomorrow to see if he won't try to influence key people in favor of decentralization." [8]

By March 25, the group had a new name, the Citizens Committee for Decentralization of the Public Schools, and an impressive executive committee, and it was ready for its first press conference. Robert Sarnoff, president of RCA, was named chairman of the group. Sarnoff was a prominent businessman, a member of the New York Urban Coalition, a trustee of the Whitney Museum, and a leader in numerous philanthropic enterprises. Other names on the list included Thomas Watson, Jr., chairman of IBM; James Linen, president of Time, Inc.; James B. Conant, former president of Harvard University; Norman B. Johnson, an NAACP attorney; the Reverend Milton A. Galamison; Mrs. Alvan Barach, president of trustees of PEA; T. Edward Hollander, coauthor with Marilyn Gittell of a critical study of six urban school systems; and Dorothy Jones, of the original IS 201 negotiating committee.

The committee, by the luster and diversity of its membership, sought to demonstrate that support for decentralization came from all races and segments of the population. In addition to issuing public statements, the committee was busy behind the scenes. In late April, a notice went out to its membership: "During the past weeks members of this committee have been meeting with representatives of the Board of Regents, the Office of the Commissioner of Education for New York State and with representatives of the Mayor of the City of New York, to bring to their attention the concerns of this committee." It urged members to go to Albany to support a strong decentralization bill.

However, at the same time that the committee was in high gear, the leadership in IS 201 and Ocean Hill-Brownsville was working actively to defeat the Bundy-Lindsay bill or any other legislation which did not incorporate what it considered sufficient community control. The rush of events and emotions had carried these districts to a point where limited decentralization was worse than no bill at all.

CHAPTER 31

Preparing for a Showdown

DURING THE FALL AND WINTER OF 1967, the governing board of Ocean Hill-Brownsville came to the conclusion that decentralization would not provide them enough power; only total community control would do. The urgency of community control was proved by the board's problems with its professional staff. While the governing board struggled to assert its authority, the supervisors' union and the teachers' union sued to void the appointments of principals in the demonstration districts who were not on the Board of Examiners' list of eligibles. This was the single most important power that the Board of Education had ceded to the projects; the governing board knew that the freedom to pick principals was vital to its capacity to influence the schools.

McCoy felt that the project was being sabotaged by uncooperative professionals. In November, seventeen of the district's twenty-one assistant principals, complaining of harassment, requested transfers to other districts. Their requests were granted as other places in the system became available. The governing board asked the Board of Education for permission to hire its own assistant principals, without regard to existing lists. Since the Board had been willing to create a new category for "demonstration principal," the governing board hoped that another new category might be established for "demonstration assistant principals," in order to circumvent the civil service list. But permission was denied by both the superintendent of schools and the state commissioner of education, pending the outcome of the court challenge to the demonstration district principals.

The governing board was convinced that the UFT's hostility was damaging the project's prospects for success. Leadership from the dis-

338

trict and the union met in November, but none of the differences between them were worked out. Shanker and McCoy continued to meet privately over the next few months; Shanker agreed, at McCoy's behest, to set up a committee of teachers within the district to prepare proposals for new programs.

Many Ocean Hill-Brownsville teachers continued to worry about the militancy and intentions of the governing board. At a meeting in December, the teachers asked questions of McCoy and the governing board which revealed the staff's anxiety and suspiciousness. Among the questions asked were: "There is a black–white split in the schools in your district. What are you doing as Unit Administrator to correct this problem?" "Do you want to have an entirely Afro-American staff in the schools in our district?" "What have you done to get parents to participate?" "There has been a rumor that your policy has been influenced by an outside militant group. Is this true?" [1]

The governing board's internal problems with its teachers were matched by its external difficulties with the Board of Education. All three governing boards refused to accept the guidelines of the Board of Education, which were issued in early December 1967. Consequently, the Board of Education refused to grant formal recognition to any of the governing boards. The Board of Education, in its guidelines, insisted that the governing boards include these four points in their proposals: "(1) provision for a fixed term in office for Project Board members; (2) recognition of the responsibility of the Project Board to (the) central Board of Education; (3) application for federal and state funds—this must be done within the framework of existing laws; (4) a statement indicating the Project Board's acceptance of the requirements of the Board of Education's guidelines." [2] The governing boards, however, rejected these requirements. They wanted their unit administrators responsible solely to them; they wanted to apply for outside funds without going through the Board of Education. And they wanted the process of evaluation to begin in September 1968, not September 1967 as the Board of Education had proposed.

As problems multiplied, the governing boards became convinced that the demonstration districts needed far more power than the Board of Education was willing to grant them, far more power than was originally contemplated in their planning proposals. In early 1968, they refined their definition of community control. The Niemeyer report stated: "They have expressed the need to control their own budget (on a number of occasions they have proposed that they have their own bank account). They want the right to hire and fire the staff and to engage in contracts and subcontracts, using local citizens, of course. They have candidly discussed all three issues with the Board of Education." The three governing boards and the Board of Education met repeatedly during the spring to try to work out an agreement, but the governing boards steadfastly refused to accept the limitations proposed by the central board. On the contrary, their demands for local power escalated.

THE IS 201 GOVERNING BOARD was slower in getting established than the Ocean Hill board, but lost no time in reaching the same deadlock in its negotiations with the central board.

Elections in the IS 201 district were delayed by internal political disputes. A parent group led by the deposed leader Helen Testamark carried protests against the planning board to the Board of Education and the City Commission on Human Rights. The dissidents alleged that parent leadership had been usurped by representatives of activist poverty agencies (MEND, in particular) and that this had been accomplished with the covert assistance of Ford Foundation personnel. The Board of Education denied the petition of the Testamark group, which then boycotted the election for the governing board.

The election, held November 10, was supervised by the Honest Ballot Association. Some 563 valid ballots were cast, 404 of them by parents. David Spencer, chairman of the IS 201 planning board, was elected parent representative at IS 201 with 72 votes; he was later elected chairman of the governing board. As in Ocean Hill-Brownsville, the leading members of the planning board became members of the governing board, if not by election as parent representatives, then by their selection as community representatives by the parent members.

Soon after the governing board was elected, it began interviewing candidates for the principal's post at IS 201 and for the unit administrator's job. There were fewer personnel conflicts at IS 201 than at Ocean Hill, because twenty-three teachers who opposed the project transferred out. The 201 board, without interference from the union, replaced them with new teachers of its choosing.

In January 1968, the governing board asked the Ford Foundation for both political and financial assistance. The board wanted a public statement from Bundy "giving his support to the demonstration projects in their present limited development." It hoped that Ford would use its influence with the mayor and the state education commissioner to help the districts get the autonomy they sought. Ford was asked to fund a substantial program for hiring parents and community workers. The

201 board viewed the schools as a community resource, not unlike the poverty program, which ought to provide jobs and training for local residents; it believed that the children and their parents would have a better attitude towards the school if it played an integral part in the economic uplifting of the community.[3]

The following month, the Ford Foundation was publicly embarrassed by a Malcolm X memorial program at IS 201, where an antiwhite play by poet Leroi Jones and antiwhite statements by Herman Ferguson (who was then on the staff of the IS 201 governing board) attracted sensational, unfavorable publicity. The Board of Education held up approval of the governing board's nominees for IS 201 principal and unit administrator until Ferguson (who was under suspension by the Board of Education because of the criminal indictment against him) was removed from the district's payroll. Not only was Ferguson paid with Ford Foundation money, but Ferguson had narrowly missed winning election by the 201 governing board as its unit administrator, despite his suspension by the Board of Education.

The Ford Foundation found itself in an unusual and unwonted position. The demands and dependency that characterized its dealings with the demonstration districts placed an unaccustomed strain on members of the foundation's staff. It was accustomed to giving money to established, ongoing institutions, not to organizations that relied on the foundation for daily, even hourly advice. Ford had helped to create the demonstration projects, had literally brought them into existence, and now was expected to see them through a succession of crises. In January 1968 the Ford Foundation allocated $38,000 to set up the Institute for Community Studies, headed by Professor Marilyn Gittell, who had worked closely with Mario Fantini in writing the Bundy report. This grant, later augmented by almost $1 million, enabled the institute to provide technical assistance to the governing boards, assistance which came to include legal, financial, administrative, and political matters. The institute became deeply involved in the affairs of the three districts, and Marilyn Gittell became known as an ardent advocate of community control.

The demonstration projects expected great sums of money from the Ford Foundation, as well as active political protection. Ocean Hill-Brownsville expected $5 million from Ford, a sum which the governing board thought would cover the hiring of 200 community people as full-time teaching assistants. In February 1968, Ford announced a total grant of only $46,000 as a supplement to the planning grants of all three districts. The meagerness of Ford's appropriation, in contrast to the high expectations, caused bitter disappointment.

IN EARLY 1968, the demonstration districts prepared a consensus document, listing the powers which were necessary for local control of the schools:

1. "The community governing board must have *total control of all money*, both capital and expense budget money. The money should be banked in a community bank with local people signing the checks. Each governing board must be empowered to seek its own state and federal funds, which will also be locally controlled."

2. The community governing board must have the power to hire and fire all personnel in the district and the power to negotiate union contracts.

3. The community governing board must have the power to contract for the building and rehabilitation of its schools.

4. The community governing board must have the power to "buy all its own textbooks and supplies *by direct purchase.*"

5. "Newly created experimental districts, where parents have been democratically elected to governing boards must be continued in existence. If they are to be expanded, new elected parents must be added to the board and no elected parent may be removed." This section reflected their concern that the legislature might pass a plan in which their small districts would be swallowed up by larger districts.[4]

The demands for control created a quandary for the Board of Education, which knew that legally it could not delegate the powers demanded by the demonstration districts. The Board's repeated commitment to gradual change made it obvious that it would not want to dispense those powers even were it legally possible. Members of the central board felt that they had more than fulfilled their commitment to the projects. They had allowed the governing boards to function without formal recognition; they had appointed the nominees of the governing boards to principalships (except for Herman Ferguson, who was suspended after his indictment on criminal charges); they had rejected challenges to the governing boards in IS 201 and Ocean Hill-Brownsville by dissident factions in their communities. When the State Supreme Court ruled in mid-March 1968 that the appointment of "demonstration principals" was an illegal attempt to bypass the civil service law, the Board of Education appealed the ruling without removing the principals. When McCoy fought for the right to hire his own staff, the Board granted him a lump sum budget for his office staff.

But from the perspective of the governing boards, the Board of Education and its staff blocked them at every turn with bureaucratic game-playing. The IS 201 board complained that it had been unable to obtain parent lists or adequate secretarial assistance during its election process; supplies were delayed, requisition slips lost; teachers who asked for assignment to the district were advised, *sotto voce,* to go to other districts instead; salary checks came late. The Ocean Hill-Brownsville board also felt that it was up against a wall of resistance at central headquarters, which was expressed silently through unnecessary mistakes, delays, and omissions. Checks for district employees who were not on the regular Board of Education payroll were delayed for months; McCoy's office was moved out of JHS 271 and into an unheated storefront, until better quarters were ready. Undoubtedly, many of these snafus grew out of hostility to the experiments, but it is likely too that the districts saw intentional rebuffs where others might have recognized ordinary bureaucratic inefficiency and ordinary resistance to new ways of operating.

The Board of Education, however, was convinced of its good intentions and continued through the spring to meet with the governing boards, hoping to come to terms and grant them formal recognition. Each meeting ended in impasse, since the governing boards were determined not to accept the Board of Education's guidelines.

After one such meeting, in March 1968, the Ocean Hill-Brownsville board left in a state of "great depression. . . . It was suggested that since we would not be recognized until we accepted the guidelines, and inasmuch as the guidelines gave the Governing Board no real power to govern the schools, it was strongly suggested that possibly the best solution would be to dissolve the Governing Board." A few days after this meeting, the mood of depression turned into rage when a State Supreme Court judge ruled that the district's six new principals had been appointed illegally.

Father John Powis, the militant white member of the local board, said that he had "never seen a racist as flagrant" as the judge. "The two week trial was a farce. The man had made his decision before the first witness testified. He is a perfect example of white America today and he is supposed to represent our system of justice. I believe that this Governing Board and this community has been 'responsible' for just too long. Maybe it's about time to become 'irresponsible.' " Another member of the governing board, a Puerto Rican antipoverty aide, said:

> I can tell you that the people did not elect this Governing Board to be a powerless puppet for the Board of Education, the Ford Foundation or Mayor Lindsay. We were selected to see to it that the black and Puerto Rican children, victims of the "white power" structure, do not continue to

be oppressed by white racism any longer. Our children will get the best education, by whatever means—peaceful or forceful—decided by the parents of this community. Our Principals will stay! If they are to be removed, the power structure will have to make their first use of the National Guard for 1968.[5]

The governing board, frustrated by its dealings with the unions and the bureaucracy and infuriated by the setback in court, decided that it was time to act. A long, vitriolic newsletter went out from the governing board, explaining to "the people of Ocean Hill-Brownsville and . . . to the entire City, State and Country . . . how the people of this community have been deceived by the power structure in their grass-roots attempt to gain control of their public schools."[6]

The newsletter explained that the original plan was approved by the Board of Education "because our school situation was so bad that it presented a threat to the establishment and because the Board of Education and the Ford Foundation needed some area to 'experiment' with the Bundy Report. . . . The very generous Ford Foundation 'gave' the community $59,000 as planning money." Their "only real power"—the selection of principals—was declared illegal, but "we go on record as supporting our principals despite this decision. . . . We the people of this community chose them and they all have State Certification as Principal, and we will not submit them to the 'impartial' Board of Examiners —We do the selecting—Not the Board of Examiners."

The powerlessness of the governing board was blamed on the Board of Education, which "never really wanted the new district or thought that it would succeed." They accused the Ford Foundation of playing "a weird role." Bundy had visited their schools and knew that they needed money "to put one or two trained parents in each classroom as teacher-assistants to prove how community participation in the schools could improve the education of our children." Proposals were submitted; parents were asked to sign up for training and classroom assignment. "Then the Ford Foundation got worried about whether we were 'responsible enough' people. And for some strange, unexplained reason," the Ford money never came through.

The schools were no better than before, "but it isn't our fault. . . . This Board has fought and taken on the whole power structure. *But we haven't been given any power at all.* . . . Come out and fight with us for full community control of our schools." The statement promised that the governing board would resign "as a body sometime in April" unless it got the powers it needed to govern the schools.

The governing board decided to boycott JHS 271 and IS 55 on April 10 and April 11 to demonstrate support for its demands. The boycott was completely effective. In its letter announcing the boycott to the parents, the governing board repeated a charge that became a refrain: "This Governing Board was promised to have the power to truly govern or run their own schools." [7]

But the governing board's insistence that it had been promised the power "to truly govern or run" its schools was a reflection of aspirations, rather than facts. The Board of Education *never* promised them any real operating authority. The Board envisioned the demonstration districts as tests of community participation and saw the "governing boards" as elected advisory groups with the power to choose a chief administrator and principals. Not even the original proposals of the various planning councils, which were all the Board had approved, sought the powers that were associated with community control.

The governing boards knew that many of their demands were not within the legal scope of the Board of Education to convey; at least, the leadership of the boards knew this. The unit administrators of each district were in daily, sometimes hourly, contact with the Institute for Community Studies at Queens College, which hired a lawyer to study the legal aspects of the dispute between the governing boards and the Board of Education. On March 12, their attorney, Howard Kalodner (who had previously been counsel to the Bundy panel), advised them that the Board of Education *could* give them more power under the existing law, but it would be substantially less than their demands, especially in the area of personnel powers. The boards were told that the law would have to be changed, if the projects were to get the degree of local control they wanted. Kalodner noted that the basic conflict between the central board and the governing boards was in their differing conceptions of the experiment. Since the Board of Education insisted that the only experiment consisted of having a locally elected board, it was not willing to give the governing boards any greater power than that held by a district superintendent. The governing boards, on the other hand, believed that their districts had been created to test community control; this belief was not based on any statement made by the Board of Education. This sharp difference of opinion, not broken promises, was the basis of the dispute between the central board and the projects. [8]

Actually, militant leaders like Rhody McCoy and David Spencer were not interested in *testing* community control, but in *having* community control. They saw no reason why their districts should serve experimental purposes, when their schools and communities were so desperately in need of improvement. They saw the school not just as a place to transmit skills and literacy, but as *the* institution which might generate a new sense of self-worth and community among blacks; where black and Puerto Rican children could receive a positive self-image by contact with adult models of their own background; where parents could gain a sense of dignity by playing a part in their children's schooling; where jobs and contracts could be consciously used to improve the economy of the surrounding neighborhood.

The rapid escalation of aspirations for black power and total community control was well portrayed in an article by Rhody McCoy, which appeared in a professional educators' journal in April 1968. McCoy declared that black educators "must become the vital force in halting the downward trend in urban education. The tokenism which we now endure must develop into a rampaging conflagration that will ultimately mean control. The policymakers and first-line implementors of educational process for black and Puerto Rican children must be black and Puerto Rican educators." He warned that

> the reform movement in education will raise the aspirations of the black community all over the country to the point where violence will be the unquestionable consequence of denial. . . . [Whites] cannot prepare fast enough to cope with the determination and commitment of those of us who are pledged to wrest from them their illegal hold on the future of ourselves and our children. If whites are not willing to relinquish this stranglehold, if they are not willing to work with blacks in resolving these Herculean problems, then the battle lines are drawn. There will be massive, persistent and even violent confrontations.[9]

Against the backdrop of passionate and revolutionary convictions, decentralization proposals were judged as tokenism. Neither the Bundy-Lindsay bill nor the Board of Education's decentralization plan was acceptable to the governing boards, because neither granted full community control. At the end of March, a third decentralization bill, sponsored by the State Board of Regents and the commissioner of education, was introduced into the legislature; the Regents bill eventually superseded the Bundy-Lindsay bill as the preferred legislation of decentralization advocates. The Regents plan recommended the division of the city school system into fifteen regular school districts, each with maximum autonomy but subject to the limited supervision of a new five-member Board of Education. The plan also provided for the creation of a limited number of small, temporary districts, called "target districts," in areas of lowest educational achievement. The bill eliminated the Board of Examiners and substituted state certification as the minimum qualification for new personnel. The Regents rejected the Bundy proposal of thirty to sixty districts on the grounds that small districts would not be able to negotiate effectively with a city-wide teachers' union, would be unable to achieve any racial integration, and would incur unnecessary duplication of costs.[10]

The Regents bill was opposed by the same groups who opposed the Bundy-Lindsay bill, most notably, the UFT, the CSA, and the Board of Education. But it seemed clear during the month of April that a climate of opinion in favor of some form of decentralization had unmistakably formed. Decentralization was supported by the mayor, the state commissioner of education, the Board of Regents, and even United States Commissioner of Education Harold Howe; in addition, there was extensive support in the press, including the *New York Times* and several small but influential journals like the *New Republic* and the *New York Review of Books.*

While political support for a strong decentralization bill appeared to be on the rise, the militant leadership in the demonstration districts was increasingly suspicious, reflexively distrusting anything which issued willingly from the "establishment." The governing boards continued to be unrecognized, at least formally, by the Board of Education, since they had agreed to accept no status less than total control. At this point, if the districts had had the political judgment to lie low, as the Catholic forces had in the 1840s when their prospects looked good in the legislature, a strong decentralization bill may well have carried. But they did not, because the momentum of their rhetoric impelled them to reject any settlement less than total community control.

On March 29, before the Regents bill replaced the Bundy-Lindsay plan, David Spencer, chairman of the IS 201 Governing Board wrote to the members of the legislature:

> Certain individuals have been actively lobbying on our behalf, presenting themselves as our representatives. We are disturbed by this, since we have authorized no one to speak for us. We are even more disturbed, because some of the views that have been presented appear to support legislation which, in our opinion, would be dangerously ineffective. . . . We are firmly opposed to the "Bundy Report" and to Mayor Lindsay's decentralization legislation. . . . Rather than see such legislation enacted, WE WOULD OPPOSE ANY LEGISLATION ON DECENTRALIZATION, for we believe that weak legislation will be worse than none at all.

Spencer explained that the only acceptable legislation would have to guarantee each district control over a lump-sum budget, control over the hiring and firing of personnel ("with appropriate safeguards to protect individuals from arbitrary actions"), the right to purchase supplies directly from suppliers, and democratic procedures for choosing local board members. Spencer concluded: "We will oppose the passage of anything less. If necessary, the legislature should postpone action this session. Public awareness of the need for strong community authority over the public schools of this City is growing. Perhaps next year the legislature will feel a greater mandate for significant change in the re-organization of the New York City Public School System." [11]

This position was shared by leaders of the Ocean Hill-Brownsville district. In early April, this view was expressed at a private meeting in the home of an employee of the Community Development Agency (the antipoverty arm of the city's Human Resources Administration), whose job was to build support for a strong decentralization plan. Fifteen people, including Rhody McCoy, David Spencer, and the Reverend Robert Nichol, met to discuss how friends of the demonstration projects could help them.

The project leaders derided any plan which retained a central board of education. Spencer said, "Decentralization gives you nothing. You must be able to control your personnel; we are hung up on this tenure nonsense." McCoy warned, "My community won't turn the schools back to the people who had them before." The spokesmen agreed that it was important for friends of the projects to understand that a plan like Bundy's was their worst enemy. Said one, "It would take twenty to thirty years for people to wake up and fight back," if a decentralization plan were passed by the legislature instead of a community-control bill. A leader from 201 described decentralization as "a mechanism created to contain guerrilla warfare. We must destroy that mechanism, encourage incidents." An Ocean Hill official agreed, "Create a crisis in every school. . . . Maybe hunting season will start early."

The demonstration district officials wanted to prevent any moderate reform and to hold out for what one described as "a simple piece of legislation: We hereby divide New York City into thirty or sixty autonomous districts to report to the Commissioner of Education." They hoped that an atmosphere of "guerrilla warfare" ("they can't afford to have 100 cops at a different school every week—the schools will 'blow' ") would lead to the complete dissolution of central authority.

Later in the meeting, the Reverend Milton Galamison arrived with a plan which he said he had worked out with Albert Shanker. He said they had agreed that parents and teachers could pass their own bill, if they could get together. They had worked out a four-point plan: (1) the central Board of Education would be abolished; (2) a three-man commission would be established to take its place; (3) community boards could not hire teachers, but they could have the power to fire; (4) local boards would be elected by the community. Galamison was hooted by the others, who called Shanker a racist; they felt that even without a bill, they were prepared to get what they wanted "the hard way," through demonstrations and pressure tactics. Galamison, in defense of his negotiations with the enemy, responded, "Al is honestly dishonest— he says 'let's slice the pie and divide it between the union and the community.' He forgives the people in the community who don't like him." [12]

But the representatives of the demonstration districts, unlike an old militant like Galamison, lacked the sophistication to deal with an "enemy" and refused to consider Galamison's proposal, just as they had refused to back any ameliorating legislation. Their intransigent spirit, their scorn for compromise and negotiation, kept them on an all-or-nothing course.

THE MILITANT ADVOCATES of community control were not without powerful friends. They had the support and sympathy of the New York Urban Coalition, which was a close approximation of the corporate power structure in the city. The National Urban Coalition had been formed in 1967 to focus the energies of the private sector on the urgent problems of the cities. The Urban Coalition wanted to bring dissidents into the "system," to preserve the system, and to restore social harmony. Members of the Urban Coalition, recognizing that social unrest should neither be ignored nor repressed, tried to place their power and prestige on the side of the poor and oppressed.

The New York Urban Coalition expressed its concern for the public school system though its Education Task Force (ETF), which was headed by Alan Pifer, president of the Carnegie Corporation. The ETF was composed primarily of representatives from the city's largest corporations, foundation executives, successful lawyers, and members of "the community," a euphemism which meant militant blacks and Puerto Ricans. There was a large gap in the composition of the task force: there was no one who spoke as a member of the broad middle class, whether white, black, or Puerto Rican. Consequently, the white businessmen on the task force, whose children attended private or suburban schools, relied on fellow members like Preston Wilcox and David Spencer to interpret the needs of the public schools for them.

The lawyers and businessmen of the ETF found the bureaucratic,

self-protective reflexes of the civil service system and the unions as repre-
hensible as the militant blacks and Puerto Ricans did. Their own commitment to private enterprise and individual initiative disposed them to
agree with the goals of community control. The New York Urban Coalition brought together a new alignment: those who sincerely wanted to
support the needs of blacks and Puerto Ricans; and those members who
never actively supported integration but could unhesitatingly support
community control, since it assured the continuation of the all-white neighborhoods where they lived.

The task force decided that the way it could be most useful was in
mediating the conflict between the Board of Education and the demonstration districts. The task force met several times with the Board of Education, to see how the stalemate might be broken. At the first meeting,
members of the Board of Education stressed that the governing boards
were never meant to be more than advisory, because of the constraints
of state law; one Board member blamed the Ford Foundation for failing
to tell the three groups what powers would be legally available to them.
Board President Giardino insisted that there were many things which
could not be granted in form, but could in different ways be permitted
in substance. Morris Iushevitz, a Board member who was also secretary
of the Central Labor Council, suggested that the Board and the three
demonstration districts go to the legislature together to request a bill
that would allow the Board to give greater powers to the projects. Mrs.
Rose Shapiro, another member, proposed that representatives of the
Board, the governing boards, and the task force sit down together as
soon as possible with the city's corporation counsel and the city budget
director to work out an agreement: "They've been struggling, we've
been struggling, and we're growing farther apart." Giardino urged that
everyone act promptly, while the legislature was still accepting new
bills.

The chairman of the task force, Alan Pifer, asked David Spencer
(who was present as a member of the task force, not as chairman of the
IS 201 Governing Board) whether the governing boards would be willing to meet with the Board of Education on that basis. Spencer "said
that the three boards would sit down if they thought there was a real
promise of something happening, and not just more talk. 'Otherwise
they'd just as soon let the pot boil, until it bubbles over.'" [13]

At another meeting a few days later on April 29, the task force
pressed the Board for more operational power for the governing boards.
One member of the task force asked "if flexibility could be built in for
transfers." He pointed out that Ocean Hill-Brownsville wanted to re-

move eight teachers, five assistant principals and one principal. Of all the problems facing Ocean Hill, "the most imminent question was how to get rid of the objectionable personnel." °

The Deputy Superintendent of Schools, Dr. Nathan Brown, told the task force that any teacher who wished to transfer out might do so if the governing board concurred. A member of the task force asked, "What happens if the governing board wants the teachers to leave but the teachers want to remain?" Dr. Brown said the governing board must bring charges "such as incompetence, insubordination, etc." When asked whether a principal might make conditions so uncomfortable for a teacher that the teacher might wish to transfer, Dr. Brown "pointed out the UFT contract and its provision for grievance procedure might not allow that." [14]

On May 7, the task force met with representatives of the three governing boards to learn their views on the Regents bill, which was the decentralization legislation that was likely to pass. The three governing boards told the task force that "the proposed legislation was inadequate and would not meet the needs of their communities." Nothing less than complete community control of personnel, school construction, and operating budgets would satisfy them. It was anticipated that the task force would convey the views of the governing boards to the state legislature.[15]

However, the task force never got the opportunity to carry these demands to Albany, because the political situation of the governing boards was altered by the unprecedented events of May 9 in Ocean Hill-Brownsville.

° During the three critical months from March to May 1968, the task force met repeatedly with the governing boards and with the Board of Education, but did not meet with UFT leaders until May 23, and then only at the union's request. The chairman of the task force admitted at the meeting that the only way the task force knew the views of the UFT was through reports in the press.

CHAPTER 32

"We Will Have To Write
Our Own Rules..."

THE CRITICAL POINT in the developing drama at Ocean Hill-Brownsville came in April 1968. The legislature was on the verge of approving a decentralization plan which did not give the governing boards total community control. The Ocean Hill governing board was increasingly embittered by its inability to have the powers it demanded and increasingly caught up in its own revolutionary aspirations. The union, defensive of its members and its contract, was dead set against community control and wary of the pending decentralization legislation. Yet, at that moment, there was still room for maneuvering, for working out problems without allowing them to turn into crises. The participants—the union, the governing board, and the Board of Education—may have been suspicious of each other's motives, but there was nonetheless communication among them.

The governing board decided that it could no longer tolerate this precarious balancing of tensions. Its decision to disrupt the status quo, like the opening scene in a Greek drama, set the antagonists on a collision course. The stakes were high: both sensed that the basis of their existence was at risk. In the struggle that followed, force counted for more than reason; passions ran high, tactics were vicious, and the politics were dirty. In such a battle the loser would be destroyed, but the victor would be tarnished by the fight itself.

IN APRIL the personnel committee of the Ocean Hill-Brownsville Governing Board recommended the "removal" of thirteen teachers, five assistant principals, and one principal (the only remaining principal in the

352

district who had not been selected by the governing board). The report explained that "the people [of the Ocean Hill-Brownsville community] were unanimous in their desire to have definite control over the Principals, Assistant Principals and teachers." The committee complained that the governing board "faced constant opposition from the Board of Education, the United Federation of Teachers, the State Commission of Education [*sic*] and the Council of Supervisory Associations. We were constantly told that our demands were 'opposed to state laws' but we found that the people in the street considered these laws written to protect the monied white power structure of this city." [1]

The report charged that "a small, militant group of teachers" opposed the project and sabotaged it in "underhanded and clever ways. . . . No set procedure has been established with this decentralized district to deal with Principals, Assistant Principals and teachers who are openly sabotaging our project." The personnel committee, hoping to establish a procedure, recommended the removal of this group of nineteen professionals. No reference was made to the removal procedure which had been detailed in the project's planning proposal. Nor was there mention of the competence or incompetence of the unwanted teachers and supervisors. The recommendation for removal was based on their alleged disloyalty to the project.

The personnel committee knew that its recommendation to "remove" the objectionable personnel was not routine: "We feel that we will be condemned by many as having to make this unpleasant recommendation. But every attempt on our part to solve the problem has met with failure. So we will have to write our own rules for our own schools. Enforcement of these rules will have to be carried out by the people of the community."

The committee also recommended:

- "Hiring of Assistant Principals with state certification whether they are on the list or not. . . ." [Both Superintendent Donovan and Commissioner Allen had already denied this proposal.]
- "Direct hiring of teachers from the south and Puerto Rico who have state certification in any state." [This recommendation violated the union contract and the state law; the minimum standard for teachers in New York, outside of New York City, was state certification, not by *any* state, but by New York State.]
- "Immediate establishment of one of our schools as a training school, where parents and community people will be trained and paid a full teacher's salary to become teachers." [This recommendation was based on an obscure provision in the state education law discovered by the

lawyer for the governing boards: if the Board of Education designated one school in the district a "training school," the governing board could staff it with people who did not meet the city's regular standards but who had potential as teachers.]

- "Immediate hiring and licensing of neighborhood musicians and artists as teachers receiving a full salary." [This recommendation was contrary to both city and state requirements for hiring teachers.]

The report of the personnel committee was approved by the governing board on May 7, and McCoy was directed to inform the nineteen unwanted professionals. The following letter, dated May 8, went out over the signatures of Rhody McCoy and the Reverend C. Herbert Oliver:

> The Governing Board of the Ocean Hill-Brownsville Demonstration School District has voted to end your employment in the schools of this District. This action was taken on the recommendation of the Personnel Committee. This termination of employment is to take effect immediately.
>
> In the event you wish to question this action, the Governing Board will receive you on Friday, May 10, 1968, at 6:00 P.M. at Intermediate School 55, 2021 Bergen Street, Brooklyn, New York.
>
> You will report Friday morning to Personnel, 110 Livingston Street, Brooklyn for reassignment.[2]

Nineteen teachers and supervisors received this directive in a registered letter which arrived at their schools in midmorning on May 9. Two of the teachers were UFT chapter chairmen. McCoy informed the press that this action was the only course open to the community, which had exhausted all other avenues of removing uncooperative, incompetent personnel.

The governing board's decision to remove these teachers and supervisors provoked a bitter struggle with the union which lasted for seven months and ultimately affected the shape of the entire school system as well as the political life of the city.

The immediate conflict centered on whether the governing board had *fired* the professionals or merely *transferred* them out of the district. The UFT insisted that its members were entitled to an impartial hearing of the charges against them. McCoy countered that the teachers were not entitled to formal charges or a hearing, because it was within his authority to transfer teachers to central headquarters for reassignment; he maintained that only teachers who were fired had the right to a hearing. The union held that the action of the governing board, whether it was construed as dismissal or as involuntary transfer, was clearly punitive, and that punitive action against teachers required submission of charges based on substantiated evidence.

The fact was that under the Board of Education's bylaws, only the superintendent of schools had the power to grant or deny requests for transfers; the governing board could not unilaterally transfer personnel out of the district. The board's bylaws required McCoy, on terminating the assignment of these teachers, to rate them either "satisfactory" or "unsatisfactory" (or in the instance of first-year probationary teachers, the rating might have been "doubtful," with supporting evidence). In view of McCoy's derogatory comments about the nineteen to the press and central officials, it was unlikely that he would have given them anything but an unsatisfactory rating. A teacher who has received an unsatisfactory rating is entitled to request a written specification of the reasons for the rating and, ultimately, a hearing.[3]

Thus it was irrelevant whether the teachers were fired or involuntarily transferred, since, in either case, they eventually had the right to receive written charges and a hearing.

If McCoy wanted to remove certain teachers without provoking a public uproar and without activating the UFT's grievance machinery, he had only to prepare a simple written request to Superintendent Donovan; no supporting charges or hearings would have been necessary, since administrative transfers were within Donovan's authority. Dr. John Niemeyer, president of Bank Street College and a supporter of the demonstration districts, stated publicly in the summer of 1968 that he had attended a meeting where both Giardino and Donovan assured McCoy of their willingness to remove unwanted personnel, quietly and in small numbers, at McCoy's request; under those circumstances, the UFT had agreed not to object. But McCoy never prepared a direct, specific request for transfers of personnel. This was later acknowledged, inadvertently, when Reverend Oliver wrote a rebuttal to Martin Mayer's book *The Teachers Strike:*

> Our whole Board had met twice with the Board of Education to get a group of personnel transferred. Donovan informed us that he would not act on a large group all at once, and that we should present cases to him one or two at a time. Mayer pretends that McCoy had failed to request the transfers, when in reality our Board had met twice with the Board of Education on the matter of transfers, but without making headway. We had been told by Giardino that if we were having difficulties with teachers, we should bring it to them (the Board of Education) and they would take appropriate action if such was necessary. We were never able to get them to see that we were talking about hundreds of teachers.[4]

Oliver did not suggest that McCoy ever prepared a written request, listing specific people, but said that the whole governing board had met with the central board "on the matter of transfers." He confirmed that McCoy and the board knew in advance that the Board of Education would not accept the transfer of a large group at one time. He also confirmed what the UFT feared: that union acquiescence in the transfer of the original group would have been the prelude to the involuntary transfer of "hundreds" of other teachers.

From their direct conversations with Donovan and Giardino and from the April 29 meeting of the Urban Coalition task force with the Board of Education, the leadership of the governing board and its Ford Foundation advisors knew that its action of May 8 would force a showdown with the Board of Education and the UFT. McCoy and Gittell confirmed this when they met in early June 1968 with representatives of major civic organizations. When asked why the governing board had not used "less inflammatory methods to transfer teachers," Gittell responded that a confrontation had to come sooner or later with the Board of Education, "and it might as well come now." McCoy concurred. Some months later Howard Kalodner, the district's lawyer, was quoted as saying: "If they had asked me, I would have probably tried to dissuade them or at least picked and chose more among those nineteen names. . . . But they were looking for a confrontation. They had to make a display with the community and with the central board." [5]

Why did the Ocean Hill-Brownsville Governing Board decide to force a confrontation? The three governing boards had reached an impasse with the Board of Education. The Board of Education refused to retreat from its guidelines, which placed limits on the authority of the projects and rejected community control. The state legislature was moving closer to a vote on a bill that did not provide total community control. Therefore, with the hope of *gaining* power by the bold act of *asserting* it, the Ocean Hill-Brownsville board decided to remove the nineteen professionals.

It was a test case. By removing the nineteen, the governing board hoped to establish the unilateral power to transfer personnel out of the district, an authority which was just as valuable as the power to fire them. There was always the possibility, given a vacillating Board of Education and a cowed union, that the local board might prevail by virtue of a fait accompli. One press release from the governing board stated bluntly that *"if these transfers were made, it would demonstrate control of the schools by the elected local Governing Board."* [6] (Italics in the original.)

Many charges and countercharges were made in the weeks and

months afterward by people seeking to apportion blame for the original contretemps. Those who supported community control, like the New York Civil Liberties Union, placed the "burden of blame" on the UFT for refusing to cooperate with the governing board and forcing it to take drastic action. Others, like Joseph Featherstone, recognized that it was the governing board that picked the time and the place for the confrontation: "It is possible to see the dismissals as routine transfers, if you squint at them, but on this issue I think the UFT had a point: the dismissals were arbitrary and punitive, they smacked of the public pillory, and, in any case, the UFT could be expected not to stand by and watch its chapter chairmen shipped out of disputed districts." [7]

In retrospect, it was clear that the Board of Education had erred in permitting the projects to get underway before establishing concise descriptions of the powers that would be granted to them and the powers that would not and could not; once more, the Board of Education had bowed to political pressures to move quickly without adequate staff preparation. A secret report of the State Education Department in February 1968 noted that the demonstration projects had been hastily conceived, without proper planning or preparation of the participants; as a result, some governing board members thought that their election entitled them to act as fully independent school boards and were simply unaware of the limitations created by state and city laws.[8]

Superintendent of Schools Bernard Donovan knew far more about the projects and the groups involved than did any of his Board members; it was his responsibility to inform the planning groups of the restraints on their experiments, although it is unlikely, given the predisposition of the militant organizations involved, that words of restraint would have been heeded. Donovan had the responsibility to advise the Board not to initiate the projects until the legal groundwork for them was complete. By the same token, the Ford Foundation funded the demonstrations precipitously, without concern for the adequacy of the plans and without regard to the consequences for the participants. Ford used its weighty influence to contribute to the districts' unreal aspirations. When the first showdown came in May 1968, Ford money and Ford-selected advisors were available to encourage the district's naive determination "to write our own rules for our own schools" regardless of the law or the consequences.

THE OCEAN HILL-BROWNSVILLE district's decision to terminate the employment of nineteen professionals was generally interpreted by the press and the public as outright dismissal. Rhody McCoy bolstered this impression by announcing to the press that unspecified vigilante action would prevent the ousted teachers from working elsewhere in the city: "Not one of these teachers will be allowed to teach anywhere in this city. The black community will see to that." [9]

The Board of Education directed the ousted employees to ignore the district's order and to report for work. Shanker warned that the UFT had recently adopted a policy of shutting down any district where teachers were denied due process, and that the union might resort to a city-wide strike if the teachers were not reinstated or granted a fair hearing.

During the next few weeks, the schools of Ocean Hill-Brownsville became the scene of a tense power struggle. The UFT was determined that its ousted members would return to their classrooms, and the governing board was equally determined that the unwanted personnel would never enter its schools again. Several hundred police were sent into the area to keep order. The disputed teachers, with police escorts, pushed their way into the schools through crowds of angry demonstrators. The governing board closed its schools in protest. When the schools reopened, 300 to 350 UFT teachers walked out to support the reinstatement of their fellow teachers. The UFT position was that if any of the dismissed teachers were denied admission, then all UFT teachers were locked out.

For a few weeks, the governing board, adhering to the unbending spirit of the personnel report which recommended the dismissal, steadfastly refused to bring charges against the nineteen. A notice to parents explaining the situation stated:

> Last week, this community—through its Governing Board—dismissed 1 principal, 5 assistant principals and 13 teachers. As usual, we were told this was against the law. We were told that we as Black and Puerto Rican parents cannot get rid of people who were messing up our children by their racism. We were told that we had to write up charges and submit them for a hearing to Shanker and Donovan. No one understands—but they really do understand—that the game is over. Decentralization means that *we* decide who will teach our children—NO DONOVAN, NO SHANKER, NO LINDSAY, NO 500 COPS—WE DECIDE!!!! [10]

Despite his initial reaction, McCoy finally brought charges on May 27. To conduct the hearing, the Board of Education appointed a special trial examiner, retired Civil Court Judge Francis E. Rivers, who was black. By the time the hearings began, only ten teachers were involved; the governing board had withdrawn its action against the one black teacher in the group, while two other teachers and the six administrators had asked to be reassigned.

While Judge Rivers began administrative hearings, Mrs. Rose Shapiro, the new president of the Board of Education, tried to work out a settlement. Mrs. Shapiro, appointed to the Board of Education by Mayor Wagner in 1964, was its first female president. Her own experience had spanned the Board's recent history. She had been a president of the United Parents Association, then was invited to join the board of PEA. She chaired the PEA's 1955 study, "The Status of the Public School Education of Negro and Puerto Rican Children in New York City," served on the Commission on Integration, and was later appointed to the New York City Commission on Human Rights. As a member of the Board of Education, Mrs. Shapiro was known as a committed integrationist, but also as a tough-minded critic of the Board's propensity to cave in under political pressure and to placate protest by launching ill-conceived programs. In a confidential memorandum to other Board members in early 1967, Mrs. Shapiro had urged the Board to resist pressures for community control, which she called a "political response to a professional problem which is plaguing educators throughout the country." She called on the Board to inform the public "on the status of integration in this city—what we have done, where we have failed, and why we have failed. The time has come when the public must be made to realize that the schools alone cannot solve a problem that is facing all large cities." Mrs. Shapiro was rare among Board members in that she was a product of the New York City schools whose children and grandchildren attended the public schools.[11]

Mrs. Shapiro believed that the dispute between Ocean Hill-Brownsville and the union could be settled rationally. She brought together representatives of the two sides and proposed binding arbitration. The union agreed, and the governing board's two representatives (Father Powis and Mrs. Clara Marshall, chairman of the personnel committee and vice-chairman of the governing board) promised to get a decision from the full governing board. Later the same day, Mrs. Marshall issued a press release in which she stated that the governing board was willing to accept binding arbitration. But two days later the governing board voted to reject binding arbitration.

Another attempt at settlement in early June was equally unsuccessful. Theodore Kheel, the city's leading labor mediator, was invited to work out a basis for ending the stalemate. Kheel's proposals required both sides to compromise. A first condition of the proposals was that teachers with serious charges against them would stay away from their jobs, pending mediation, and those with minor charges would continue to teach. The UFT accepted Kheel's terms. The Ocean Hill-Brownsville board refused to allow any of the disputed teachers to return, dooming Kheel's mediation efforts.

Throughout this stage of the controversy, Mayor Lindsay tried to hew to a middle-ground, with an eye on the fate of decentralization legislation in Albany. He deplored Ocean Hill-Brownsville's ouster of professionals, but blamed the situation on the Board of Education's failure to define the duties and status of the local board. Mrs. Shapiro promptly responded that the local board had refused to accept the Board of Education's guidelines for the past six months and that the real issue was the local board's rejection of limits on its powers. Her statement had little effect, because shortly afterwards the *New York Times* editorially echoed Lindsay's position, criticizing the local board's action while blaming the Board of Education for the confusion over the local board's powers.

The turbulence in Ocean Hill-Brownsville had an immediate effect in Albany, where the legislature was close to an agreement on the Regents' decentralization bill. After the ousters at Ocean Hill, the UFT launched a massive lobbying effort to defeat the bill. Two weeks after Rhody McCoy sent out his registered letters, the accord on a strong decentralization bill collapsed. The governing board (which included the local assemblyman, Samuel D. Wright) knew that the legislature was about to act on decentralization. Perhaps they thought that dramatizing their demands would cause the legislators to strengthen the bill. Or perhaps, as David Spencer had suggested, they wanted to kill decentralization and let pressure build for total community control. In any case, the board's action of May 9 and the ensuing turmoil gave the opponents of decentralization the ammunition to succeed.

In place of the Regents plan, a bill was enacted which postponed decentralization for a year. The Marchi bill, named for its sponsor, Republican State Senator John Marchi of Staten Island, temporarily enlarged the Board of Education to thirteen members, which enabled Mayor Lindsay to add pro-decentralization members; empowered the Board of Education to delegate to local school boards "any or all of its functions, powers, obligations and duties"; and recognized the three demonstration districts as equivalent to regular local school boards. Shanker agreed not to oppose the Marchi bill after he was assured that it would not change teachers' rights or hiring practices.[12]

The demonstration districts received a new lease on life. Though they were still embroiled in conflict with the union, the new legislation promised them a sympathetic Board of Education with the authority to grant many of their demands. Their position was improved even further when the Ford Foundation made a grant of almost $1 million to the Institute for Community Studies; $275,000 was earmarked for the Ocean Hill-Brownsville Governing Board. The projects had the backing of the Urban Coalition, the City Commission on Human Rights, the New York City Council Against Poverty, State Commissioner Allen, Mayor Lindsay, and major civil rights groups. The leadership of the districts looked to the fall term secure in its conviction that community control, as they defined it, was the wave of the future.

CHAPTER 33

Confrontations and Strikes

THE SUMMER OF 1968 provided a brief respite from the turmoil of
May and June in Ocean Hill-Brownsville. By the time school closed for
the summer, the children in the district had missed almost seven weeks
of school, the dispute between the union and the governing board was
unresolved, and Judge Rivers was hearing charges against the ten
ousted teachers.

A current of upheaval was in the air. On a number of college campuses, most notably Columbia University in New York City, radical students staged forcible takeovers of buildings and won concessions by pre-
senting nonnegotiable demands. A sense of bitterness and alienation
affected many political activists after the slayings of the Reverend Martin Luther King, Jr., in April and Senator Robert Kennedy in June. It
was a year of crisis and uncertainty; as the presidential election approached, the highly charged atmosphere reflected acrimony among
Democrats, massive demonstrations against the Vietnam war, and the
violent reaction in ghetto areas to the assassination of Martin Luther
King, Jr.

The 1968 report of the National Advisory Commission on Civil Disorders (the Kerner Commission) discerned a "national climate of tension
and fear," largely attributable to the urban ghetto riots, which began in
1964 and reached an apogee in 1967 in Newark and Detroit. The Kerner
Commission warned that overreaction to an apparently insignificant "initial incident" could ignite a full-scale confrontation between ghetto-
dwellers and police. Mayor Lindsay, who had been vice-chairman of the
Kerner Commission, did not want to be responsible for allowing any "incident" in New York to escalate into a riot; a large part of his future

appeal as a candidate for re-election was his ability to avert racial disorders. This desire—necessity even—to avoid a showdown which might lead to a conflagration made officials of the Lindsay administration eager to head off any trouble before it started, especially in ghetto neighborhoods.[1]

In the demonstration districts, particularly Ocean Hill-Brownsville and IS 201, the city and the Board of Education were determined not to provide the fuel for a ghetto explosion. The conviction that Ocean Hill-Brownsville was a "tinderbox" or a "powderkeg" was reiterated frequently in the months ahead. In hopes of calming passions, several conciliatory moves were made during the summer of 1968:

- The Board of Education rejected a petition for a new election by an antigoverning board group which collected some 2,000 signatures in Ocean Hill-Brownsville; Reverend Oliver called the dissidents "Pied Pipers of educational genocide" who were stabbing the children of the community in the back.[2]
- There were no repercussions, in the press or at the Board of Education, when the governing board turned over JHS 271 for a three-day "black power caucus" in July, which featured prominent black separatists.
- Lindsay, in accordance with the Marchi Law, appointed five new members to the Board of Education in July, all committed to school decentralization; among the new members was the Reverend Milton A. Galamison, one-time scourge of the Board of Education. The expanded board began to draft a new city-wide decentralization plan which would immediately increase the power of local school boards.
- The city's Community Development Agency, the community-action arm of the antipoverty program, pledged its active support to community control of the schools and organized meetings in ghetto areas to work towards that goal.

A month after the expansion of the Board of Education, its new decentralization plan was ready. It went further than any previous plan in assigning powers to local boards, but still not far enough to satisfy the districts that wanted complete community control. The union was not pleased with the plan, either, for it appeared to abridge the union contract. Shanker warned that the union would strike the schools in September unless the plan were amended to conform with the teachers' contract.

With the beginning of the fall term only days away, Rhody McCoy announced that he had hired 350 teachers to replace the UFT teachers who had struck the district in the spring. He and his board were now ready to accomplish in one fell swoop what the union contract and the

Board of Education's bylaws would never permit under normal circumstances.

The predictable outrage of the UFT was bolstered by the findings of Judge Rivers, released a few days before the opening of school. In all ten cases, Judge Rivers denied McCoy's requests for transfer. Rivers noted that, "Perhaps if the Unit Administrator had sent to the Superintendent of Schools a simple request to transfer the teachers, without assigning any supporting charges, he (the Superintendent) may have been able to do so without a hearing." Clearly, McCoy had not sent such a request to Donovan, or it would have been submitted to Rivers as evidence supporting McCoy.[3]

At the hearings, Howard Kalodner, counsel for Ocean Hill-Brownsville, asked Rivers to presume the legitimacy of the transfers "so as to put the burden of proof on the accused teachers." Rivers, however, maintained that each teacher had the right to be informed of the nature and cause of the charges, to be confronted with witnesses, and to be able to cross-examine them; to be represented by counsel and to call witnesses in his defense; and "to be presumed innocent of the charges until his accuser has proved them by a fair preponderance of the credible evidence."

Supporters of the governing board at the New York Urban Coalition, the Ford Foundation, and elsewhere had urged McCoy to bring charges because McCoy had assured them that this group of teachers was so incompetent and/or insubordinate that it was vital to remove them even though the school year was almost finished. They were sure, from McCoy's vehement denunciations of these teachers, that there was ample evidence to back up the governing board's drastic action. But the Rivers hearings proved this confidence ill-founded.

Three of the ten teachers were charged with having "expressed opposition to the project and contributed to the growing hostility between the 'Negro and White teachers,'" but neither witnesses nor evidence was presented as substantiation, and the charges were dismissed. Five teachers were charged with incompetence. Their defense counsel showed that some had recently been given good ratings or letters of commendation by their supervisors in Ocean Hill-Brownsville. In other instances, Rivers pointed out that the principal had failed to give any assistance to the teacher or otherwise to exercise "the guidance and leadership required of a principal" by the Board of Education's bylaws. One teacher, accused of permitting students to throw chairs at each other, produced evidence that the furniture in his classroom was attached to the floor at the time of the alleged incident.

Another teacher, the membership chairman of the UFT chapter at IS 271, was charged with being "hostile to the program of the Ocean Hill-Brownsville Demonstration School District" and attempting "to instill in the minds of his colleagues anxieties about their participation in the program." McCoy's attorney stated that the governing board "made no complaint against him on the basis of competency." The UFT introduced several exhibits attesting to the teacher's "outstanding ability." The sole witness to testify on behalf of the charges was another teacher, who reported a private conversation at a school Christmas party in 1967 when the accused criticized conditions in the school. Judge Rivers concluded that the accused teacher was "actually attempting to give constructive criticism of the project," and even if he were not, his comments were protected by the First Amendment "and also by the privileges adhering in a Union official under the Fair Labor Practice Provisions."

On receiving the Rivers report in September, Superintendent Donovan ordered Rhody McCoy to return the ten teachers to their jobs. Ocean Hill was legally obliged to comply with Donovan's order. But the governing board ignored Donovan's directive; according to Reverend Oliver, they "never formally discussed the Rivers findings, to say nothing of voting on them." The board's reasoning was amplified in a sympathetic article by Random House editor Jason Epstein: "The governing board has ignored these findings on the grounds that the judge refused to hear important evidence and that the right to decide who should teach in the schools of the district belongs to the governing board and not to a retired judge." The excluded evidence was defined by Judge Rivers as hearsay testimony and the testimony of nonprofessionals about a teacher's competence as a teacher.[4]

Had this controversy been an ordinary dispute between the teachers' union and a local school board, the Rivers report would have settled the issue: the teachers were removed for insufficient reason, and the superintendent of schools ordered the district to take them back. But this was no ordinary dispute. Through its rhetoric, the Ocean Hill-Brownsville district had become the symbol of black and Puerto Rican people struggling for self-determination. McCoy and the governing board urged their supporters to disregard anachronistic rules and regulations which, in their view, stood in the path of the aspirations of oppressed peoples. They sought to polarize the issue in such a way as to make it appear that anyone who opposed them abetted, even if unconsciously, white racism.

By drawing the issue in this fashion, the governing board won wide support. The New York Civil Liberties Union, for example, might have been expected to defend the teachers' academic freedom and right to due process. Instead, the NYCLU became a prime defender of the governing board, because of its desire to support the interests of minority-group children. In October, the NYCLU blamed the controversy on the UFT's desire to wreck decentralization and prevent community control. The NYCLU disregarded Rivers' hearing report and charged that "the entire due process issue has been from the beginning a myth created by the UFT." The NYCLU referred to the ousted teachers as the "most uncooperative" staff, repeating charges which Judge Rivers had already invalidated.[5]

BY THE BEGINNING of the fall 1968 term, the governing board was ready for a confrontation with the UFT, having hired a full teaching staff to replace the UFT teachers who walked out in the spring.

The union was ready too. Its delegate assembly authorized a city-wide strike unless the Board of Education extended the protection of previous contracts to all decentralized districts under its new plan and agreed to return all UFT teachers to Ocean Hill-Brownsville. Some members of the Board of Education thought they could convince the UFT to restrict its strike to Ocean Hill, which was patently impossible since those schools were now fully staffed with teachers loyal to the governing board.

The day before school was to open, September 8, Mayor Lindsay announced that a strike had been averted and that Ocean Hill-Brownsville would not prevent the return of the UFT teachers. However, McCoy stated publicly that the UFT teachers would not resume their normal duties, but would be reassigned within the district. It was now no longer a question of the ten who had been dismissed originally, but of the district's other UFT teachers who had been replaced by new teachers. Reassignment was not acceptable to the union; Shanker declared that school would not open on September 9.

The first teachers' strike lasted only two days. Almost 54,000 of the city's 57,000 teachers joined the strike. It ended to the satisfaction of the union, which won the assurances that it sought from the Board of Education. A "memorandum of understanding" was drawn up in which the Board of Education agreed to extend the protection of the union contract and city-wide bylaws to all local boards; to pay the more than 300 Ocean Hill teachers for the time they were out on strike in the spring; to order the hundreds of UFT teachers still assigned to Ocean Hill to "resume their professional duties"; to require local boards who wanted to dismiss teachers to submit charges to binding arbitration; and to confer "superseniority" (protection from involuntary transfer) on union officials, protecting them against antiunion reprisals. The teachers' strike, like all public-employee strikes, was a violation of the state's Taylor Law; Shanker and the union were ultimately prosecuted for their strike action, but the contract which they signed with the Board of Education was nonetheless valid. The Board of Education was now legally bound to return the UFT teachers to the schools of Ocean Hill-Brownsville. From this point on, future disputes were a question of how to get the governing board, whether by persuasion or by force, to accept the terms agreed to by the union and the Board of Education. This was not understood by the governing board, the press, or the public.[6]

The UFT membership voted to accept the agreement with the Board of Education and to end the strike, but cautiously authorized a new strike on forty-eight hours' notice if the Ocean Hill portion of the agreement were not fulfilled. Their caution proved necessary.

The tone of the governing board's acquiescence in taking back the ousted teachers made clear that it would bear no responsibility for the success of the agreement that had been imposed on it: "a. We will no longer act as a buffer between this community and the establishment" —this phrase warned that the governing board could no longer contain the wrath of its outraged community. "b. This community will control its schools and who teaches in them. c. We do not want the 210 teachers to return to this district. Since the legal machinery of this sick society are forcing these teachers on us under threat of closing our schools and dissolving our district, the Board of Education should return to our district any of the teachers who wish to return. Our original decision remains as before. We refuse to sell out. If the Board of Education and the Superintendent of Schools forces them to return to a community who does not want them, so be it." [7]

The UFT teachers who reported to Ocean Hill on September 11 had to force their way through angry crowds to enter the schools; they were verbally assailed both outside and inside the schools. Once they were inside, there were no teaching assignments for them. Instead they were subjected to a day of harassment, calculated to discourage them from returning to their assignments. They were assembled together in the modern auditorium of IS 55 and told to wait for McCoy. While they waited, about fifty people entered the room and taunted them with anti-white invective, as two members of the governing board stood by in silence. When McCoy arrived an hour late, he announced that he could not address the teachers because of the community's hostility to them.

They returned to their schools with police escorts. One principal locked his UFT teachers in a guarded room for their physical safety.[8]

The uprising of the community was not entirely spontaneous. A few years after the events, one of McCoy's chief advisors admitted that McCoy had an understanding with militants like Robert "Sonny" Carson of Brooklyn CORE: he could call them in and he could call them off. Carson and his followers were on the front lines of the action in Ocean Hill throughout the strikes of 1968. When the governing board was compelled by threat of legal action to make agreements, the militants emerged to make compliance impossible.[9]

Since none of the reinstated UFT teachers was permitted to resume teaching, the executive board of the UFT called for a new city-wide strike.

THE SECOND STRIKE BEGAN on Thursday, September 12. It did not end until September 30. During this strike, Lindsay appointed three new members of the Board of Education, which gave his appointees a majority. The new president of the Board was John Doar, previously a United States assistant attorney general under Robert Kennedy and for several years the federal government's chief civil rights prosecutor. Doar had only recently arrived in New York to take over as president of the Bedford-Stuyvesant Restoration Corporation, a mammoth antipoverty effort initiated by Robert Kennedy.

Now not only Mayor Lindsay and State Commissioner Allen but also the Board of Education sided with the Ocean Hill governing board. This lineup made the UFT unsure whether there was anyone in a position of authority willing to enforce its previous agreement with the Board of Education. During the second strike, the UFT knew that the nonunion schools of Ocean Hill would continue to function; the union's only hope of attaining in fact what had already been awarded on paper was to create sufficient public pressure on school and city authorities to enforce the "memorandum of understanding." Though Board members privately discussed drastic ways of requiring compliance by Ocean Hill during the second strike, like taking control of the district or cutting off its funds, their fear of putting a match to the "tinderbox" prevented them from adopting any strong punitive course.

An attempt by State Commissioner Allen to settle the second strike failed. Allen proposed the temporary suspension of both the governing board and the ten UFT teachers who had been the center of the original controversy. The Lindsay Board of Education, against its own

inclinations, promptly suspended the governing board. The UFT, however, refused to accept the proposal without guarantees that the governing board would not be reinstated while the ten teachers remained suspended, that the other UFT teachers in the district would be restored
to their regular teaching jobs, and that neutral observers would
be placed in the district's schools to prevent intimidation of UFT teachers. McCoy, who was not suspended, would not give an inch either.
When Superintendent Donovan ordered him to allow the return of
more than 100 UFT teachers, McCoy flatly refused. A few days later,
the plan a failure, the Board of Education lifted the suspension of the
governing board.

The UFT called a mass rally in front of City Hall, and more than
15,000 persons, mostly teachers, demonstrated their support for the
union. The appearance of civil rights leader Bayard Rustin alongside Albert Shanker was an attempt to minimize the racial aspect of the strike
and to focus the strike on the issues of due process and teachers' rights.
It was a vain effort.

The fact remained that the union was overwhelmingly white, and
the Ocean Hill board was overwhelmingly black: It was on the issue of
racism that the governing board took its ground. From the beginning of
the project, its leaders viewed every obstacle as an instance of racism.
Whenever anything went wrong, whenever a request was denied, whenever they had to compromise for less than they demanded, they detected
racism.

The racial and religious antagonisms which had simmered in the
schools for several years burst to the surface during the second strike.
Picketing teachers claimed that they were subjected to antiwhite, anti-
Semitic invective. Governing board partisans charged the teacher pickets with using antiblack invective. Tension increased each day. Parents
were angry because the schools were closed; blacks were angry because
a small, black school board was being stepped on by a powerful union;
union members were angry because it appeared that the mayor was
union-busting; Jews were angry because Jewish teachers were pushed out
of their jobs without cause while the Board of Education complacently
tolerated outbursts of anti-Semitism at its public meetings.

Both the governing board and the UFT contributed to the hate-
filled atmosphere. A governing board official could usually be found on
television each evening, tossing off a fiery statement with anti-Semitic or
antiwhite overtones. The UFT, trying to make the public see why all the
schools had to stay closed, distributed 500,000 copies of anti-Semitic
sheets that had been handed out anonymously or placed in teachers'

mailboxes by little-known militants. These sheets, broadly circulated, stirred Jewish fears of black anti-Semitism.°

Having rubbed raw the sensitivities of the city's ethnic and religious groups, the second strike was finally settled on terms which embittered the governing board. The Board of Education and the UFT agreed that the UFT teachers, now reduced in number to 110, would return to Ocean Hill with as much police protection as necessary; their return would be safeguarded by teams of observers, containing representatives of the UFT, the Board of Education, and the mayor; the observer teams would report to a committee of the Board of Education which was empowered to close schools, if necessary, to protect UFT teachers. In hopes of making the settlement palatable to the governing board, the Board of Education granted enough extra teaching positions to Ocean Hill to allow them to retain all their newly hired staff as well as the returning UFT teachers.

The union insisted on impartial observers over the objections of the governing board, which wanted only its own observers and UFT observers. During an all-night negotiating session, Shanker's chief aide, Sandra Feldman, kept a personal record of the discussions and noted: "We turn down—got to have impartial even without power [to close schools], to have credibility in case need to take further strike or other action. Just our people, public would think carping. . . ." Many commentators, in choosing the side of the black board over the white union, became so vehemently antiunion that they denied any validity to the union's position. Thus, while Shanker was holding out for impartial observers, Murray Kempton, columnist for the *New York Post,* wrote that "the only such observer satisfactory to Shanker is probably the UFT chapter chairman." Kempton called Shanker a "goon" of Donovan, aiming "to break the Ocean Hill governing board and make certain that no other community body raises its head to suggest that a teacher or supervisor do an honest day's work again." [11]

° After the strikes, the New York Civil Liberties Union accused the UFT of knowingly fanning racial fears in order to discredit community control. However, there was ample evidence that anti-Semitism was on the rise. The Anti-Defamation League concluded in early 1969 that it had reached a crisis level in 1967–1969, "unchecked by public authorities." The ADL study documented the repetition of the black separatist theme that Jewish teachers and supervisors sought the "mental genocide" of ghetto children and willfully prevented the advancement of blacks in the school system. Tolerance of anti-Semitism was so general in this period that even the Metropolitan Museum of Art published and defended a catalogue containing an essay which was anti-Semitic in tone.[10] The relationship between black and Jewish organizations, as well as their respective communities, was seriously damaged during the strikes, which had the effect of weakening the city's liberal political coalition.

More than 1,000 policemen were on duty in and around the eight schools of Ocean Hill-Brownsville on Monday, September 30. The number of UFT teachers wishing to return had dropped now to eighty-three. On the first day, few had classroom assignments. On the morning of the second day, Reverend Oliver issued a call to parents and "supporters of Ocean Hill-Brownsville throughout New York City," seeking their help in getting rid of "Shanker's phonies." Oliver announced that "on Tuesday morning, October 1st, the Governing Board met and ordered Mr. McCoy, the Unit Administrator, to remove the unwanted teachers because they were disrupting our educational program." He said that after the teachers were "forced in. . . . A few taught quietly, but in most cases, they began interfering and sabotaging our education program again. The police came to protect the very same teachers who left our children and who have now returned to destroy our new educational program." At the time this was written, most UFT teachers had not yet been assigned to a class.[12]

That day, disorders broke out at several of the district's schools, starting at JHS 271, where nonunion teachers told the UFT teachers they were not wanted and then walked out, taking the students and shutting down the school. In clashes outside JHS 271, nine persons were arrested and ten policemen injured. A contingent from JHS 271, teachers and pupils, marched to IS 55; after their arrival, order broke down, children ran through the halls, and the principal dismissed the school. The demonstrators then marched to two elementary schools, which also closed.

Inside the schools, while they were open, relationships between the UFT teachers and the governing board teachers (whom the unionists called "scabs") were abrasive. The UFT teachers resented the strikebreaking role of the new teachers and disdained their competence as teachers, since many were just out of college and were teaching to gain deferment from the military draft. Many of the union teachers were physically frightened by their situation, but were sustained by their belief that they were defending important principles: academic freedom, due process, the right of a teacher to work without harassment or political dictation from his superiors, the validity of the union contract, and the supremacy of city and state laws over the will of a renegade school board. They believed that they were standing up for the rights of every present and future teacher in the city. Any one of them could have transferred to another job in the school system, but their commitment to professionalism made them hang on stubbornly, despite the dangers.

Meanwhile, the new teachers were motivated by a different set of ideals, which they adhered to with equal passion. Many of them were recent graduates of leading colleges (at PS 178, thirty out of thirty-nine teachers were under twenty-five). Seventy percent of them were white; fifty percent of the white teachers were Jewish. They blamed the UFT for cynically stirring up white and Jewish hostility to the governing board and community control. They were proud to be scabs; they saw themselves in the vanguard of a grassroots movement in the ghetto which would end the exploitation of the poor and oppressed. Their devotion was increased, not diminished, by external pressures on the district. They cared little for professionalism and unionism, perceiving both as a cover for maintaining an unjust status quo. They were scornful of the union teachers, believing that they valued job security over the welfare of the children. They shared a conviction that working for ideals (as they saw themselves) was morally superior to working for a salary (as they saw the UFT teachers).

On Wednesday, October 2, the day after the disorders around JHS 271, Superintendent Donovan closed the school for a one-day cooling-off period. Though McCoy then had the governing board's order to remove UFT teachers from classroom assignments, he delayed taking action. The situation was precarious when JHS 271 reopened. On Thursday and Friday community control activists from IS 201 and other parts of the city entered the building as "observers." Between one-third and two-thirds of the union teachers had classroom assignments, but few had any children assigned to their classes. Some UFT teachers were made "assistants" to inexperienced teachers, others were assigned to watch hallways or the lunchroom.

On Friday field trips were planned for many classes, and UFT teachers were not invited to go along. (Early in the morning, a worried mayoral aide called to find out if a field trip was headed for City Hall.) In a last-minute switch, the UFT teachers were asked to go, but without assignments; their field representatives, fearful for their physical safety beyond the eyes of the police, advised them to stay behind. In one school, the children who remained were gathered in the auditorium while UFT teachers sat in empty classrooms.

Over the weekend, the tenuous peace disintegrated. The governing board instructed McCoy to relieve the UFT teachers of their classroom assignments, which caused the Board of Education again to suspend the governing board, this time for thirty days. It made little difference. McCoy announced that he would follow the instructions of the governing board even though it was suspended: he was removing the UFT teachers from the classrooms. When the union teachers reported to school on Tuesday morning, October 8, their principals informed them that they had been relieved of their assignments and would be given other "professional" duties. They were given busy-work, like textbook inventory, or placed in empty classrooms.

Superintendent Donovan declared that he was reassigning McCoy and seven of his principals (the eighth decided to transfer out of the district) to central headquarters for defying his orders. The principals, with McCoy's approval, reported to 110 Livingston Street, but McCoy remained at his desk, ignoring Donovan's directive.

When the delegate assembly of the UFT learned that Donovan had finally acted to curb the defiance of the district leadership, it agreed to defer action on yet a third teachers' strike. However, the next day, October 9, the district was again shaken by disorders and raucous demonstrations. Some 200 demonstrators battled the police outside JHS 271.

Inside JHS 271, the tension was electric. The staff at 271 included Albert Vann and Leslie Campbell, leaders of the Afro-American Teachers Association, whose newspaper frequently printed antiwhite, anti-Semitic diatribes on the theme of "educational genocide." A member of Donovan's staff declared that the school was "unsafe" because of the level of animosity among the teachers. In one classroom in the school, the names of the UFT teachers were posted on a blackboard for identification. With JHS 271 again at the center of the storm, Superintendent Donovan ordered it closed. Donovan reported to the Board of Education that

> there was almost continual harassment and intimidation of the returning UFT teachers by members of the staff and by persons other than staff who entered the building. The constant state of tension within the building kept attendance at a low level and prevented effective instruction within the building. In addition, there were threats made against various teachers, individually and by groups. . . . There was also a threat to the safety and life of at least one of my observers in the schools.[13]

At this critical juncture, while the city was at the brink of a third strike, the suspended governing board picked up additional support for its defiance of the Board of Education's authority: Mrs. Mamie Clark, wife of State Regent Kenneth Clark, joined with other well-known blacks and whites to form an "Emergency Citizens Committee to Save School Decentralization and Community Control." Its purpose was to work on behalf of the preservation of the Ocean Hill-Brownsville district, as well as the broader principles of decentralization and community control. Other members included Whitney Young, executive director of the National Urban League; the Reverend Donald Harrington, state chairman of the Liberal party; and Trude Lash, executive director of the Citizens Committee for Children. The group declared that the destruction of Ocean Hill would mean "a betrayal of the rights of minority groups to play a vigorous role in the education of their own children." 14

Further backing for Ocean Hill-Brownsville came from the Public Education Association, whose prestigious board included Mrs. John B. Oakes, the wife of the editor of the *New York Times* editorial page. The board issued "a call to sanity and action," lambasting the UFT. The entire crisis, they held, had been "manufactured" by the UFT, which had resorted to "militant and lawless tactics." The PEA trustees averred that the issue of job security "was never at stake," but was nothing more than "a naked power grab by the Union to restrict administrative discretion, thus resulting in a virtual Union takeover of the public schools." 15

Superintendent Donovan, at the urging of Doar and Galamison, announced that JHS 271 would reopen on Monday October 14 and that all seven principals would return to their jobs. To the union, this meant that Donovan was retracting the only reproach ever administered to the district, despite its flagrant insubordination. What was more, JHS 271, the worst trouble spot in the district, would reopen without any changes made. The UFT teachers still had no assurance that they would be allowed to teach. Furthermore, Donovan stated that the principals had a "legitimate right" to assign teachers to a "variety of professional activities." Within the context of the previous week, the "variety" might include hall duty, textbook counting, or other nonteaching duties.16

Shanker asked his membership to authorize a third strike if Donovan reopened JHS 271. In a letter to his members, he explained, "We wanted to wait, but we could have done so only on the basis of concrete action, because we've had too many empty words and promises. They held up no better than the written agreements the Board of Education, the Superintendent and the Mayor find so simple to violate." Shanker declared that

> if it becomes necessary to close down Ocean Hill-Brownsville . . . because of its refusal to abide by decency and due process, we stand ready to turn to any community in the city that wants to conduct an honest experiment in innovation . . . [but] on only the minimum basis that due process, free speech, free thought, and academic freedom will be maintained. . . . The issue is what it has been all along—will we have a school system in which justice, due process and dignity for teachers is possible, or will we have a system in which any group of vigilantes can

enter a school and take it over with intimidation and threats of violence.
. . . This may be a long one—for this time we are staying out and not
going back until we are sure we still have a school system in the City of
New York.[17]

On Monday, October 14, JHS 271 reopened. UFT picket lines
closed down almost every other school in the city outside the demonstration districts.

Shanker was right, the strike was a long one. Five weeks passed before a settlement was reached. Having twice signed agreements which
were not carried out, the union now demanded the dissolution of the
Ocean Hill-Brownsville project. The Board of Education refused to dis-
mantle a district which had become a symbol for black and Puerto
Rican aspirations.

As days turned into weeks with no sign of a settlement, desperate parents sought alternatives. Working mothers made makeshift arrangements for their children. Some teachers cooperated with parent associations and held classes in private homes. The recently elected
vice-president of the Board of Education, Reverend Galamison, urged
parents to sleep in school buildings to keep them open. Many schools
were broken into and occupied by angry parents. Some school custodi-
ans changed door-locks to prevent parent break-ins. Only in the demonstration districts did schooling proceed normally (although in Ocean
Hill-Brownsville, attendance dropped by as much as 70 percent during
the strike).

Nonetheless, the strike was successful in closing about 85 percent of
the schools. Mayor Lindsay, Commissioner Allen, the Board of Education, the Ocean Hill-Brownsville Governing Board, and the UFT were
engaged in almost continuous negotiations throughout the strike. A
dozen or more plans were set forth to end the strike. None of them
worked, for the same reason: no matter what guarantees were offered
the union, the union did not trust the Board of Education to enforce
them.

As the strike dragged on, the frustration and rage of parents and
the press grew. Whatever the provocation in one district, there seemed
to be no justification for a general strike which halted the education of
all public school children. It seemed to some community-control advocates that the continuation of the strike might lead ultimately to a public demand to break the power of the UFT.

This view was espoused at a meeting called November 7 by then
Manhattan Assemblyman Jerome Kretchmer and political activist Bella

Abzug. About twenty people attended, coming from groups such as the United Bronx Parents, the New York Civil Liberties Union, the Citizens Committee for Children, the United Parents Association, the Field Foundation, and the Ad Hoc Committee of Teachers for Community Control.

Kretchmer wanted to create a lobby "to get our kind of decentralization from Albany." He anticipated that the greatest resource of the community control movement would be the antiwar, reform Democrats who had campaigned to "dump Johnson" and worked in the unsuccessful campaigns of Eugene McCarthy and Robert Kennedy. He maintained that the UFT's only support came from other labor unions and that "the longer the strike goes on, the more parents, teachers and students are being brought around to open the schools. . . . Maybe we don't want the schools officially opened. We're making more points in this battle in the minority communities, chaos for a year is not such a bad thing; no one was getting taught anything anyway." Out of this meeting came the nucleus of the group which led the legislative and public campaign for community control of the schools.[18]

ALTHOUGH MILITANT SUPPORTERS of community control like the Kretchmer group saw advantage to their position in the continuation of the strike, city and state officials and the union were under tremendous pressure to end the strike and reopen all the schools. Lindsay and Commissioner Allen wanted the crisis resolved as quickly as possible on terms that preserved the district.

At last, an agreement was reached which brought the strike to a close on November 17, 1968. Enforcement of the settlement was personally guaranteed by State Commissioner Allen. The conditions were the following:

1. The 79 remaining UFT teachers would return to teach in the district.
2. The Commissioner would appoint a three-man State Supervisory Committee to protect teachers' rights city-wide.
3. The Ocean Hill-Brownsville district would be governed by a state trustee and the governing board would be suspended until it was prepared to comply with the terms of the agreement.
4. The three principals in the district whose appointments had been challenged in the courts (and at that date ruled invalid) would be assigned to central headquarters.
5. Three of McCoy's teachers charged with misconduct would be assigned to the district office pending their hearings (they were subsequently cleared).

6. Salary lost by teachers and time lost by pupils during the strike would be made up by temporarily lengthening the school day and by eliminating certain holidays.

While most of the city greeted the end of the strike with relief, the members of the governing board were furious. They continued to think of themselves as a legal entity and could not understand how an agreement made without their assent could be imposed on them. They went to court to try to have their suspension lifted and were told by the judge that their governing board was "no more than an unofficial body of citizen advisors without power to transfer and suspend . . . (having) no authority to countermand any orders of the Board of Education on any subject." [19]

In the weeks that followed the establishment of the state trusteeship for their district, they attempted to undermine and overturn the agreement by openly violating it, the same strategy which had voided the two previous strike pacts. Suspended teachers and governing board members, specifically barred by the agreement from entering the schools, insisted on their right to do so. Boycotts, demonstrations, and disorders erupted periodically, usually in response to a call from governing board members. Within a month, three successive state trustees had been appointed by Commissioner Allen, the first two pleading exhaustion, after bearing up to a continuing round of demands, threats, and confrontations. The second trustee, William Firman, had a twenty-four-hour state police guard at his home in Albany for weeks after his departure from the district, because of threats against the lives of him and his family.

The third trustee, Wilbur R. Nordos, was able to work with McCoy and the governing board. They may have learned of his speculation that state education officials "had gone so far as to select McCoy's replacement, and were prepared to operate the district with new staff, even to the point of possibly dismissing or reassigning the entire staff of 271 and recruiting a brand-new staff, from the principal right down through all the teachers." The district settled down after Nordos took charge, and three months later, on March 7, 1969, the governing board was reinstated by the Board of Education at Nordos' recommendation.[20]

IN THE END, the Ocean Hill-Brownsville board lost far more than was necessary; at any one of several different points in the struggle, the board had the moral and political power to effect a favorable compromise. By refusing to accept any terms less than total victory, the board was its own worst enemy. Ultimately, because of its symbolic power, the governing board did not serve the interests of blacks and Puerto Ricans well. In the same way that Bishop Hughes has been criticized by historians for failing to perceive a viable alternative to separatism and for putting the interests of his own group above the interests of the whole community, so must the leaders of Ocean Hill-Brownsville be faulted for failing to understand that the interests of their people would be best advanced by breaking down the barriers of race and class and by overcoming the self-limitations of separatism.

The union won the right to return its members to the classrooms of Ocean Hill-Brownsville, as well as strong procedural protection for the future. But its image as an idealistic and socially progressive union was tarnished among the liberal intelligentsia and many black leaders. The union's victory established it as a political power in the city and state, but the price of victory was high.

EPILOGUE

CHAPTER 34

The New Law

AFTER THE STRIKES, the struggle shifted to the New York State Legislature, where a decentralization bill for the system as a whole would be framed. It was the task of the legislature to decide how the city school system was to be restructured, whether there would continue to be a central authority, how to divide power between the central authority and the local boards, how these boards would be chosen, whether to retain the Board of Examiners, how many districts to establish, and whether to retain the three controversial demonstration districts.

The militant advocates of total community control wanted the city carved into at least thirty autonomous school districts, with no central administration. The United Bronx Parents, for example, proposed "The People's Plan," which was less a plan than a series of hypotheses. It presumed that the city was composed of "natural communities," which were capable of acting without any predetermined ground rules. Each of these communities would decide for itself how it wanted to elect its community board and which schools to include in its district ("the district may include the number of schools deemed educationally sound to the community"). The plan did not suggest how to resolve disputes between conflicting groups claiming to be "the community" or between communities who wanted to include the same schools in their districts.[1]

The major community-control legislation was prepared by the Board of Education. It proposed to divide the city into thirty community school districts with broad powers, to retain the three demonstration districts, to abolish the Board of Examiners, and to retain the Doar board as a central authority to establish minimum standards and fiscal controls.

There were two dissents filed by members of the Board of Education. A minority report was submitted by three members of the pre-Doar board, including former president Rose Shapiro. They proposed legislation to make existing local school boards operational with in-creased powers, retain the Board of Examiners, eliminate the demonstra-tion districts, and provide for parent participation by setting up "school policy committees" in each school. They contended that poverty was "*the* major factor in the low achievement of children" and that the only meaningful participation of parents occurs in individual schools.[2]

Another dissent came from the Board's vice-president, Reverend Galamison, who attacked the Board's plan for not going far enough in delegating powers to community school boards. Galamison believed that the local boards should have greater fiscal and budgetary autonomy than the central board's plan allowed, as well as substantial participa-tion in school construction.

The UFT submitted its proposed legislation, calling for no more than fifteen school districts of equivalent size, elimination of the demon-stration districts, an elected central board, elected local boards, and re-tention of the Board of Examiners (a reversal of its position of the year before, when it had agreed to the elimination of the Examiners). The UFT held that large districts would maximize the possibilities for integration while minimizing the impact of small, well-organized groups in local school-board elections.

On one side of the issue, supporting a strong community-control bill, were the Board of Education, the state commissioner of education, Mayor Lindsay, the Board of Regents, and an assortment of organiza-tions that comprised the city's philanthropic–social welfare establish-ment.

On the other, opposing community control and supporting gradual administrative decentralization, were the UFT, the supervisors' union, and most of organized labor. One of the few people outside the labor movement to criticize community control was Louis Yavner, coauthor of a 1951 management survey of the schools which was often cited by com-munity-control proponents. Yavner, on behalf of the Citizens Union, sharply criticized the assumptions that undergirded the drive for com-munity control. Yavner wrote: "I believe that most of us have fallen in love with matters of form rather than substance during the debates and crises that have taken place since Bundy. . . . Government by gim-mickry is just giving the people bread and circuses."

Yavner did not agree with the Doar board's plan to make profes-sionals accountable by giving local boards direct power over them:

> The majority's community control plan will not help materially. Assume it
> is installed. The local board will be able to hold the teacher responsible

for some things, but not for the lack of certain necessities—ranging from class size to square meals for kids—then must look to the principal, then to the district superintendent. Then where? . . . Money is the root of the school problem. Even if all money now appropriated were spent as well as it should be, and it isn't, we would still need much more money to enable our schools to do all that should reasonably be expected of schools in a poverty society.

The great weakness of the Board's plan, Yavner held, was that the Doar board did not know either the school system or the city: "Given superhumans to administer it, their plan like any other could be made to succeed. It would more likely fail. It has built-in conflicts among religious and other groups, it has built-in political patronage, it has built-in inefficiency and squandering of public funds in procurement of construction, it has built-in destruction of a merit system that could be improved. . . . In New York City or elsewhere, the spoils system and favoritism can come back awfully fast." [3]

BUOYED BY A STRONG antiunion public reaction to the humbling of the Ocean Hill-Brownsville district, the idea of community control commanded an influential following. Its supporters held some or all of the following to be true:

- That the chaos of the fall was due to the legislature's failure to pass a strong decentralization plan in 1968.
- That the strikes and disorders were caused by the UFT's desire to discredit community control.
- That the eight schools of Ocean Hill-Brownsville proved the value of community control by making great educational gains during the brief period when UFT teachers were barred from the district.
- That black and Puerto Rican communities would be angry and disruptive until granted control over their neighborhood schools.
- That opponents of community control were motivated by either racism or selfish economic reasons.
- That community control would provide militant blacks an opportunity to participate constructively in the system, thereby reducing disorderly protests and social instability.

The campaign for community control reflected these views. A leader of the community control lobby, Ira Glasser of the New York Civil Liberties Union, wrote legislators that, "Everyone is now agreed that the chaos in the three demonstration districts was primarily due to the ab-

sence of clearly drawn lines of authority." David Seeley, speaking on behalf of the PEA to the City Commission on Human Rights, said that,
"The public schools have left large numbers of Negroes and Puerto Ricans with a sense of frustration and anger. As the city's educational
problems have remained unsolved, the frustration and anger have become ever more explosive. The number of violent incidents and outbursts in the schools have increased year by year—all, be it noted,
within the present school structure." Mayor Lindsay's key aides, called
into the school controversy after it had reached the crisis stage in 1968,
were convinced that the UFT schemed to sabotage Ocean Hill and decentralization; they were unaware that the union had participated in the
creation of the original experiment.[4]

The treatment of the issue in the media was decidedly favorable to
community control. While the news columns of the *New York Times*
provoked the ire of both sides, its editorials backed community control.
The *Wall Street Journal* gave front-page coverage to the educational
progress made in Ocean Hill while under community control. The noted
critic Alfred Kazin visited the Ocean Hill schools during the third teachers' strike and reported in the *New York Times* (on the editorial page)
that "the holy flame of learning" was burning "hotter than ever" there.
In the same period, *Time* magazine ran an admiring piece on Ocean
Hill-Brownsville, called "Teachers Who Give a Damn," which presented
the governing board's interpretation of the events.[5]

The Ocean Hill-Brownsville Governing Board fought for its survival, while ridiculing the Board of Education bill for not giving full autonomy to local districts. In its newsletter, the governing board asserted
that it had achieved educational success, despite the crisis, but was
being relentlessly destroyed by the racist power structure:

> From September 9, till the teacher's strike was settled on November 19,
> the Ocean Hill community proved to the entire world that, with 580
> Principals, Assistant Principals and teachers who respected us and our
> children under real community control, our children could learn as never
> before. Forget all the so-called arguments of "due-process." Forget all the
> tales of whether "decentralization" was dead or alive. . . . Remember only
> that our 9,000 children really started learning.[6]

The lobby for a strong community-control bill consisted of two
groups. The Committee for Citizen Participation, organized in January
1969 by Assemblyman Jerome Kretchmer, carried out most of the day-
to-day lobbying activities, arranging meetings, coffee-klatches, petitions,
advertisements, and visits to legislators in Albany. Its cochairmen were
Ira Glasser of the New York Civil Liberties Union and Anna Lou Pick-

ett, a political district leader from Kretchmer's area on the West Side of Manhattan. The other lobby was the Coalition for Community Control, an assemblage of high-level names and civic organizations that maintained a lobbyist in Albany and issued statements on behalf of community control. It was formed in March and headed by Francis Keppel, the respected former United States Commissioner of Education. Under the slogan "Community Control is the Hallmark of American Public Education," the Coalition for Community Control brought together twenty-six civic groups and civil rights organizations, including the PEA, the Citizens Committee for Children, United Neighborhood Houses (the coordinating agency for the city's settlement houses), the NAACP, the New York Urban League, and the Women's City Club. The two lobbies worked closely together, jointly sponsoring a public rally in Albany and distributing statements to legislators.

The social class make-up of the opposing forces was as different as their views on the issue of community control. The UFT position was supported by most of the organized labor movement, who defended the tenets of unionism; by those conservative Jews who saw an impartial merit system as a protection against discrimination; and by other white ethnics who felt that their jobs and neighborhoods were threatened by the growing black population. These groups were broadly middle class and lower-middle class. The community control forces represented a more complex roll of participants: militant spokesmen for lower-class blacks and Puerto Ricans; radical intellectuals in the foundations, universities, and small journals; certain political activists from the West Side of Manhattan; wealthy patricians on the boards of the city's leading social-welfare organizations; and representatives of the city's corporate elite. There is no precedent in New York school history for this particular combination from the upper and the lower strata of society; the former, however, appear to be the direct descendants of the upper-class school reformers who fought for centralization in the 1890s and the Gary plan some twenty years later.

As the legislature began serious deliberations on reorganization of the New York City school system, the political climate appeared to favor the proponents of community control.

In March the State Board of Regents submitted its decentralization bill, which was virtually the same as the Board of Education plan, with but one major change: it legislated the Doar board out of office. The Regents believed that the new system, having elected local school boards, should have either a one-man or three-man commission to administer the central agency. The Regents bill quickly gained the endorsement of the community control lobby, Mayor Lindsay and the *New York Times.*

The Republican leadership, which controlled both the state senate and the assembly, intimated that it would support any plan which had the consent of the Board of Regents, the mayor of New York, and the Board of Education. However, to the surprise of both the mayor and the Regents, the Board of Education refused to endorse the Regents bill. Despite their commitment to decentralization, the majority on the board would not agree to a bill which removed them from office. Afterwards, though board members continued to lobby for their own bill, the Doar board was, in the words of one of Lindsay's closest political operatives, "politically dead." [7]

The search for a bill that would placate both the UFT and the black legislators went on for days, as one proposal after another collapsed. Finally, Alton Marshall, representing both the Republican leadership and Governor Rockefeller, brought the opposing groups into round-the-clock negotiations and at last produced a bill that had enough protection to satisfy the union and enough local autonomy to satisfy the minority lawmakers. Ironically, the decisive votes for the bill were mustered among conservatives, led by Rosemary Gunning of Queens, who came to realize that decentralization was beneficial to the neighborhood school movement. Once the conservatives saw the uses of decentralization, the measure quickly passed both houses of the legislature with huge majorities. On April 30, 1969, less than an hour after its passage, it was signed into law by Governor Rockefeller.

Though the Board of Regents supported the new law, others were certain that any bill which satisfied the union could not be good. Lindsay called the new law "unworkable and unconstitutional." The *New York Times* denounced the bill as "a bad measure patched together in an outrageous travesty of the legislative process . . . a monstrosity" which affronted everyone except the unions; in its editorial view, the bill was a victory for "reactionary forces." Rhody McCoy called the bill a "prelude to the destruction of public education," and his counterpart in the IS 201 district called it "a disgrace." [8]

First reactions, however, prejudged the new law. The governance of education is a political question, and the political process shaped a compromise system, one which attempted to respond to the demands of the greatest number of interested parties. Most opposition to the bill stemmed from its scuttling of the demonstration districts; the projects would have been preserved if Lindsay's Board of Education had agreed to back the Regents bill. Once the issue was opened up, it turned out that the militant districts' strident ideological posturing was an embarrassment even to some black and Puerto Rican legislators. Like John Purroy Mitchel, the mayor who forty-two years earlier had tried to save his political career by wrapping himself in the American flag and denouncing his opponents as anti-American, the Ocean Hill-Brownsville district tried to survive by wrapping its cause in the black liberation flag and calling its detractors racists. One black legislator checked to be certain that the bill would have no difficulty passing, then cast his vote against it. In public, he deplored the bill, but in private he was glad to be rid of the raucous Ocean Hill-Brownsville board, relieved that he was once again free to live by his own definition of the responsibilities of a black person, rather than by the terms set by the governing board.[9]

THE NEW LAW enacted the following measures:

1. It removed the Doar board and replaced it with a paid, five-member interim Board of Education (one member appointed by each borough president); this board was supposed to be replaced a year later by a board composed of five elected members (one from each borough) and two members appointed by the mayor. (The election of a permanent board never occurred because of a court ruling that it would give equal representation to boroughs of vastly different population).

2. It empowered the interim board to divide the city into thirty to thirty-three school districts, each containing no less than an average of 20,000 pupils; this provision eliminated the three small demonstration districts as regular districts. (The number was reduced to 15,000 in 1973.)

3. It provided for election of community school boards by proportional representation in 1970.

4. It directed the Board of Education to appoint a chancellor in place of the superintendent of schools, with wide-ranging powers including the authority to suspend or remove a community board or any of its members for failing to comply with the law, bylaws, rules, regulations, or directives of the city board.

5. It granted substantial operating powers to community boards over all education in their districts except high schools, subject to the ultimate authority of the chancellor and city board.

6. It retained the Board of Examiners as the central agency to examine prospective teachers and supervisors, but allowed the bottom 45 percent of the schools—compared by reading scores—to hire teachers who had passed the National Teachers Examination regardless of whether they were on the examiners' list.[10]

CHAPTER 35

Aftermath

THE INTERIM BOARD OF EDUCATION was appointed soon after passage of the decentralization law. The Board's first president was Joseph Monserrat, a Puerto Rican; one of his successors as president was Isaiah Robinson, the first black president of the Board of Education. Instead of going out of office after a year in favor of an elected central board, the interim board was continued in office by the state legislature, because of the political difficulty of finding a satisfactory formula for an elected board, one which would be properly representative without being large and unwieldy. In 1973 the legislature made the interim board—one member appointed by each borough president—a permanent board and gave the mayor the power to add two additional board members, making a total of seven members.

More than a year passed before the central board located a new chancellor. He was Harvey Scribner, commissioner of education in Vermont, and a blunt-spoken educational reformer. Scribner quickly set himself at odds with the system's professional organizations. In 1971, a court ruling enjoined the Board of Examiners from issuing lists of supervisors, holding that the low proportion of black and Puerto Rican supervisors in the school system was evidence that the entry tests were discriminatory; Scribner, agreeing with the decision, declined to contest it. During his tenure in office, Scribner alienated a majority of the Board of Education as well. In early 1973, months before his three-year contract expired, the Board of Education accepted Scribner's resignation and soon afterward appointed Irving Anker, Scribner's deputy and a member of the system since 1937, as the new chancellor.

Before the first school board election in 1970 proponents of decentralization and community control anticipated that blacks and Puerto Ricans would take control of most of the new boards, since more than

60 percent of the public school register was by then black and Puerto Rican. It was forgotten that the boards were elected by all the city's voters, not just public school parents. It was forgotten, too, that low-income voters do not usually participate politically as much as middle- and upper-income voters.

Militant advocates of community control, angry because the new law did not provide total community control and did not retain the demonstration districts, boycotted the community school board elections and urged minority groups to do the same. The UFT, church groups, and civic organizations formed slates of candidates in most school districts. The average voter turnout was 15 percent, ranging from a low of 4.9 percent in Brownsville to a high of 22 percent in a Queens district. Of those elected, about 15 percent were black and 10 percent were Puerto Rican (their proportions in the population were, respectively, 21 percent and 12 percent). Community control proponents complained that too few blacks and Puerto Ricans were elected, although the same groups had encouraged minority voters to boycott the election.

In the second community school board elections in 1973, voter participation declined to less than 11 percent, although there was no voter boycott and there was an intensive publicity campaign. UFT-endorsed slates won control of most local boards, an outcome which no one had expected or wanted when the law was framed. The rationale for establishing locally elected boards had been to increase the representativeness of the boards and to stimulate public participation. But the absence of an information system in most districts left many voters confused about the identity of candidates and strengthened the appeal of slate-voting, which gave organized blocs like the UFT and the Catholic Church more power than they had had when local boards were appointed. The low level of participation in the special elections recalled the nineteenth-century precedent when, after six elections marked by voter apathy, the school elections were merged with the regular elections.

EACH OF THE DEMONSTRATION DISTRICTS in its waning days demonstrated a different facet of the problem of community control in a city as diverse as New York.

In Ocean Hill-Brownsville, the local school-board elections in 1970 became a bitter political contest. It was a contest, on one level, for jobs and power, and on another level, for control of the schools' ideology. As a result of the decentralization law, the eight schools of the district were scheduled to be absorbed into Brooklyn's District 23. The governing board, showing again the tenacious intransigence that had characterized its past performance, decided to boycott the election, in hopes of pressuring the city and state to preserve the independence of the district. The district's most persistent local critic, Assemblyman Samuel D. Wright, entered the election; Wright was an original member of the governing board and leader of the local Democratic club.

Rhody McCoy used the schools to fight Wright and to rouse support for the boycott. In September 1969, McCoy established an "Office of Educational Awareness and Development," one of whose purposes was "to motivate and assist the community in distinguishing between leadership which is solely self-seeking and that which seeks the greatest good for the greatest number." In December, McCoy and Reverend Oliver distributed a warning to parents that "some forces in this city and community stand with the bloody dagger in the backs of our children." Shortly before the March 1970 election, Wright was vilified by flyers which called him an "Enemy of Black and Puerto Rican People" (Wright was black). Within days of the election, McCoy sent parents a progress report which claimed that 64 percent of the children in the district, not including those already on grade level, had made "substantial progress" and listed ten reasons why parents should boycott the school-board election, concluding: "I urge you NOT TO VOTE—DO NOT be responsible for the failure of your child by legalizing this fraud." [1]

Fewer than 1,000 of 23,000 eligible voters participated in the school election, but Wright and a board supporting him were elected. McCoy called the election a farce and directed his staff not to cooperate with the new district superintendent, but the Ocean Hill-Brownsville experiment was concluded. Its supporters believed that entrenched interests killed the experiment for fear of being embarrassed by its educational success. People who wanted the project to succeed believed and publicized unverified claims of dramatic gains, especially in the area of reading, where the centralized system had conspicuously failed. The legislative leader of the community-control forces, Assemblyman Jerome Kretchmer, quoted McCoy that "by February 1, 1970, every youngster in that school system will be classified as a reader." *New York* magazine reported that McCoy had "wrought, in slightly over a year, deep organic change in the eight schools of the district," and that the district's reading program "has succeeded in raising the reading levels of many children . . . in a remarkably short time." Since Ocean Hill-Brownsville was the only district in the city that refused to give its children standardized reading tests, the claim that the district had conquered the reading problem went unchallenged. Then in 1971, less than a year after the dissolution of the governing board, the eight schools of Ocean Hill-Brownsville took a citywide reading test and registered scores lower than most ghetto schools in the borough. All eight schools recorded lower reading scores than in 1967, when the experiment began.[2]

The dangers of politicizing the schools should be evident. If school officials are held accountable at election time, they either have to prove they have succeeded or appear to have succeeded. In the absence of any clear gauge of accountability, school officials may be tempted to cover up failure with rhetoric, to search out scapegoats, to use communications with parents to lobby for their own survival. In the nineteenth century when there were elected local boards in New York, the election of a new local board sometimes meant the dismissal of entire teaching staffs, though the reason then was not ideological but was a function of the spoils system. Ocean Hill-Brownsville was the first instance in New York's history where a local board and its staff aggressively employed its schools, its students, and teachers in a campaign for self-preservation.

Horace Mann predicted, 120 years earlier, what would happen should "hostile partisans" fight for control of the public school:

> A preliminary advantage, indispensable to ultimate success, will be the appointment of a teacher of the true faith. . . . Why should not political hostility cause the dismemberment of districts, already too small. . . . What better could be expected, than that one set of school books should be expelled, and another introduced. . . . And who could rely upon the reports, or even the statistics of a committee, chosen by partisan votes, goaded on by partisan impulses, and responsible to partisan domination; and this, too, without any opportunity of control or check from the minority.

As first one side, then its opponents, wins control of the school board, "truth" is redefined: "Right and wrong have changed sides. The children must now join in chorus to denounce what they had been taught to reverence before, and to reverence what they had been taught to denounce."[3]

The Ocean Hill-Brownsville Governing Board, instead of providing freedom for different views, became a partisan advocate of one view only. Dissent from its position was seen as error, or worse, traitorous sabotage. When the school board cannot tolerate dissidence, when the nonconforming teacher risks dismissal, then the classroom becomes a partisan arena whose doctrines are determined at the polling place.

IS 201 DEMONSTRATED what can happen when an inexperienced board in a poor community suddenly finds itself responsible for the success of its educational program and for the disposition of millions of dollars in public funds. The struggle with a hostile union and a sluggish bureaucracy unified the governing board. But once the fight was over, once IS 201 ceased to be a revolutionary beachhead, "a knife poised at the jugular vein of the system," factionalism rent the governing board. The governing board had authority to fill hundreds of jobs, to let valuable contracts and to hire space; this was power, political power and dollar power, in a community where few people earned the kind of salaries that school officials were paid.[4]

The IS 201 Governing Board hired Charles Wilson as its first unit administrator; bright, articulate, scholarly in manner, Wilson had a masters degree in educational psychology and public administration. His commitment to educational reform sometimes conflicted with his commitment to community control. In one instance, Wilson recommended an innovative educator to be principal of an elementary school; the governing board turned him down because he was not a strong disciplinarian. At Wilson's insistence, the board relented and appointed him. Wilson learned "that the attitudes of community people are not as change oriented as speeches of the activists would seem to imply. In many instances, parents and community groups seemed reluctant about the simplest change in routines and wished to reinstitute a punitive repressive system."[5]

Wilson resigned in the summer of 1969. In his last official message to the governing board, he warned against the emergence of a "new elite" from within: "New elites are opposed to the liberation and empowering of the people in much the same way that other elites have been opposed to the liberation of the people. The protest style distracts attention from this ominous fact and creates the illusion of masses of people in motion when, in fact, only a few are in the know."[6]

After Wilson left the district, he reflected on his experiences. He concluded that the old "colonialism" was replaced by the local leadership's "neo-colonialism," which continued to serve the old order in return for money and power. Ambition and self-seeking opportunism, Wilson wrote, made it apparent that the argument "was really not about new education for the colonized but new educational *leaders* for the colonized, and the hustlers in the rebel camp envisioned an endless gravy train." When millions of dollars became available to the district for a government-funded Community Education Center, Wilson observed that "the Poverty Program mentality of some of the rebels took over" leading to "poor programming (at high salaries) . . . nepotism [and] the resurrection of every debunked Poverty scheme developed in the past ten years for East and Central Harlem." [7]

Wilson described the local leadership as "rebels without a long term strategic approach [who are in turn] led by individuals frequently unencumbered by a sense of history, limited by a grim anti-intellectualism, hobbled by lack of capacity to trust anyone—even themselves—and driven to extremism by underdeveloped egos paired with overdeveloped vocal chords." The project's stability, he wrote, was continually undermined by ethnic conflict between blacks and Puerto Ricans. Internal frictions were intensified by the leaders' "oppositional mentality" and habitual lack of trust:

> Puerto Ricans were suspicious of Blacks and vice versa, some whites by their mere presence in the rebel camp were objects of mistrust. Substantive changes in the schools were suspect. For example, changes involving humanizing the penal quality of the public schools were resisted by some of the very victims. . . . Knowledgeable opinions or technical expertise were also suspect. . . . In such an environment, no one . . . was to be fully trusted. This deep-rooted suspicion helped keep many capable people away. . . . There were, of course, interminable debates on all kinds of issues. . . . In most such debates, reason took a back seat to flamboyant rhetoric and to, of all things, *old school traditions!* By turning up the volume of their discontent or by adopting petulant postures, those with personal ambitions to gratify could gain as much consideration as those putting forward sound proposals.[8]

The strains within the governing board and between members of the governing board and its professional staff finally erupted into public view in early 1971. Two veteran members of the governing board resigned with a vitriolic blast at the district's black educators and at the "leaderless, listless" governing board. They declared that educational conditions were worse than ever,

> because the cripplers in the past (largely white) have now been joined by destructive opportunistic education pimps (largely black), who prey on the Harlem community, sucking its life blood, the community's future, which is embodied in their children. . . . We have seen little achievement and accountability, mainly because the so-called black professionals who talk that talk about community control and accountability . . . display the same contempt as their white counterparts.[9]

A few days later, a statement signed "MEMBERS OF THE STAFF" was sent to Ronald Evans, the principal of IS 201, agreeing with the resigning board members that children were being miseducated: "Black power slogans, Free Angela posters, Black studies programs and a predominantly black staff and administration have made little difference to the children. Granted, they are somewhat more proud of being black; but that isn't enough." The anonymous teachers complained of a lack of innovation, of insufficient remedial reading, of the irrelevance of many courses, of an inadequate guidance program and an absence of constructive supervision.[10]

Evans was defended by governing board members who held that he had thwarted the "ruling clique" on the board, who were accustomed to making decisions "capriciously and unilaterally" and had used their positions "to place their friends and relatives in Community Education Center jobs." [11]

Part of the reason for the discontent was that there continued to be serious educational problems in the district, and there was no longer an outside bureaucracy or union to blame. It was, after all, at 201 itself that the movement for community control and professional accountability had begun. Less than 8 percent of the children at IS 201 were reading on grade level, and the parents expected accountability from someone. Good progress in some of the elementary schools, where reading scores had gone up since the experimental district was created, was not enough, since expectations had been raised so high. In the three years that the project had functioned, there had not been time enough to work miracles. Still, some things had been learned: the teachers were dedicated; no one could claim that the staff was racist; black culture was respected; and the problems were not yet vanquished. Those who had come to power as revolutionaries were confronted with the realization that children are not educated by slogans. Perhaps the greatest benefit of the experiment was to enable those who were outsiders to get inside the system, to free themselves of the oppositional mentality and to take responsibility for the consequences of their actions.

THE TWO BRIDGES MODEL SCHOOL DISTRICT, the third of the demonstration districts, has been neglected to this point in the narrative, because its problems and its origins were entirely different from those of the other two demonstration districts. And yet Two Bridges, perhaps more than the other two districts, demonstrated the problems which inhere in particularistic community control, that is, a system in which the controlling group is free to promote its own interests without concern for the whole. In Two Bridges, ethnic groups competed to capture the local board in order to promote their interests without regard to the interests of other groups in the school district. The result of this competition demonstrated that unrestrained community control in an ethnically mixed district carries a high potential for ethnic conflict.

According to a report published by the Institute for Community Studies, which was funded by the Ford Foundation to assist and evaluate the three demonstration districts, Two Bridges was selected as a demonstration area for reasons different from those that led to the designation of IS 201 and Ocean Hill-Brownsville. The choice of the latter two was "as much a political response as an educational experiment. . . . Their unit administrators, Charles Wilson and Rhody McCoy, were looked upon as black leaders at the height of the development of black consciousness. Black power became the leverage for changing the personnel, programs and policies in these two districts. . . ." But, wrote the institute's participant-observer at Two Bridges, "educationists at the Ford Foundation chose this community as the crucible for school reform for what one Ford program director described as 'its middle class social and political views.' " Unlike the other two districts, the students of Two Bridges were racially and ethnically mixed: 40 percent Puerto Rican, 12 percent black, 35 percent Chinese, and 12 percent non–Puerto Rican white. The Ford Foundation grant of $40,000 went to the Two Bridges Neighborhood Council, which dominated the planning process as well as the elected governing board. An antipoverty agency associated with this organization ran a slate in the election and won all but two seats, which went to two middle-class Chinese who received the bloc vote of Chinese parents.[12]

Within six months, the professionals on the governing board resigned, because of the board's increasing militancy. Ethnic conflicts emerged over the allocation of Federal program money. Whether money was to go for English remedial classes (which would benefit blacks and Puerto Ricans) or for Chinese language classes was bitterly disputed in the schools and the community; Chinese parents organized and successfully pressed their demands for Chinese classes over the protests of other ethnic groups. White and Chinese parents combined to frustrate the governing board's attempt to eliminate ability tracking (their children were in the highest tracks).

After months of interracial and interethnic combat, a new election was held to fill three vacancies on the board. Parents opposed to the board campaigned with the support of the professional staff. This group won all three seats, electing two Chinese and one black. These new members, combined with moderates on the existing board, voted to abandon the demonstration project and to ask the Board of Education to incorporate their schools into a larger district.

The participant-observer from the Institute for Community Studies concluded: "In the end, the Two Bridges experience raises grave doubts about the political feasibility of organizing the disadvantaged in an ethnically mixed, heterogeneous community." Ethnic politics became the determining factor in establishing educational priorities; this situation encouraged groups to compete for personnel and policies which benefited *their* interests. Particularistic community control led not to cooperative pluralism but to antagonistic separatism. It polarized the community into hostile factions and, in doing so, failed to nurture a responsible majority with respect for minority rights. In the years immediately following decentralization, District 1 on the Lower East Side was the city's most troubled school district.

The danger of this kind of divisiveness was recognized as early as 1851, less than a decade after the initiation of community control; the president of the Board of Education stated then that "the present system could never have existed, nor could it now be maintained if it did not properly respect the many and various shades of political, religious and social opinions and practices which prevail in particular localities." [13] If the majority advances its interests to the detriment of the minority, then the ability of the public school to accord equal treatment to all its pupils is sacrificed. Unequal treatment, based on ethnic considerations, is certain to result in conflict within a single district or within the city system as a whole. The consequence for the public school is disastrous.

WITHIN A FEW YEARS after the Great School War of the 1960s ended, the combatants, with few exceptions, dispersed. Rhody McCoy went to the University of Massachusetts to prepare an oral history of Ocean Hill-Brownsville. Preston Wilcox opened a consulting firm. The Reverend Milton Galamison was invited to teach at Harvard University. John Doar returned to antipoverty work in Bedford-Stuyvesant. Superintendent Donovan retired to work for the Center for Urban Redevelopment in Education and to advise parochial schools. Commissioner Allen for a brief time was United States Commissioner of Education under President Richard Nixon; he died in a plane crash in 1971. Charles Wilson became assistant to John Niemeyer, the president of Bank Street College of Education. Marilyn Gittell continued to teach at the City University of New York and to direct the Institute for Community Studies. Mario Fantini left the Ford Foundation to become dean of the School of Education at New Paltz State College. David Spencer moved outside of New York City to work for a prisoner rehabilitation agency. Dr. Kenneth Clark denounced decentralization in 1972, declaring that the new system had failed to improve education and that local boards were too concerned with politics. Samuel Wright, as chairman of the District 23 community school board, continued to draw fire from supporters of the disbanded governing board. McGeorge Bundy, still president of the Ford Foundation, made his peace with Shanker and the UFT. In part because of the Ford Foundation's involvement in the New York school upheavals, the United States Congress passed legislation which taxed foundations' income and placed checks on controversial grants. As punishment for violating the state's Taylor Law, forbidding public employee strikes, Albert Shanker received a token jail sentence, and the UFT was fined $250,000. The UFT was strengthened in 1969 by winning a close election to represent the public school paraprofessionals, almost all of whom were black or Puerto Rican; the UFT's merger with the state National Education Association in 1972 made Shanker the head of the nation's largest local and one of the most powerful labor leaders in the state and nation.

THE DRIVE FOR COMMUNITY CONTROL was a direct assault on the idea of the common school, that is, a school which is supported by all, controlled by all, and which propagates no particular religious, ideological, or political views. The advocates of community control wanted public schools supported by all but controlled and ideologically directed by whoever won the local school-board elections.

When decentralization went into effect and locally elected boards took office, the community-control movement dissipated. Few, if any, of its substantive complaints about the quality of public education had been acted on. The system was shaken up, though not as much as it had been in the 1840s and 1890s. Once again, ethnic politics took precedence over educational issues. The activists' ability to rouse public support was vitiated by the fulfillment of their demand for public participation in school policy. Now that local boards were accountable to the public at election time, educational policy ceased to be a charged issue; education officials once again had political authority, derived directly from the electorate.

The diminution of militancy around the public schools concurred with a national trend away from strident demonstrations. With the onset of an economic recession, and a general cutback in the funds available for public education, the new local boards found themselves deciding how to cut their budgets rather than how to reach out in new directions.

The decentralized school system proved to be neither the disaster that its enemies had feared nor the panacea that its proponents had anticipated. There were districts where very little change resulted in personnel or in policies, and districts whose boards undertook to censor books and teachers or to create ethnic quotas for the staff. But there were benefits, too. Recruitment to the teaching staff, once tightly regulated by the Board of Examiners, was broadened. District superintendents were no longer chosen by test scores which bore no relation to ability to lead or to administer a district; under the new law, local school boards hired their own district superintendents on a contractual, rather than a tenured, basis. Black and Puerto Rican professionals were quickly elevated to supervisory posts in some districts, a step which would have required a wait of many years under the old methods.

Some community school boards welcomed innovative approaches. School reformers, such as Lillian Weber, director of the Workshop Center for Open Education at City College, continued to pour their energies into improving classroom practices, teacher training, and teacher-student relationships. School reform in the 1960s focused on who would control the schools; school reform in the 1970s was directed towards bettering the climate of the school and overcoming the relationship between education and social class, that is, discovering how to teach children of all backgrounds with equal effectiveness.

CHAPTER 36

The Search for Community

Some of New York's problems are peculiar to the metropolis, partly because of its size and partly because of the shifting nature of its population. As people move in and out of the city, neighborhoods change within a brief period from residential to commercial (and even, as happened on the Lower East Side, from commercial to residential); overutilized schools become underutilized schools and vice versa. In a city with almost 1,000 schools, the need for replacement and renovation is constant, placing a steady drain on public resources.

One solution to this problem which was pressed by school reformers at the turn of the century was to build huge schools which drew students from a large area, as a way of minimizing the impact of population shifts. Efficiency experts admired this strategy, though it saddled the city with heavy investments in big school buildings; the size of these buildings made it more appropriate to call them school plants rather than schools. Not until the 1960s did reformers point out that a school for 5,000 students has the environment of an impersonal factory rather than that of a school, where personal relationships are important.

But not all of New York's problems are specific to the city. Its mixed record in educating first- and second-generation immigrant youth is shared by other school systems. The social dislocations of the mid-twentieth-century migrations from Puerto Rico and the American South were in many ways similar to the earlier transatlantic migrations; the difficulties of adjusting to urban life have been inevitably reflected in school statistics. Still, the situations were not identical; there was a component of racism in the schools' hiring practices and attitudes toward dark-skinned children which diminished their opportunity to utilize the school to escape from poverty. The public schools have been almost as slow as other major institutions to eradicate racist practices.

The poor performance of large numbers of black and Puerto Rican children was the chief cause of criticism of the school system in the 1960s. It was not providing basic literacy to a sizable proportion of its students; it was not abating crime and poverty; it was not molding good citizens; it was not preparing young people for useful work; it was not eliminating social and economic inequity; it was not spearheading the reform of society; it was not producing happy, fulfilled human beings. Reformers did not realize that the same charges had been leveled periodically over the previous 150 years, or that the schools had never fully accomplished any of these ends. But by the 1960s, an almost mystical belief in the power of the public school to change society and to save individuals had become so ingrained that the system was certain to disappoint even the gentlest of critics.

Attacks on the schools, as in the past, were aimed at the system of governance. In the 1890s, the reformers said, "If the *experts* ran the system instead of petty ward politicians, then the problems could be solved." In the 1960s, the reformers said, "If the *people* controlled the schools instead of the bureaucrats, then the problems could be solved." In both instances, the schools were an easy target. In the 1890s, the reformers could point to blatant examples of inefficiency and arbitrariness under the ward system. In the 1960s, the reformers could justly complain of the inefficiency and cumbersomeness of the system's massive central bureaucracy.

One of the persistent ironies of reform is the impossibility of predicting the full consequences of change; every school war has had outcomes which were unintended, and, in many cases, unwanted. Nicholas Murray Butler, for example, imagined that centralization would remove capriciousness and error; he expected his reforms to empower visionary experts, not bureaucratic functionaries.

The controversy over black control of black schools recalled the Catholic campaign against the Public School Society in the 1840s. In neither case did those who began the struggle get what they wanted. Catholics had wanted either public subsidy of Catholic schools or the possibility that Catholic neighborhoods could control their local school boards. The resolution of the issue established community control, but barred the introduction of any teachings specific to a particular religion. As in 1842, the settlement of the school controversy in 1969 created elected local boards, but on terms that did not permit the inculcation of a particular ideology. Though this determination was not made explicit in the law, it was implicit in the legislature's rejection of total community control and in its elimination of the three demonstration districts.

The history of the public schools in New York City is inextricably related to the city's social and political history. Each major controversy was resolved politically, but *resolution* has not been *solution*. Every important issue remains and recurs:

The question of separation of church and state continues to be a lively and unsettled dispute. What once appeared to be absolute separation has been gradually abandoned in favor of limited public support for parochial schools. In the 1970s, parochial schools, feeling the pressure of rising costs, renewed their pursuit of greater or even full public subsidy; despite an adverse ruling by the United States Supreme Court in 1973, New York State and several other states continued to search for ways to assist parochial schools.

Neither centralization nor local control has solved the problems of the school system. Each has its advantages and disadvantages, which cause a pendulum movement over the years from one form to the other. When school officials have known what they wanted to do and how to do it, then faith in centralization was strong, as in the early nineteenth century and in the 1890s. But when both the means and the ends of schooling seemed confused and uncertain, and when the political legitimacy of the educational authorities appeared doubtful, there has been a trend to decentralize control of the schools, as in the 1840s and 1960s.

The education of lower-class children has been from 1805 until the present the most vexing dilemma of the New York public schools. New York, like other major metropolises, attracts low-income people with the lure of economic advancement. In the course of a generation or two, those who succeed move away from the slum to the outer reaches of the city, then to the suburbs, as members of the middle class. Thus, New York has a constantly replenished low-income population and a steady middle-income exodus. The proportion of poor children in the public schools is greater than the proportion of poor people in the city, because of the large number of private schools, a condition which is characteristic of many other large cities and which was true of New York in the nineteenth and early twentieth centuries. A journalist in 1905, for example, complained that the public schools in New York were handicapped by "the lack of active interest and support on the part of our well-to-do citizens, who do not send their children to the public schools and therefore have no immediate and vital stake in them." [1]

THE CITY'S RICH MELANGE of classes, races, religions, and ethnic groups guarantees an ever-present potential for conflict over the distribution of power. It is not accidental that each major school war coincided with the arrival in the city of a large new immigration. The thrust of public school history has been to reject different manifestations of separatism —whether religious or racial—and to evolve, however fitfully, in the direction of pluralistic, multi-cultural participation and control.

The common school idea, for all the buffeting it has taken in the past 150 years, has survived because it is appropriate to a democratic, heterogeneous society. It presumes that children should be taught those values which are basic to a free and just society, including respect for the individual's rights, a sense of social responsibility, and above all, perhaps, a devotion to comity, that precious value of a democratic society which grants the legitimacy of opposing views and permits groups to compete without seeking to crush one another. It presumes that schooling is a learning process, not an indoctrination process—a time for debate and discussion, not a time for instilling received opinion; it presumes that society needs and wants men and women who are capable of voting, deciding, and acting as free agents. As Robert M. Hutchins recently wrote, the public schools are "the common schools of the commonwealth, the political community. They may do many things for the young . . . but they are not public schools unless they start their pupils toward an understanding of what it means to be a self-governing citizen of a self-governing political community." [2] However good or poor any individual public school may be, and whatever its ethnic composition may be, this notion of community has always been central to its purpose.

The public school operates on behalf of the community, but how "community" is defined is the source of political and ideological controversy. A child lives simultaneously in many communities: his neighborhood, city, state, and nation; his ethnic group, race, and/or religion; his parents' occupation and interests may place his family in other communities as well. To suggest that the school serve one community and reject others is to create a partial vision, to limit children's potentialities instead of expanding them. The school that exalts only one race or class or locality denies the common humanity of its pupils, denies the diversity and mobility that is characteristic of democratic society. Respecting common values and common humanity need not imply the pursuit of homogeneity; no one wants to be a faceless figure in a mass society. The school can applaud individual and cultural diversity without resorting to the extremes of separatism and chauvinism.

The child's nonschool education begins in the family and continues while he is in school, through exposure to religion, television, comic books, movies, voluntary organizations, summer camps, and numerous other institutions and influences. Recognizing the importance of these informal educational networks, some contemporary writers on education would disband the school or limit it strictly to skill-training. But to do so would strip the school of the unique role that it can play in a mass society where sophisticated propaganda bombards the average citizen. For the school is the one educating institution whose purpose it is systematically to equip its students with the analytical tools of reasoning and judgment, in order that they may evaluate, criticize, and make choices.

Critics of the public schools in each generation have emphasized failure and inefficiency. What is inevitably lost sight of is the monumental accomplishments of the public school system of New York City. It has provided free, unlimited educational opportunities for millions, regardless of language, race, class, or religion. It has pioneered in the creation of programs for children with special gifts or special handicaps. It has willingly accepted the responsibility for solving problems which were national in scope, the result of major demographic shifts. The descendants of the miserably poor European immigrants who overflowed the city schools in the nineteenth and early twentieth century are today the prosperous middle class of the city and its suburbs. Without the public schools, despite their obvious faults, this unprecedented social and economic mobility would be inconceivable.

The introduction in 1970 of a non-competitive "open admissions" policy, guaranteeing all city high school graduates the right to a tuition-free college education at the City University of New York, was intended to boost sharply and quickly the educational qualifications of large numbers of blacks and Puerto Ricans, as well as children of white working-class background. The steady rise in educational level among blacks and Puerto Ricans in the sixties and seventies has been accompanied by the growth of an ambitious and energetic middle class— managers, professionals, artists, and businessmen—who are making their own contributions to the life of the city.

Whether control is centralized or decentralized, the problems of managing a school system for one million children are staggering. Other areas of the public sector are as crisis-prone as the schools, but none is as vulnerable to political struggle, particularly at the local level. The school is the principal public institution, beyond the government itself, intentionally designed to influence the values, habits, and behavior of the rising generation. Since people do not agree on which values, habits, and behaviors should be encouraged, school policy will always be controversial, especially when traditional attitudes are undergoing change. But what makes the public school unusually vulnerable to attack is that it is directly subject to—and ideally responsive to—public control. Child-rearing practices or television programming may have more impact on children than how they learn to read, but neither families nor the mass media are publicly controlled. Because of the American conception of lay control of public education, the school is likely to remain at the center of social conflict. Struggles for control of education will shift in emphasis as different groups seek to influence the schools for their own purposes; the politics of education is as valid as any other kind of politics and involves as many participants.

While the language of school wars relates to educational issues, the underlying contest will continue to reflect fundamental value clashes among discordant ethnic, cultural, racial, and religious groups. And this very fact underlines the importance of comity in the politics of education—comity, that basic recognition of differences in values and interests and of the desirability of reconciling those differences peacefully which the school itself aims to teach. The effort to advance comity, in educational affairs and in the affairs of the larger society, has always been at the heart of public education. Whatever their failings, whatever their accomplishments, the public schools have been and will be inescapably involved in the American search for a viable definition of community.

APPENDIX

New York City: Population and Pupils, 1800–1970

| YEAR | TOTAL POPULATION | PUBLIC SCHOOL ENROLLMENT | | NONPUBLIC SCHOOL ENROLLMENT |
		AVERAGE DAILY ATTENDANCE	ENTIRE REGISTER	
1800	60,489			5,249 (I, Ch/Ch)
1810	96,373	400 (AQA)		
1820	123,706	2,146 (AQA)		
1830	202,589	6,178 (AQA)		16,000 (I)
				3,000 (Ch/Ch)
1840	312,710	13,189	38,000	3,000–5,000 (C)
				16,000 (I)
				3,500 (Ch/Ch)
1850	515,547	18,153	53,546 (PSS)	9,500 (I, OS)
		15,805	45,872 (W)	8,500 (C)
1860	813,669	58,000	153,000	14,000 (C)
1870	942,292	103,679	270,000	23,000 (C)
				16,000 (I)
				9,000 (OS)
1880	1,206,299	133,000	268,000	30,000 (C)
				15,000 (I, OS)
1890	1,515,301	158,000	307,809	52,000 (C)
				15,000 (I, OS)
1900	3,437,202	380,000	420,000	75,000 (C)
				20,000 (I, OS)
1910	4,766,883	603,455	770,243	82,000 (C)
				600 (J)
				15,000 (I, OS)
1920	5,620,048	861,751	941,881	90,000 (C)
				1,000 (J)
				12,000 (I, OS)
1930	6,930,446	1,063,860	1,113,254	100,000 (C)
				4,000 (J)
				20,000 (I, OS)
1940	7,454,995	1,029,019	1,177,886	210,000 (C)
				5,000 (J)
				28,000 (I, OS)
1950	7,891,957	858,883	879,315	280,000 (C)
				16,000 (J)
				27,000 (I, OS)
1960	7,781,984	973,771	986,697	360,000 (C)
				36,000 (J)
				27,000 (I)
1970	7,894,862	927,766	1,143,853	330,000 (C)
				47,000 (J)
				34,000 (I)
				15,000 (OS)

KEY: (C) Catholic; (J) Jewish; (OS) Other Sects; (W) Ward Schools; (PSS) Public School Society; (AQA) Average Quarterly Attendance; (Ch/Ch) Church/Charity Schools; (I) Independent, Nonsectarian Schools.

SOURCES: United States Census Reports; Annual Reports, New York State Department of Education; Annual Reports of the Superintendent of Schools, New York City Board of Education; Carl Kaestle, *The Evolution of an Urban School System*, p. 52; Thomas Boese, *Public Education in the City of New York*, pp. 57, 129; James A. Burns, *A History of Catholic Education in the United States* (New York: Benziger Bros., 1937); Vincent P. Lannie, "Archbishop John Hughes," p. 570; John Talbot Smith, *The Catholic Church in New York* (New York: Hall & Locke, 1905), pp. 155, 285, 286, 314, 332, 488; Works Progress Administration, *Historical Records Survey* (New York City), Roman Catholic Church, Archdiocese of New York, Volume 2, 1941, pp. 16–22; Bureau of Education of the Jewish Community, *A Survey of the Financial Status of the Jewish Religious Schools of New York City*, 1911, p. 1; New York City Planning Commission, *Three Out of Ten: The Nonpublic Schools of New York City*, 1972, pp. 29–38; and Alexander M. Dushkin, *Jewish Education in New York City* (New York: Bureau of Jewish Education, 1918), p. 430.

A word of caution about these figures: the numbers representing independent schools are taken from New York State Department of Education estimates, which appear to be based on guesswork rather than on a careful census. Up until 1950, the number of pupils in independent schools is probably underestimated, with the exception of the years 1829 and 1867, when a school census was taken. Figures for Catholic school enrollment are in some instances interpolated from those years for which there was a count made (1854, 1857, 1862, 1864, 1867, 1880, 1900, 1920, 1932, 1940, 1950, 1960, 1970). When there were conflicting public school counts by the city and state departments of education (and there usually were), I chose those of the city, which I found to be more reliable and most frequently in agreement with other sources.

Notes

CHAPTER 1   EARLY NEW YORK: SOCIAL CONDITIONS AND SCHOOLS

1. Samuel L. Mitchill, *The Picture of New York* (New York: I. Riley, 1807), p. 2.
2. A. Emerson Palmer, *The New York Public School* (New York: Macmillan, 1905), p. 13.
3. Carl F. Kaestle, *The Evolution of an Urban School System* (Cambridge, Mass.: Harvard, 1973), p. 53.
4. Lawrence A. Cremin, *The American Common School* (New York: Teachers College Press, 1951), pp. 29–30.
5. William Oland Bourne, *History of the Public School Society* (New York: William Wood & Co., 1870), pp. 6–8.
6. Palmer, *New York Public School*, pp. 26–27.
7. David Salmon, *Joseph Lancaster* (London: Longmans, Green & Co., 1904), p. 7.
8. Ibid., pp. 7, 9.
9. *Manual of the Lancasterian System of Teaching Reading, Writing, Arithmetic, and Needlework, as Practised in the Schools of the Free School Society of New York* (New York, 1820), pp. 21–22, 27–28, 62–63.
10. Public School Society of New York, *Annual Report* (New York, 1833).
11. *Manual of the Lancasterian System*, p. 60.
12. Bourne, *Public School Society*, pp. 36–39.
13. Ibid., pp. 642–44.

CHAPTER 2   THE SOCIETY EXPANDS

1. Robert Dale Owen, *Free Enquirer*, vol. 2, no. 2, November 7, 1829, p. 14.
2. Bourne, *Public School Society*, pp. 110–18.
3. New York City Board of Education, *Annual Report for 1884*, pp. 52–53; Palmer, *New York Public School*, pp. 177, 292.

CHAPTER 3   THE IRISH ARRIVE

1. Robert Ernst, *Immigrant Life in New York City, 1825–1863* (New York: King's Crown Press, 1949), p. 67; Ernst writes that while the Irish and the Negroes shared the same oppressive poverty and discrimination, the Irish were the Negroes' worst enemy; commenting on this antagonism, a contemporary observer noted that "some of these people would shoot a black man with as little regard to moral consequences as they would a wild hog," p. 104.
2. Ibid., pp. 136, 178.

CHAPTER 4   THE CATHOLICS CHALLENGE THE SYSTEM

1. Vincent Peter Lannie, "Archbishop John Hughes and the Common School Controversy: 1840–1842" (Ph.D. dissertation, Teachers College, Columbia University, 1963), p. 29; published as *Public Money and Parochial Education* (Cleveland: Case Western Reserve, 1968).
2. Bourne, *Public School Society*, pp. 160–62.
3. Ibid., pp. 162–63.
4. John A. G. Hassard, *Life of the Most Reverend John Hughes* (New York: Appleton & Co., 1866), p. 338. Theodore Maynard, *The Story of American Catholicism* (New York: Macmillan, 1941), pp. 290–91.
5. Hassard, *Reverend John Hughes*, pp. 276, 278.
6. Andrew M. Greeley, *The Catholic Experience* (New York: Doubleday, 1967), p. 107; Maynard, *American Catholicism*, p. 298.
7. Bourne, *Public School Society*, p. 179.
8. Lannie, "Archbishop John Hughes," p. 61; George E. Baker, *The Works of William Seward*, vol. 3 (Boston: Houghton Mifflin, 1853), p. 210.
9. B. A. Hinsdale, *Horace Mann and the Common School Revival in the United States* (New York: Scribner's, 1900), p. 74.
10. Glyndon Van Deusen, *William Henry Seward* (New York: Oxford University Press, 1967), p. 70; Frederick Seward, *Autobiography of William H. Seward from 1831 to 1834 with a Memoir of His Life and Selections from His Letters from 1831 to 1846* (New York: D. Appleton, 1877), p. 502; Baker, *Works of Seward*, vol. 3, p. 480.
11. *Catholic Register*, February 20, 1840; *New York Truth Teller*, February 15 and 22, 1840.
12. Bourne, *Public School Society*, pp. 324–28.
13. Report of the Committee on Arts and Sciences, Document No. 80, Board of Assistant Aldermen, April 27, 1840; Bourne, *Public School Society*, pp. 722–34.
14. Lannie, "Archbishop John Hughes," p. 150.
15. *Freeman's Journal*, July 4, 1840.
16. Ibid., July 11, 1840.
17. Lannie, "Archbishop John Hughes," pp. 41–42.

CHAPTER 5   THE BISHOP TAKES COMMAND

1. Hassard, *Reverend John Hughes*, p. 228.
2. Ibid., p. 230.
3. Lannie, "Archbishop John Hughes," p. 232 (letter dated August 27, 1840).
4. Bourne, *Public School Society*, pp. 331–38.
5. Ibid., pp. 338–44.
6. Lannie, "Archbishop John Hughes," p. 206 (Hughes to Seward, August 29, 1840; Seward to Hughes, September 1, 1840).
7. Bourne, *Public School Society*, pp. 344–45.
8. *Freeman's Journal*, September 26, 1840.
9. Lannie, "Archbishop John Hughes," pp. 239–63.
10. Ibid., p. 259.
11. Bourne, *Public School Society*, pp. 189–95.
12. *New York Herald*, October 1, 1840.
13. Hassard, *Reverend John Hughes*, p. 234.

14. "The Important and Interesting Debate on the Claim of the Catholics to a Portion of the Common School Fund with the Arguments of Counsel, Before the Board of Aldermen of the City of New York," *Freeman's Journal*, October 29 and 30, 1840. Pp. 9–10, 14, 19, 43.
15. Bourne, *Public School Society*, pp. 316–20.

CHAPTER 6   FIRST ROUND IN ALBANY

1. Frederick Seward, *William H. Seward*, p. 502.
2. Lannie, "Archbishop John Hughes," p. 354.
3. Bourne, *Public School Society*, pp. 353–56.
4. Van Deusen, *William Henry Seward*, p. 68; Frederick Seward, *William H. Seward*, pp. 502–3.
5. Lannie, "Archbishop John Hughes," p. 360.
6. Ibid., p. 363; *Freeman's Journal*, April 3, 1841.
7. Lannie, "Archbishop John Hughes," p. 369.
8. Bourne, *Public School Society*, pp. 356–73.
9. Bishop John Hughes to Secretary of State John C. Spencer, May 10, 1841, Researches of American Catholic Historical Society of Philadelphia (New York Historical Society).
10. Bourne, *Public School Society*, pp. 373–402.
11. Ibid., p. 426; Van Deusen, *William H. Seward*, p. 70.

CHAPTER 7   "NO FLINCHING!"

1. *New York Tribune*, October 20 and 30, 1841; *New York Post*, October 28 and 30, 1841.
2. *Freeman's Journal*, October 30, 1841.
3. All quoted newspaper accounts are from the period October 30–November 3, 1841.
4. Bourne, *Public School Society*, pp. 498–500.
5. Ibid., pp. 501–16.
6. Ibid., citing *New York Evangelist*, March 3, 1842, pp. 506–16; citing report of the Commissioners of Common Schools of Brooklyn, p. 511.
7. Lannie, "Archbishop John Hughes," p. 523.
8. Ibid., pp. 511, 529.
9. Hughes to Seward, March 22, 1842, Records of the American Catholic Historical Society of Philadelphia (New York Historical Society).
10. *Freeman's Journal*, April 9, 1842.

CHAPTER 8   THE WARD SCHOOL SYSTEM TAKES ROOT

1. Seward to Hughes, Records of the American Catholic Historical Society of Philadelphia, 1842; cited in Hassard, *Reverend John Hughes*, pp. 250–51.
2. *Freeman's Journal*, May 7, 1842.
3. Ibid., April 3, 1841.
4. Greeley, *Catholic Experience*, pp. 103, 123, 126; David J. O'Brien, "American Catholicism and the Diaspora," *Cross Currents*, Summer, 1966, pp. 307–23; Ray Allen Billington, *The Protestant Crusade* (New York: Macmillan, 1938), p. 290.

5. Report of William L. Stone, Esq., Deputy Superintendent of Common Schools to the Board of Education of the City and County of New York, January 10, 1843.
6. John Franklin Reigart, *The Lancasterian System of Instruction in the Schools of New York City* (New York: Teachers College, 1916), p. 100; Edward H. Reisner, *The Evolution of the Common School* (New York: Macmillan, 1930), pp. 427–28.
7. Thomas Boese, *Public Education in the City of New York* (New York: Harper Brothers, 1869), pp. 71, 84.
8. Ibid., p. 86.
9. Ibid., pp. 93–94.
10. Ibid., p. 69.
11. Documents of the New York Board of Education, Collection of the New York Public Library, June 6, 1851.
12. Documents of the New York Board of Education, Collection of the New York Public Library, Hearings of the Board of Education, May 18, 1864; see also, Cephas Brainerd, Testimony before the Assembly Committee on Colleges, Academies and Schools, Collection of the New York Public Library, February 6, 1867; M. R. Werner, *Tammany Hall* (New York: Doubleday, 1928), p. 109.
13. Palmer, *New York Public School*, p. 144.
14. Laws of New York State (1851), chapter 386; Laws of New York State (1854), chapter 101.
15. *New York Times*, January 7, 1857.
16. *New York Sun*, April 5, 1864; *New York Herald*, April 2, 1864; Laws of New York State (1864), chapter 351; see also, Report of the New York State Commission appointed "to secure the more perfect establishment, government, regulation and economy of common schools in New York City," established in 1857, report submitted to state assembly on February 11, 1858.
17. Junius Henri Browne, *The Great Metropolis; A Mirror of New York* (Hartford, Conn.: American Publishing Company, 1869), p. 680.
18. Kate Holladay Claghorn, "The Foreign Immigrant in New York City," United States Industrial Commission, Report on Immigration, vol. 15, 1900, pp. 449–92. (New York Public Library).
19. *New York Evening Post*, March 15, 1869.
20. Ibid., February 2, 1867; February 12, 1867; *New York Tribune*, January 29, February 2, 1867; "The 'Ring' in the Board of Education of New York City," *The Nation*, March 7, 1867.
21. *New York Times*, February 11, 1867.

CHAPTER 9   THE TWEED RING IN CHARGE

1. *New York Times*, January 13, 1869.
2. John Pratt, "Boss Tweed's Public Welfare Program," *New York Historical Society Quarterly*, October 1961.
3. *New York Times*, March 19, 26, 1869, April 8, 1869; Charles Wingate, "An Episode in Municipal Government," *North American Review*, January 1875, pp. 119–74, and July 1875, pp. 113–175.
4. *New York Tribune*, March 19, 1869; *New York Sun*, March 20, 1869; *New York Evening Post*, March 19, 1869.
5. *New York Times*, May 6, 1869; Statement of Mayor A. Oakey Hall, Mayoral Papers, New York Municipal Archives, May 1869.

6. *New York Times,* May 13, 1869.
7. Ibid., November 11, 1871, December 20, 1871, October 20, 1870, May 20, 1870. Denis Tilden Lynch, *"Boss" Tweed* (New York: Boni & Liveright, 1927), pp. 335–37.
8. *New York Times,* March 25, 1871.
9. Ibid., April 30, 1871; "Speech of William Wood at Organizing Meeting of Department of Public Instruction," Documents of the New York Board of Education, Collection of the New York Public Library, April 29, 1871.
10. *New York Times,* June 27, 1871.
11. Ibid., December 20, 1871.

CHAPTER 10   INSIDE THE CLASSROOM

1. Report of a Committee of the Board of Education, Documents of the Board of Education, Collection of the New York Public Library, 1853.
2. "Report of Committee of Baltimore School Officials Sent to Visit the Public Schools of Philadelphia, New York, Brooklyn and Boston," Documents of the New York Board of Education, Collection of the New York Public Library, 1867.
3. Browne, *The Great Metropolis,* pp. 680–1.
4. Howard Ralph Weisz, "Irish-American and Italian-American Educational Views and Activities, 1870–1900: A Comparison" (Ph.D. dissertation, Columbia University, 1968), pp. 70, 78–79, 81–83.
5. "Report of a Committee of the New York Municipal Society Appointed to Investigate the System of Public Education with a View to Retrenchment in Expenditures," February 4, 1878; Teachers College Library. Weisz, "Irish-American and Italian-American," p. 215.
6. William Wood, Remarks to the Board of Education, Documents of the New York Board of Education, Collection of the New York Public Library, January 9, 1878.

CHAPTER 11   THE BIRTH OF A REFORM MOVEMENT

1. Annual Report of the Superintendent of Schools, New York Board of Education, 1890.
2. Stephen A. Walker, ex-president of the Board of Education, to Mayor Hugh Grant, November 1, 1892, New York City Municipal Archives, Mayoral Papers.
3. Mayor Franklin Edson to Stephen A. Walker, president of the Board of Education, response by Walker, April 15, 1884, New York City Municipal Archives, Mayoral Papers.
4. "Report of the Special Committee of Eight on Changes in the School System, New York Board of Education," 1889, Teachers College Library.
5. "Memorial to the Board of Education," Public Education Society of the City of New York, February 6, 1889, Teachers College Library.
6. *School,* February 12, 1891; January 14, 1892.
7. "Report of the Special Committee of Seven," New York Board of Education, November 1891, Teachers College Library.
8. Journal of the New York Board of Education, April 15, 1891, pp. 482–86; March 2, 1892, pp. 242–44.
9. Letters in the New York Municipal Archives, Mayoral Papers.
10. *School,* February 2, 1893.

11. "Report of the Commissioners of Accounts to Mayor Hugh Grant," December 31, 1892, New York Municipal Archives, Mayoral Papers.

CHAPTER 12   "SAVE, SAVE THE MINUTES"

1. Jacob A. Riis, "The Problem of the Children," *The Children of the Poor* (1892), reprinted in *Jacob Riis Revisited,* ed. Francesco Cordasco (New York: Doubleday Anchor, 1968), p. 125.
2. Riis, "Children," p. 131.
3. *School,* September 8, 1892.
4. Ibid., September 29, 1892.
5. Ibid., October 13, 1892.
6. Ibid., December 1, 1892, quoting the *New York Tribune.*
7. Ibid., January 26, 1893, February 2, 1893.
8. Joseph Mayer Rice, "The Public School System of New York City," *Forum,* January 1893, pp. 616–30.
9. *New York Times,* April 11, 1893.
10. "Report of the Commission to Revise the Laws Affecting Common Schools and Public Education in the City of New York," 1893, Teachers College Library.
11. *School,* October 26, 1893.

CHAPTER 13   THE BEST MEN TO THE RESCUE

1. *The Triumph of Reform, A History of the Great Political Revolution, November 6, 1894,* (New York: Souvenir Publishing, 1895), p. 157.
2. *The Triumph of Reform,* p. 126.
3. Stephen Olin, "Public School Reform in New York," *Educational Review,* June 1894, pp. 1–6.
4. Charles C. Wehrum, "The Common Schools of the City of New York," 1894, Documents of the New York Board of Education, Collection of the New York Public Library.
5. *The Triumph of Reform,* pp. 8, 11.
6. Ibid., pp. 5–6.
7. "Report of Sub-Committee on Public School System," Committee of Seventy, December 15, 1894, Teachers College Library.
8. *School,* January 3 and 31, 1895.

CHAPTER 14   PROFESSOR NICHOLAS "MIRACULOUS" BUTLER

1. David Hammach, "The Centralization of New York City's School System, 1896," (Master's thesis, Columbia University, 1970).
2. *School,* February 21, 1895.
3. Ibid., April 4, 1895.
4. *School,* April 25, 1895.
5. *Educational Review,* May 1895, p. 98.
6. Ibid., June 1895, p. 102.
7. *New York Evening Post,* December 28, 1894.
8. Nicholas Murray Butler to Robert Maclay, December 3, 1895; Butler to N. Prentiss, December 5, 1895, Butler Collection, Columbia University.

(All Butler correspondence cited here is in the Butler Collection, Columbia University.)

9. Mrs. Schuyler van Renssalaer to Butler, December 4, 1895.
10. Butler to Stephen H. Olin, December 20, 1895.
11. Papers of the Citizens Committee for Public School Reform, Teachers College Library.
12. Butler to State Senator Frank Pavey, January 30, 1896.
13. James W. Pryor to Butler, February 20, 1896.
14. Susan Abbot Mead to Butler, n.d. [1896].
15. Hamilton Fish to Butler, January 20, 1896; Charles B. Page to Butler, February 11, 1896; George C. Austin to Butler, January 16, 1896.
16. Butler to Charles B. Page, February 6, 1896; Butler to Page, February 16, 1896.
17. Butler to James Milne, principal of the State Normal School at Oneonta, New York, March 9, 1896.
18. Butler to president of Packer Collegiate Institute, February 10, 1896.
19. Butler to J. B. Bishop, March 4, 1896; Butler to John E. Milholland, February 4, 1896.
20. Butler to State Senator Stranahan, March 6, 1896; Butler to Charles Skinner, State Superintendent of Public Instruction, 1896; *School,* February 20, 1896.
21. *School,* February 27, 1896.
22. Ibid.
23. Ibid., April 9, 1896; Sol Cohen, *Progressives and Urban School Reform* (New York: Teachers College, 1963), p. 42.
24. "Public Schools in Danger!" New York Municipal Archives, Mayoral Papers, undated [1896]; Butler to Stranahan, March 27, 1896; *New York Herald,* March 27, 1896.
25. *School,* February 20, 1896.
26. Ibid., April 9, 1896.
27. Ibid., March 26, 1896.
28. F. R. Coudert to Mayor Strong, New York Municipal Archives, Mayoral Papers, 1896.
29. Petition from several inspectors and others to Mayor Strong, New York Municipal Archives, Mayoral Papers, 1896.
30. Staff member to Mayor Strong, New York Municipal Archives, Mayoral Papers, April 17, 1896.

CHAPTER 15   NEW EDUCATION FOR THE NEW IMMIGRATION

1. Butler to N. Prentiss, May 9, 1896, Butler Collection, Columbia University Library.
2. *School,* January 23, 1899.
3. Ibid., January 18, 1900.
4. Lawrence Cremin, *The Transformation of the School* (New York: Knopf, 1961), p. viii.
5. *New York Times,* January 17, 1905.
6. Ibid., March 20, 1905; for descriptions of the post-centralization changes, see Selma Berrol, "Immigrants at School, 1898–1914" (Ph.D. dissertation, City University of New York, 1967).
7. Richard Hofstadter, *Anti-Intellectualism in American Life* (New York: Knopf, 1962), p. 328.

8. Annual Report of the Superintendent of Schools, New York City Board of Education, 1912.
9. Samuel Abelow, *Dr. William H. Maxwell* (New York: Scheba Publishing Company, 1930), pp. 105–6.
10. Ibid., pp. 121–22.
11. Robert Hunter, *Poverty* (New York: Macmillan, 1904), p. 210; Annual Report of the Superintendent of Schools, New York City Board of Education, 1904.
12. Annual Report of the Superintendent of Schools, Brooklyn Board of Education, 1887.
13. H. Thiselton Mack, "Moral Education in American Schools," *Special Reports on Educational Subjects,* Vol. X, *Education in the United States of America,* English Ministry of Education, 1900, Part I, pp. 74, 84.
14. Adele M. Shaw, "True Character of New York Public Schools," *World's Work,* December 1903, pp. 4204–21.
15. Roy L. Garis, *Immigration Restriction* (New York: Macmillan, 1927), p. 203.
16. John R. Commons, *Races and Immigrants in America* (New York: Macmillan, 1907), pp. 69–70.
17. Oscar Handlin, *Race and Nationality in American Life* (New York: Doubleday Anchor, 1957), p. 74.
18. Berrol, "Immigrants at School," p. 51; *New York Times,* October 24, 1904.
19. "The Public Schools of New York," *New York Tribune,* March 1896.
20. Leonard P. Ayres, *Laggards in Our Schools* (New York: Russell Sage Foundation, 1909); David K. Cohen, "Immigrants and the Schools," *Review of Educational Research,* February 1970, pp. 13–27.
21. Nathan Glazer and Daniel Patrick Moynihan, *Beyond the Melting Pot* (Cambridge, Mass.: MIT, 1963), pp. 184, 199; Weisz, "Irish-American and Italian-American," pp. 96, 395.
22. "The Public Schools of New York," describing GS 42, 30 Allen Street; Moses Rischin, *The Promised City* (Cambridge, Mass.: Harvard, 1962); *New York Times,* September 15, 1897.
23. Claghorn, "The Foreign Immigrant," p. 478.

CHAPTER 16   PROBLEMS OF CENTRALIZATION

1. Report of Committee on School Inquiry, New York City Board of Estimate and Apportionment, 1911–1913, pp. 19–21, 140–41, 193–94, 547–662. Teachers College Library.
2. Abelow, *Dr. William H. Maxwell,* p. 147.
3. William Henry Maxwell to Nicholas Murray Butler, February 7, 1918, Butler Collection, Columbia Library.
4. *Current Biography* (New York: H. W. Wilson, 1940), pp. 130–34.

CHAPTER 17   DISILLUSIONED PROGRESSIVES

1. George B. McClellan, Jr., *The Gentleman and the Tiger, The Autobiography of George B. McClellan Jr.,* ed. Harold Syrett (New York: Lippincott, 1956), p. 292.
2. Fourteenth Annual Report of the Superintendent of Schools, New York City Board of Education (New York, 1912).

CHAPTER 18   THE SOLUTION TO NEW YORK'S PROBLEMS

1. *New York Times,* June 15, 1915.
2. *New York Globe,* April 8, 1915; *New York Times,* January 29, 1915.
3. *New York Times,* February 1, 1915.
4. Sol Cohen, *Progressives and Urban School Reform,* pp. 89–90.
5. Randolph S. Bourne, *The Gary Schools* (Boston: Houghton Mifflin, 1916), p. 14 (see also, pp. 15, 17, 19, 35, 39, 57–64, 117); the original Gary series appeared in the *New Republic* on March 27, April 3, April 10, April 24, May 1, 1915.
6. R. Bourne, *Gary Schools,* p. 85.
7. "The Originator of the Gary Plan," *American Review of Reviews,* November 1915, pp. 588–89.
8. John and Evelyn Dewey, *Schools of Tomorrow* (New York: Dutton, 1915).
9. *New York Times,* March 25, 1915.
10. Murray and Adeline Levine, "The Gary Schools: A Socio-Historical Analysis of the Process of Change," *California Elementary Administrator,* Spring 1970; this article appears, with slight changes, as the introduction to an annotated new edition of Randolph S. Bourne's *The Gary Schools* (Cambridge, Mass.: MIT Press, 1970), pp. xxxiii–xxxiv.
11. *New York Times,* June 9, 1915.
12. Ibid., June 23 and July 4, 1915.
13. Ibid., September 17, 1915.
14. Ibid., September 24, 1915.
15. William A. Prendergast, "Why New York City Needs a New School Plan," *American Review of Reviews,* November 1915, pp. 584, 586, 587.
16. R. Bourne, *The Gary Schools,* p. 184–85.
17. *New York Times,* October 22, 1915.
18. Thomas S. Baker, "What the Gary School Plan Proposes and Does," *New York Times,* October 17, 1915.
19. *New York Times,* December 6, 1915.
20. Ibid., December 7, 1915.

CHAPTER 19   TAMMANY AND THE BUREAUCRACY AS ALLIES

1. *Seventeenth Annual Report of the Superintendent of Schools,* New York City Board of Education (New York, 1916); *School and Society,* III, pp. 628–29.
2. Howard W. Nudd, "The Buckingham Tests of the Gary Schools in New York City," *Journal of Education,* April 20, 1916, editorial, p. 435.
3. William G. Willcox, "What Modern Education Means," *Journal of Education,* June 22, 1916, pp. 677–78.
4. *New York Globe,* November 23, 1915; *New York Times,* February 6–9, 1916, May 18, 1917.
5. *New York Times,* January 30, 1915, February 5, 1915.
6. Willcox, "Modern Education," p. 677.
7. Woman's Municipal League, "Modern Schools for New York City," Report of the Education Committee (New York, 1916), Teachers College Library.
8. *New York Times,* January 9, 1916.
9. Ibid., July 4, 1917.
10. Ibid., August 8, 1917.

CHAPTER 20   "MR. MAYOR, HANDS OFF OUR PUBLIC SCHOOLS!"

1. Cleveland Rodgers, *Robert Moses* (New York: Henry Holt, 1952), p. 327.
2. *New York Times,* September 1, 1917; see also, W. A. Swanberg, *Citizen Hearst* (New York: Scribner, 1961), pp. 365–66.
3. Bayrd Still, *Mirror for Gotham* (New York: New York University Press, 1956), p. 287; Edwin R. Lewinson, *John Purroy Mitchel* (New York: Astra, 1965), pp. 200–2.
4. *New York American,* March 28, 1917.
5. *New York Times,* October 2, 1917.
6. Ibid., September 21, 1917, October 5, 1917.
7. Ibid., September 8, 1917.
8. Ibid., September 13, 1917.
9. Ibid., September 22, 1917.
10. Ibid., September 22, 1917.
11. Ibid., October 5, 1917.
12. Ibid., October 14, 15, 17, 1917.
13. Ibid., October 2, 16, 1917.
14. Ibid., October 17, 18, 19, 20, 22, 23, 1917.
15. Ibid., October 31, 1917.
16. *Journal of Education,* November 15, 1917, p. 491.
17. Allan Nevins and John A. Krout, eds., *The Greater City* (New York: Columbia University Press, 1948), pp. 84, 86, 87. Rodgers, *Robert Moses,* p. 19.
18. Abraham Flexner and Frank P. Bachman, *The Gary Schools, A General Account* (New York: General Education Board, 1918), pp. 200–4.
19. Abraham Flexner, *I Remember* (New York: Simon & Schuster, 1940), p. 255; Raymond E. Callahan, *Education and the Cult of Efficiency* (Chicago: University of Chicago Press, 1962), p. 143.
20. "William A. Wirt," *Dictionary of American Biography,* XIII, Supplement Two (1934), p. 728.
21. *Journal of Education,* November 29, 1917, p. 545.
22. *New York Times,* January 2, 1918; *Fiftieth Annual Report of the Superintendent of Schools,* New York City Board of Education (New York, 1949), p. 62.
23. *Sixteenth Annual Report of the Superintendent of Schools,* New York City Board of Education (New York, 1915), pp. 178–79; *Twenty-first/Twenty-second Annual Report of the Superintendent of Schools,* New York City Board of Education (New York, 1920), pp. 279–80.

CHAPTER 21   DIVIDENDS OF THE DEPRESSION

1. Lawrence Cremin, *The Transformation of the School* (New York: Knopf, 1961), p. 181.
2. Sol Cohen, *Progressives and Urban School Reform,* pp. 124–26.
3. *Fiftieth Annual Report of the Superintendent of Schools,* p. 88.
4. Sol Cohen, *Progressives and Urban School Reform,* pp. 161–62.
5. *Fiftieth Annual Report of the Superintendent of Schools,* p. 126.
6. Bureau of School and Cultural Research, "Historical Review of Studies and Proposals Relative to Decentralization of Administration in the New

York City Public School System," New York State Education Department, June 1967; Special Collection, Politics and Education Program, Teachers College.

CHAPTER 22   FROM AMERICANIZATION TO INTEGRATION

1. *New York Times,* April 23, 1956.
2. *Brown* v. *Board of Education of Topeka,* 347 U.S. 483 (1954).

CHAPTER 23   THE DISCOVERY OF SEGREGATION AND SCANDALS

1. Frances Blascoer, *Colored School Children in New York* (New York: Public Education Association, 1915). This survey found that Negro children had a high rate of truancy and 60 percent were overage for their grade. The author held that poor home environment and society's racism sapped the children's ambition. She noted that Negro children from good homes progressed "on a par with the white children in the same schools." Pp. 16–17.
2. *New York Times,* October 26, 1954.
3. Ibid., December 24, 1954.
4. "The Status of the Public School Education of Negro and Puerto Rican Children in New York City." Presented to the Board of Education Commission on Integration. Prepared by the Public Education Association, assisted by the New York University Research Center for Human Relations, October 1955.
5. *New York Times,* July 24, 1956.
6. Ibid., February 26, 28, March 1, March 4, 1957.
7. Ibid., March 23, 1957.
8. Ibid., July 27, 1957.
9. Ibid., August 5, October 31, 1957.
10. Ibid., September 2, 1957; "Toward the Integration of Our Schools," Final Report of the Commission on Integration, New York City Board of Education.
11. Ibid., July 8, 1959.
12. Ibid., June 29, August 15, 1959.
13. Ibid., January 29, 1959; Henry T. Hillson and Florence C. Myers, "The Demonstration Guidance Project, 1957–1962," New York City Board of Education, May 1963.
14. "Toward Greater Opportunity," a Progress Report from the Superintendent of Schools to the Board of Education dealing with Implementation of Recommendations of the Commission on Integration, June 1960, p. 156.
15. *New York Times,* September 2, 1960; Jacob Landers, "Improving Ethnic Distribution of New York City Pupils," New York City Board of Education, May 1966, p. 28.
16. *New York Times,* August 13, 1961; September 12, 1958.
17. Mark Schinnerer, "Report on New York City School Reorganization," submitted to New York State Commissioner of Education James E. Allen, Jr., December 26, 1961.
18. *New York Times,* January 4, 7, 8, 15, 1962.

CHAPTER 24   BOYCOTTS AND DEMONSTRATIONS

1. *New York Times,* September 2, 1962.
2. Ibid., August 20, August 23, 1963.
3. Ibid., August 26, 28, September 3, 1963.
4. *New York Times,* September 7, 1963.
5. *New York Times,* November 11, 1963; Kenneth B. Clark, *Dark Ghetto* (New York: Harper & Row, 1965).
6. *New York Times,* December 10, 1963.
7. Ibid., December 17, 1963.
8. Ibid., December 23, 1963.
9. Ibid., January 14, 20, 30, 1964.
10. Ibid., January 26, 29, February 3, 1964.
11. Ibid., February 4, 1964.
12. Ibid., December 23, 1963.
13. Ibid., March 9, 1964.
14. Ibid., March 11, 1964.
15. Ibid., March 13, 17, 1964.
16. Ibid., March 27, 1964; Rose Shapiro, "A Review of the Steps the Board of Education has Taken to Integrate the New York City Schools," a confidential memorandum circulated among members of the Board of Education, February 6, 1967; Special Collection, Politics and Education Program, Teachers College.

CHAPTER 25   THE ALLEN REPORT

1. "Desegregating the Public Schools of New York City," a Report Prepared for the Board of Education of the City of New York by the State Education Commissioner's Advisory Committee on Human Relations and Community Tensions (the Allen Report), May 12, 1964. The research director of the Allen Report was Dr. Robert Dentler.
2. *New York Times,* May 13, 14, 1964.
3. Allen Report, pp. 11–13.
4. Ibid., pp. 13–14.
5. Ibid., pp. 14–16.
6. *New York Times,* May 17, 1964.
7. Ibid., May 29, June 2, 3, 1964.
8. Bert Swanson, *The Struggle for Equality* (New York: Hobbs, Dorman & Co., 1966), p. 83.
9. *New York Times,* March 9, 1965.

CHAPTER 26   ENTER LINDSAY

1. *New York Times,* February 22, 1966.
2. Ibid., October 22, 1965.
3. Martin Mayer, "Close to Midnight for the New York Schools," *New York Times Magazine,* May 2, 1965.
4. *New York Times,* April 30, 1966.

Chapter 27 IS 201: An End and a Beginning

1. Preston Wilcox, "Releasing Human Potential: A Study of the East Harlem-Yorkville School Bus Transfers," New York City Commission on Human Rights, 1961; Special Collection, Politics and Education Program, Teachers College.
2. "Questions and Answers on New York's Schools," City-Wide Committee for Integrated Schools, leaflet circulated prior to February 1964 school boycott: reprinted in *Black Protest*, ed. Joanne Grant (New York: Fawcett, 1968), pp. 408–10.
3. Carolyn Eisenberg, "The Parents Movement at IS 201: From Integration to Black Power, 1958–1966." (Ph.D. dissertation, Teachers College, Columbia University, 1971), p. 33.
4. Ibid., p. 71.
5. "Sequence of Events Surrounding Community Involvement with Public School 201," unpublished, privately circulated; Special Collection, Politics and Education Program, Teachers College.
6. Preston Wilcox, "To Be Black and to Be Successful" (New York: MUST, March 1, 1966); Special Collection of the Politics and Education Program, Teachers College.
7. Preston Wilcox, "One View and a Proposal," *Urban Review*, July 1966, pp. 12–16.
8. Robert Dentler, "In Reply to Preston Wilcox," *Urban Review*, July 1966, pp. 16–17.
9. *New York Times*, July 10, 1966.
10. Ibid., September 2, 1966.
11. Statement by Lloyd Garrison, Office of Public Information, New York City Board of Education, September 8, 1966.
12. Eisenberg, "Parents Movement at IS 201," pp. 173–74; "Sequence of Events."
13. *New York Times*, September 9, 1966.
14. "Sequence of Events," p. 13.
15. *New York Times*, September 19, 1966; Miriam Wasserman, *The School Fix* (New York: Outerbridge & Dienstfry, 1970), p. 224.
16. *New York Times*, September 20, 1966.
17. Ibid., September 21, 1966.
18. *New York Times*, September 24, October 2, 3, 1966.
19. Letter to Albert Shanker, Parent/Community Negotiating Committee for IS 201, October 3, 1966; also *New York Times*, October 4, 1966.
20. *New York Times*, October 18, 19, 1966.
21. Ibid., December 21, 22, 28, 31, 1966.
22. "Real Public Hearing," Ad Hoc Board of Education, December 20, 1966; Special Collection, Politics and Education Program, Teachers College.
23. Kenneth B. Clark, *Dark Ghetto*, pp. 112, 131; see also, James S. Coleman, *Equality of Educational Opportunity* (United States Office of Education, 1966), pp. 21–22, 325; Frederick Mosteller and Daniel P. Moynihan, eds., *On Equality of Educational Opportunity* (New York: Random House, 1972); Christopher Jencks, *Inequality* (New York: Basic Books, 1972).
24. Ayres, *Laggards in Schools;* David Cohen, "Immigrants and Schools"; Katherine Murdoch, "A Study of Race Differences in New York City," *School and Society*, January 1920, pp. 147–50; Colin Greer, *The Great School Legend* (New York: Basic Books, 1972).

CHAPTER 28   DECENTRALIZATION EMERGES

1. Staff Paper No. 9, New York City Temporary Commission on City Finances, *Governing the Public Schools* (New York: The Commission, 1966).
2. Harry Gottesfeld and Sol Gordon, "Academic Excellence: Community and Teachers Assume Responsibility for the Education of the Ghetto Child," unpublished, undated; Special Collection, Politics and Education Program, Teachers College.
3. "Decentralization: Statement of Policy," New York City Board of Education, April 19, 1967. [Originally Superintendent Donovan recommended seven decentralization proposals for approval; notable among them was an experimental school unit consisting of the Joan of Arc Junior High School and its elementary school feeders, located on the upper West Side of Manhattan, a racially, socially, and economically diverse area. The Joan of Arc project never was launched because of conflict within the community; one of the elementary schools voted not to participate in the experiment, and competing parent leaders could not unite "on the fundamental issues of the character of the elected board and on establishing criteria for community participation." Arthur Tobier, "Decentralization: Waiting for Something to Turn Up," *The Center Forum*, August 28, 1967, pp. 1, 2.]
4. Mario D. Fantini, "A Strategy for Initiating Change from within Large Bureaucratic School Systems," n.d.; "How Can We Best Educate the Economically Disadvantaged Student," a Paper Prepared for the New Jersey Governor's Conference on Education, April 2, 1966; Special Collection, Politics and Education Program, Teachers College.
5. *New York Times*, July 7, 1967.
6. Ibid., May 22, 1967.

CHAPTER 29   THE MAKING OF A POWER STRUGGLE

1. Ocean Hill-Brownsville Demonstration District, "The Independent School Board," n.d., Special Collection, Politics and Education Program, Teachers College (all documents in this location will be referred to as *Special Collection*).
2. "A Plan for an Experimental School District in District 17, Brooklyn, An Ocean Hill-Brownsville Experiment," submitted to the New York City Board of Education, May 27, 1967, Special Collection.
3. "Anti-Semitism in the New York City School Controversy," Anti-Defamation League of B'nai B'rith, January 1969 [hereafter referred to as ADL report], p. 213.
4. "An Evaluative Study of the Process of School Decentralization in New York City," Final Report by the Advisory and Evaluation Committee on Decentralization to the Board of Education of the City of New York, July 30, 1968 [hereafter referred to as the Niemeyer report], p. 37, Special Collection.
5. Ocean Hill-Brownsville Demonstration District, "Information Sheet Re: Samuel D. Wright's charges against the Ocean Hill Governing Board as enumerated in an August 9th letter to State Commissioner Allen," August 1968. Special Collection.
6. Maurice J. Goldbloom, "The New York School Crisis," *Commentary*, Janu-

ary 1969, wrote: "Interestingly, those paid to get out the vote were also candidates for the governing board. Not surprisingly, they turned out to be the successful candidates. No formal procedures for the election or qualifications for candidates were set by the planning board, which also took charge of counting the vote." Marilyn Gittell, director of the Institute for Community Studies (which was administering the Ford Foundation's grant to the demonstration districts), responded ("The School Strike," *Commentary*, April 1969): "Many canvassers in Ocean Hill who were paid to get out the vote ran for office—some were elected, others were not. Mr. Goldbloom's suggestion that those who were paid were the only ones elected is not accurate." The facts were not a matter of perspective; while it was true that only six out of twenty paid canvassers won governing board seats, six of seven governing board seats were won by paid canvassers.

7. "Statement by the Teachers of the Ocean Hill-Brownsville Experimental District," September 1967, Special Collection.
8. "Progress Report for Month of July," from Rhody McCoy, summer coordinator of Ocean Hill-Brownsville School Project, July 1967, Special Collection; Rhody McCoy, "The Year of the Dragon," *Confrontation at Ocean Hill-Brownsville*, Maurice R. Berube and Marilyn Gittell, eds. (New York: Praeger, 1969), p. 56; "A Plan for an Experimental School District: Ocean Hill-Brownsville," July 29, 1967, Special Collection.
9. McCoy, "The Year of the Dragon," p. 56.
10. McCoy, "Progress Report for the Month of July."
11. "A Plan for an Experimental School District: Ocean Hill-Brownsville," July 29, 1967.
12. *New York Times*, September 3, 1967.
13. Rhody McCoy, "Progress Report on Negotiations with the Board of Education," August 30, 1967, Special Collection.

CHAPTER 30   THE BUNDY REPORT

1. Staff Paper No. 9, Temporary Commission on City Finances; Marilyn Gittell, *Participants and Participation* (New York: Praeger, 1967), p. 20; Marilyn Gittell and T. Edward Hollander, *Six Urban School Districts: A Comparative Study of Institutional Response* (New York: Praeger, 1967).
2. Gittell, *Participants and Participation*, pp. 42–44.
3. Mayor's Advisory Panel on Decentralization of the New York City Schools, *Reconnection for Learning: A Community School System for New York City* (New York: Ford Foundation 1967), pp. 1, 3, 8, 13.
4. Mario Fantini, Marilyn Gittell and Richard Magat, *Community Control and the Urban School* (New York: Praeger, 1970).
5. "The United Federation of Teachers looks at school decentralization: a critical analysis of the Bundy report with UFT proposals, policy positions adopted by the Executive Board on November 28, 1967" (New York: UFT, 1967).
6. *New York Times*, February 1, 1968; Morton Inger, Harris Dienstfry, David Outerbridge and Arthur Tobier, "The Bundy Report: A Critical Analysis," *The Center Forum*, January 26, 1968, pp. 7–9.
7. "Resolution passed by Metropolitan Council of American Jewish Congress supporting Bundy plan," in file of Center for Urban Education, New York City, November 29, 1967, U.S. Office of Education Library, Washington,

D.C. However, in 1969, two key officials of the American Jewish Congress wrote a book which was critical of decentralization: Naomi Levine (with Richard Cohen), *Schools in Crisis* (New York: Popular Library, 1969).

8. Materials from Citizens Committee for Decentralization of the Public Schools in file at Center for Urban Education, dated March–June 1968.

CHAPTER 31   PREPARING FOR A SHOWDOWN

1. "Questions by the teachers who are teaching in the Ocean Hill-Brownsville Demonstration Project, addressed to Mr. McCoy, the Unit Administrator, and members of the Governing Board," December 1967, Special Collection, Politics and Education Program, Teachers College. (All documents in this location will be referred to as Special Collection.) See also Eugenia Kemble, "New York's Experiments in School Decentralization," *The United Teacher,* December 20, 1967, for the UFT point of view.
2. Niemeyer report, pp. 41, 44.
3. Letter from IS 201 Governing Board to Ford Foundation, "Presentation to Ford Foundation," January 1968, Special Collection.
4. "Recommendations of a joint meeting between representatives of Ocean Hill-Brownsville and IS 201 Governing Boards," February 9, 1968, Special Collection.
5. Ocean Hill-Brownsville Demonstration School District, news release, March 5, 1968, Special Collection.
6. Ocean Hill-Brownsville Demonstration School District, "What could be— What the truth really is! Telling it the way it is!" Newsletter, March 1968, Special Collection.
7. Letter to parents from Parent Association presidents of JHS 271 and IS 55, April 7, 1968; also, Ocean Hill-Brownsville School District, flyer announcing "We Demand Total Control of Our Schools," and calling for March 28, 1968, meeting, Special Collection.
8. Howard Kalodner, "Legal Aspects of Grant of Authority to Demonstration Decentralization Projects," March 12, 1968, Special Collection.
9. Rhody McCoy, "A Black Educator Assails the 'White' System," *Phi Delta Kappan,* April 1968, pp. 448–449.
10. *New York Times,* March 30, 1968.
11. Letter from David Spencer, chairman of the IS 201 Governing Board, to State Legislators, March 29, 1968, Special Collection.
12. Author's notes, Special Collection.
13. Minutes of meeting of Education Task Force of the New York Urban Coalition with the New York City Board of Education, April 25, 1968, Special Collection.
14. Minutes of meeting of Education Task Force of the New York Urban Coalition with the New York City Board of Education, April 29, 1968, Special Collection.
15. Summary of meeting of Education Task Force of the New York Urban Coalition with representatives of the demonstration districts, May 9, 1968, Special Collection.

CHAPTER 32   "WE WILL HAVE TO WRITE OUR OWN RULES . . ."

1. Report of the Personnel Committee to the Ocean Hill-Brownsville Governing Board, Mrs. Clara Marshall, chairman, n.d., Special Collection, Politics and Education Program, Teachers College (hereafter referred to as *Special Collection*).
2. Letter from Reverend C. Herbert Oliver and Rhody A. McCoy to various teachers and supervisors, May 8, 1968, Special Collection.
3. Bylaws of the New York City Board of Education, July 1, 1964. Article II, Section 101 (1): "Transfers of members of the teaching and supervising staff from one school to another shall be made by the Superintendent of Schools, who shall report immediately such transfers to the Board of Education for its consideration and action." Article III, Section 105 (1): "No member of the teaching and supervising staff . . . shall be removed, except for cause after a hearing, by the affirmative vote of a majority of all the members of the Board of Education." Article X, Section 89 (7a): "In the case of a teacher or supervisor whose assignment to a given school has covered at least 20 school days since the beginning of the school year and is terminated at some time other than the last day of the school year, the principal of such school shall give to such teacher or supervisor not later than four school days following the last day of service in such school, a signed statement characterizing his work as Satisfactory or Unsatisfactory . . . This statement shall cover the period of service in such school immediately preceding such termination of service in such school. The certification of unsatisfactory or doubtful work shall be accompanied by appropriate supporting data. A report of such rating with accompanying data, as indicated, shall be submitted to the Superintendent of Schools . . ." Article III, Section 105a: Any employee of the Board of Education who wishes to appeal a rating "is entitled to appear in person, to be accompanied and advised by an employee of the Board of Education, to be confronted by witnesses, to call witnesses and to introduce any relevant evidence." Such person is also entitled to receive "a written statement of the reasons, facts and conditions," and to have his appeal for a change in rating heard by a committee designated by the Superintendent of Schools.
4. C. Herbert Oliver, "A Response in part to the Martin Mayer Article in the February 2, 1969, Magazine Section of the *New York Times*," undated, Special Collection.
5. Naomi Levine, *Schools in Crisis* (New York: Popular Library, 1969), pp. 56–57; ADL report, p. 83; Lillian S. Calhoun, "New York Schools and Power—Whose," *Integrated Education*, January–February 1969, p. 16; Niemeyer Report, p. 5.
6. Ocean Hill-Brownsville Community Schools, "Fact Sheet: To Parents, Students and Community People," June 1968. [The full paragraph reads: "*This move was made to stop the constant lack of co-operation and to improve the quality of teaching. It also meant that if these transfers were made, it would demonstrate control of the schools by the elected local Governing Board.*" (Italics in original.)], Special Collection.
7. Joseph Featherstone, "Off to a Bad Start," *New Republic*, March 29, 1969, pp. 19–22.
8. "Experimental Decentralization in New York City, A Report to the Commissioner of Education," State Education Department, February 1968. [The report noted that the Ford Foundation appeared convinced that de-

centralization promised success and should be extended throughout the city, but "the evidence to support this conviction is not apparent to this departmental team at this time."], Special Collection.

9. *New York Times,* May 16, 1968.
10. Ocean Hill-Brownsville Governing Board, "Fact Sheet: To the People of Our Community," May 1968, Special Collection.
11. Shapiro memorandum, "Steps Taken to Integrate."
12. *New York Times,* June 6, 1968.

CHAPTER 33   CONFRONTATIONS AND STRIKES

1. *Report of the National Advisory Commission on Civil Disorders* (New York: Bantam Books, 1968), p. 334.
2. *New York Times,* June 19, 20, 1968.
3. "Report and Recommendations of Francis E. Rivers, Esq., Special Trial Examiner; Administrative Hearing into Complaints of Rhody A. McCoy, Unit Administrator of Ocean Hill-Brownsville, Requesting Transfer of Teachers," August 26, 1968, pp. 4–8, 21–23, 28–31, 32–37, Special Collection.
4. C. Herbert Oliver, "Response to Martin Mayer"; Jason Epstein, "The Brooklyn Dodgers," *New York Review of Books,* October 10, 1968.
5. New York Civil Liberties Union, "The Burden of Blame," October 9, 1968, Special Collection.
6. *New York Times,* September 11, 1968.
7. Ocean Hill-Brownsville Governing Board, press release, September 8, 1968, Special Collection.
8. Naomi Levine, *Schools in Crisis,* pp. 75–78; ADL report, appendix, p. 9.
9. Author's interview.
10. ADL report, pp. 1–4.
11. Sandra Feldman, "Strike Log," 1968, Special Collection; Murray Kempton, "Time to Tell Off Shanker," *New York Post,* September 26, 1968.
12. C. Herbert Oliver, "A Fact Sheet to Our Parents with Children in JHS 271, IS 55, PS 144, PS 178, PS 73, PS 137, PS 155, and PS 87—and to Supporters of Ocean Hill-Brownsville Throughout New York City," October 1, 1968, Special Collection.
13. ADL report, appendix, p. 12.
14. *New York Times,* October 4, 1968.
15. Public Education Association, "A Call to Sanity and Action," October 1968, Special Collection.
16. *New York Times,* October 12, 1968.
17. Albert Shanker, Letter to UFT membership announcing initation of third strike, undated.
18. Minutes of a meeting convened by Jerome Kretchmer and Bella Abzug, November 7, 1968, Special Collection.
19. *New York Times,* November 30, 1968.
20. Interview with Wilbur Nordos, *Why Teachers Strike,* ed. by Melvin I. Urofsky (Garden City: Anchor Books, 1970), p. 282.

CHAPTER 34   THE NEW LAW

1. United Bronx Parents, "The People's Plan for School Decentralization," 1969, Special Collection.
2. New York City Board of Education, "Plan for Development of a Community School District System for the City of New York," adopted by the Board of Education January 29, 1969. See dissents, pp. 32–43.
3. Louis E. Yavner, "Review of decentralization plans," Report to the Citizens Union School Committee, 1969, Special Collection.
4. Ira Glasser, "Why We Oppose the Marchi Bill," Committee on Citizen Participation, March 25, 1969, Special Collection; David Seeley, "Statement of the Public Education Association on the Reduction of Racial and Religious Antagonisms," New York City Commission on Human Rights, December 4, 1968.
5. *Wall Street Journal*, April 10, 1969; *New York Times*, November 9, 1968; *Time*, October 4, 1968.
6. Ocean Hill-Brownsville School District, Newsletter, January 1969, Special Collection.
7. Interview by author.
8. *New York Times*, May 1, 2, 1969.
9. Author's interview.
10. Ibid., May 1, 1969.

CHAPTER 35   AFTERMATH

1. Rhody McCoy, "Creation of an Office of Educational Awareness and Development," memorandum to all staff members, September 10, 1968; "To the Parents of Ocean Hill-Brownsville, From: The Governing Board and Unit Administrator," December 11, 1969; Ocean Hill-Brownsville Demonstration District, "Important Message to Parents of Ocean Hill-Brownsville from the Unit Administrator," March 1970; Anonymous leaflet, "Enemy of Black and Puerto Rican People, Samuel D. Wright, Assemblyman," n.d., Special Collection.
2. Carol Wielk, "Educational Progress at Ocean Hill," *Community*, May 1969, p. 3; WCBS-TV's *Public Hearing*, "Albany's Decision on School Decentralization," a debate between Albert Shanker and Jerome Kretchmer, March 30, 1969 (transcript); Fred Ferretti, "Who's to Blame in the School Strike," *New York*, November 18, 1968, p. 27; "School Ranks Report 1971," Bureau of Educational Research, New York City Board of Education.
3. Horace Mann, *Twelfth Annual Report* (1848), reprinted in *The Republic and the School*, ed. by Lawrence A. Cremin (New York: Teachers College, 1957), pp. 94–101.
4. Charles Wilson and David Spencer, "The Case for Community Control #2," IS 201 Governing Board, 1969, Special Collection.
5. Charles Wilson, "Final Report by Unit Administrator to IS 201 Governing Board," August 1969, p. 68, Special Collection.
6. Wilson, "Final Report."
7. Charles Wilson, "IS 201 in Perspective," *Harlem USA*, ed. by John Henrik Clarke (New York: Collier, 1971), pp. 236, 238.
8. Wilson, "IS 201 in Perspective," pp. 235–37.

9. Babette Edwards and Hannah Brockington to David Spencer, chairman of the IS 201 Governing Board, filed with Secretary of the New York City Board of Education, February 5, 1971, Special Collection.
10. "Members of the Staff" to Ronald Evans, principal of IS 201, March 1971, Special Collection.
11. *New York Times*, March 23, 1971.
12. Adele Spier, "Why Two Bridges Failed," *Community*, September/October 1970, pp. 1–3.
13. Palmer, *The New York Public School*, p. 145.

CHAPTER 36   THE SEARCH FOR COMMUNITY

1. "The School Situation in New York," *Outlook*, May 1905, p. 214.
2. Robert M. Hutchins, "The Schools Must Stay," *The Center Magazine*, January/February 1973, p. 20.

Bibliography

My original intention was to prepare a conventional bibliography which would list all the materials consulted. However, I decided that it might be more valuable to mention only those works which I found to be especially useful, and I hope the chapter footnotes will serve as additional reference sources.

First School War: Public Schools or Catholic Schools?

This is the best documented period in school history, because the Public School Society commissioned William O. Bourne to write a thorough history, reproducing many of the original debates, statements, and correspondence. Bourne's *History of the Public School Society of the City of New York* (New York, William Wood: 1870) is excellent. Another good source is Thomas Boese's *Public Education in the City of New York* (New York, Harper Bros.: 1869). Vincent P. Lannie's *Public Money and Parochial Education* (Cleveland: Case Western Reserve, 1968) is a first-rate rendering of the controversy. John R. G. Hassard's *Life of the Most Reverend John Hughes* (New York, Appleton & Co., 1866) is an admiring and valuable biography. The reports and manuals of the Public School Society are themselves the best illustration of the attitudes and practices of the institution. Julia Agnes Duffy's "The Proper Objects of a Gratuitous Education: The Free School Society of the City of New York, 1805 to 1826" (Ph.D. dissertation, Teachers College, 1968) provides useful insights into the Quakers who founded the early free schools. Carl Kaestle's *The Origins of an Urban School System: New York City, 1750–1850* (Cambridge, Mass.: Harvard, 1973) documents the half-century before and after the establishment of New York's first school system. An exceptionally fine book about immigration and social conditions in the first half of the nineteenth century is Robert Ernst's *Immigrant Life in New York City, 1825–1863* (New York: King's Crown Press, 1949). Bayrd Still, in *Mirror for Gotham* (New York: New York University, 1956), has collected visitors' impressions of the city from Dutch days to the present. (Still's book reminded me that "Gotham," which is popularly identified with New York City, is a village in England whose inhabitants were proverbial for their foolishness.)

Between the Wars: 1842–1888

There is no general work for this period in the life of the public schools. Since this was an era in which the public schools were community-controlled, such a study would be extremely useful. Putting together a composite picture requires detective work. Boese has some material on the ward schools. A. Emerson Palmer's *The New York Public School* (New York: Macmillan, 1905) outlines major changes on a decade-by-decade basis, but without filling in the social or political context; for example, he never explains *why* the schools were reorganized in 1864, 1869, 1871, and 1873. The annual reports of the Board of Education and the superintendent of schools are excellent sources, both as statistical summaries and as renderings of the concerns and perceptions of contemporary school officials. The standard works on Boss Tweed are helpful in recreating the period, though they usually ignore the schools (see especially M. R. Werner's *Tammany Hall* [New York: Doubleday, 1928] and Gustavus Myers' *The History of Tammany Hall* [New York: Boni & Liveright, 1917]). Howard Ralph Weisz's "Irish-American and Italian-American Educational Views and Activities, 1870–1900: A Comparison" (Ph.D. dissertation, Columbia University, 1968) conveys an unconventional, extrainstitutional perspective on the public schools.

SECOND SCHOOL WAR: THE RISE OF THE EXPERT

There is no adequate single source for the story of centralization. A brief overview of the battle for reform is contained in Sol Cohen's *Progressives and Urban School Reform* (New York: Teachers College, 1963); besides being a history of the Public Education Association, Cohen's book is valuable as background for later periods in public school history. The weekly journal *School* provides a good account of the professionals' distaste for centralization and reform. The documents of Good Government Club E can be found in the New York Public Library; the papers of the Citizens Committee for Public School Reform are in the Teachers College Library. The newspapers of this period reflect a strong anti-Tammany, anti-local-control bias. For a critical view of the reform movement, see David Hammach's "The Centralization of New York City's Public School System, 1896" (Master's thesis, Columbia University, 1970); Hammach stresses the class identification of the reformers and underestimates the objective conditions that made a reform movement possible. The reform movement cannot be fully comprehended without reference to the works of Jacob Riis, who focused attention on the public school as the key weapon in the war against the slum. Herbert Shapiro's "Reorganizations of the New York City Public School System, 1890–1910" (Ph.D. dissertation, Yeshiva University, 1967) surveys the issues and includes a good bibliography.

BETWEEN THE WARS: 1896–1913

Superintendent Maxwell's annual reports are good reading as well as frank descriptions of the problems of the public schools; they might well serve as models for present-day school officials. Selma Berrol's "Immigrants at School, 1898–1914" (Ph.D. dissertation, City University of New York, 1967) ably depicts both the immigrants' impact on the public schools and the changes which followed centralization. Moses Rischin's *The Promised City* (Cambridge, Mass.: Harvard University, 1962) is a fine portrait of New York's Jews from 1870 until 1914. Nathan Glazer and Daniel Patrick Moynihan's *Beyond the Melting Pot* (Cambridge: MIT, 1963) contains useful information about the city's major ethnic and racial groups. Howard Weisz's previously cited thesis is an important source of documentation about Irish and Italian attitudes towards education. Lawrence Cremin's *The Transformation of the School* (New York: Knopf, 1961) is the best work on the progressive education movement.

THIRD SCHOOL WAR: THE CRUSADE FOR EFFICIENCY

Randolph Bourne's *The Gary Schools* (Boston: Houghton Mifflin, 1916) admiringly describes the Gary plan and illustrates why progressives found it fascinating. John and Evelyn Dewey's *Schools of Tomorrow* (New York, 1915) is a significant account of the Gary schools. An excellent account of the Gary controversy in New York and of the Mitchel administration is in Edwin R. Lewinson's *John Purroy Mitchel: Boy Mayor of New York* (New York: Astra, 1965). The national impact of the efficiency movement is ably documented in Raymond E. Callahan's *Education and the Cult of Efficiency* (Chicago: University of Chicago Press, 1962).

BETWEEN THE WARS: 1920–1954

Sol Cohen's previously cited book contains interesting material about this period. The suggestion that the high ideals of progressive educators were watered down as a "life-adjustment" curriculum is argued by Richard Hofstadter in *Anti-Intellectualism in American Life* (New York: Vintage, 1962). The Board of Education's fiftieth annual report (New York, 1949) is a very good official history of the years from 1898 to 1948; the board's annual report of 1942 includes a charming portrayal of the everyday life of the New York City public school student. Though I have found participants in the "activity program" of the late 1930s and early 1940s who insist that it was excellent, I have not been able to find a scholarly appraisal of the program or of the subsequent effort to make it citywide.

FOURTH SCHOOL WAR: RACISM AND REACTION

The history of the schools from 1954 through 1973 presents special problems because, while there have been thousands of pages published about the subject, very little has been written with detachment and objectivity. There is a mountain of material that was written in the heat of battle—either as a bloodless, defensive account by a bureaucratic functionary or as a revolutionary call to action. It is a toss-up as to which of the two versions is the more distorted, although of course the distortions themselves are interesting.

Newspaper stories serve as a rough gauge of events, though I am aware of the familiar complaints about inaccuracy and misquoting. What I found remarkable was that news stories themselves tended to become events which people reacted to, regardless of accuracy. Many participants in unfolding crises received their information from newspaper accounts, which makes them more significant than they would be if they were simply a record. For this reason, I found the tone and content of newspaper coverage to be an important historical sequence in itself.

Interviews were a different problem. After interviewing several dozen people, I realized that events were subject to unpredictable distortions when seen through the perspective of a single participant; in retrospect, people often magnify their own role and belittle or ignore that of others. I did not consider recollections to be reliable unless I could get independent verification from others. I discovered a similar problem in oral histories; in one instance, a man in his nineties had recorded his memory of events that took place fifty years before, and from other research, I spotted several glaring errors in his reconstruction of history.

David Rogers' *110 Livingston Street* (New York: Random House, 1968) has a good bibliography, which can be supplemented by the citations in the Bundy report. Carolyn Eisenberg, in "The Parents' Movement at IS 201: From Integration to Black Power, 1958–1966" (Ph.D. dissertation, Teachers College, Columbia University, 1971), ably documented that crisis, although I would disagree with many of her interpretations. Mario Fantini, Marilyn Gittell, and Richard Magat's *Community Control and the Urban School* (New York: Praeger, 1970) presents the case for community control, as does *Community Control of the Schools*, ed. Henry Levin (Washington, D.C.: Brookings Institution, 1970). *Confrontation at Ocean Hill-Brownsville*, ed. Maurice R. Berube and Marilyn Gittell (New York: Praeger, 1969), includes some of the most important documents of the period, including the New York Civil Liberties Union report, Maurice J. Goldbloom's "The New York School Crisis," Jason Epstein's "The Brooklyn Dodgers," the report of Judge Rivers, and others. Naomi Levine's *Schools in Crisis* (New York: Popular Library, 1969) is an excellent, judicious account of the controversy.

The materials that were most valuable in piecing together an account of the last great school war, especially during the years 1966–1970, were the letters, flyers, press releases, and so on which were prepared by participants in the events. A good many of the footnotes in this section come from these ephemeral sources. These papers have been placed in files of the Politics and Education Program at Teachers College and are referred to as the Special Collection.

Index

Fish, Hamilton, 151
Flexner, Abraham, 212, 214, 223, 227, 228
Flower, Roswell P., 136, 141
Forced transfers, 281, 284
Ford Foundation, 397
  and demonstration districts, 350, 357, 395
  and Mario Fantini, 315–316
  and Herman Ferguson, 341
  and the Institute for Community Studies, 361
  and Rhody McCoy, 364
  and the Ocean Hill-Brownsville district, 320, 322, 344
  and Helen Testamark, 340
  and the United Federation of Teachers, 328
Fordham University, 37
Forum, 126
Fosdick, Raymond B., 212–213, 214
Free Academy, 103
Free Choice Transfer policy, 280
Free School Society, 9–11, 16–19, 20–22
  changes name to Public School Society, 22
Free-transfer plan, 273
Freeman's Journal, 44, 45, 61, 68, 79
Friedenberg, Edgar, 329
Friends of Peace Society, 226
Froebel, Friedrich, 111
Fuller, H. S., 145, 146, 147
Fusion, 192, 194, 195, 222
Fusion-Republican, 220

G

Galamison, Milton A., 255, 261–262, 396
  and the Allen plan, 280, 284
  and boycotts, 273–274, 276–277, 278–279, 285
  and decentralization, 363
  and integration, 257, 269
  and local school boards, 382
  and Rhody McCoy, 322

and the Ocean Hill-Brownsville district, 374, 375
and the People's Board of Education, 308
and Albert Shanker, 348–349
Garrison, Lloyd, 264, 289–290, 299–301, 305
Gary, Elbert, 225–226
Gary plan, 197, 207, 215–218, 224
  antagonism toward the, 208–209
  attacked, 226–228
  description of the, 198–203
  evaluation of the, 210–211
  introduction of the, 204–209
  lessons learned from the, 233
  similarity to the Lancasterian system, 216
Gary School League, 216
Gaynor, Mayor William, 192, 194
General Education Board, 212, 227
George, Henry, 110
Gerrymandering of districts, 246
Ghettos, *see* Slums
Giardino, Alfred, 334, 350, 355, 356
Gilroy, Mayor Thomas, 119, 120, 122, 132, 139
Gittell, Marilyn, 330–331, 337, 356, 396
  and community control, 341
  and decentralization, 332, 334
Glasser, Ira, 383, 384
Glendale, 259, 331
Godkin, E. L., 90, 141
Goff, John, 136
Good Government Clubs, 134, 151
  Club A, 135–136
  Club E, 134, 142, 149
Goodman, Paul, 329
Gordon, Sol, 313
Gottesfeld, Harry, 313
Gottesfeld-Gordon plan, *see* Yeshiva plan
Grace, Mayor William R., 113
Grading in school, 100
Grant, Mayor Hugh J., 119–120, 121, 141
Greeley, Andrew M., 80–81
Greeley, Horace, 69